RESEARCH METHODS
FOR POSTGRADUATES

RESEARCH METHODS FOR POSTGRADUATES

Third Edition

Edited by

Tony Greenfield with Sue Greener

WILEY

Library of Congress Cataloging-in-Publication Data

Names: Greenfield, Tony, editor. | Greener, Sue, editor.
Title: Research methods for postgraduates / edited by Tony Greenfield with
   Sue Greener.
Description: Third edition. | Chichester, UK ; Hoboken, NJ : John Wiley &
   Sons, 2016. | Includes index.
Identifiers: LCCN 2016011607 (print) | LCCN 2016016287 (ebook) | ISBN
   9781118341469 (pbk.) | ISBN 9781118763001 (pdf) | ISBN 9781118762998
   (epub)
Subjects: LCSH: Research–Methodology.
Classification: LCC Q180.55.M4 R473 2016 (print) | LCC Q180.55.M4 (ebook) |
   DDC 001.4/2–dc23
LC record available at https://lccn.loc.gov/2016011607

A catalogue record for this book is available from the British Library.

ISBN: 9781118341469

Set in 10/12pt Times by Aptara Inc., New Delhi, India
Printed in Singapore by C.O.S. Printers Pte Ltd

1   2016

Contents

About the Author

Tony Greenfield was born in Chapeltown, South Yorkshire on 26 April 1931 to Geoffrey James Greenfield (1900–1978) and Hilda Aynsley (1903–1976).

Tony worked in a Cumbrian iron mine when he left Bedford School at the age of 17. He later worked in coal mines, a brass tube factory and a copper mine and studied mining engineering at Imperial College London. He received the diploma in journalism from the Regent Street Polytechnic, worked technical journals and on the *Sunday Express* and *Sunday Mirror* before turning to technical journalism, in Sheffield, for 10 years. He was an active member of the Sheffield Junior Chamber of Commerce of which he was chairman of the Local Affairs, Business Affairs and Public Speaking committees and editor of *The Hub,* the chamber's monthly magazine. At the 1963 conference in Tel Aviv of Junior Chamber International, he was acknowledged as the editor of the best junior chamber magazine in the world.

He moved into the steel industry to write technical reports for Operations Research (OR) scientists. There he found satisfaction in solving production problems, and studied OR, mathematics, statistics and computing, leading to an external degree from University College London. He moved into steel research and became head of process computing and statistics. Much of his work was in design and analysis of experiments for which he received his PhD. He co-authored the first interactive statistics package to be written in Fortran. When the laboratories closed, he joined the medical faculty of University of Sheffield where he was statistician to a multi-centre study of cot death. He taught medical statistics to undergraduates, supported post-graduates and medical staff with consultancy. Tony moved to Belfast as professor of medical computing and statistics at Queen's University. Early retirement enabled him to work as a research consultant.

Tony's passion is to persuade all scientists and engineers to write, speak and present their work in language that other people understand well enough to use. And, like W.B. Yeats, he asks scientists to "think like a wise man but communicate in the language of the people".

Like Isaac Asimov, he is "on fire to explain and doesn't indulge in scholarly depth". He believes strongly that the economic fortune of Europe depends on the success in the world markets of our manufacturing industries.

"Statisticians and statistical practitioners across Europe know that statistical methods have improved business and industrial peformance – and can do so in the future", he says. *"Our national quality of life will be improved and secured if we can communicate the philosophy, as well as the methods, of statistics to engineers and others in the manufacturing and the service industries. Businessmen and engineers need to understand the benefits of applied probability and*

statistics; they need to understand how the methods are applied to their own work; they need to be fully converted to a frame of mind that will make them automatically question sources of variability in everything that they do and, without outside prompting, adopt the statistical approach".

He and others founded ENBIS to stimulate the application of statistical methods to economic and technical development and to business and industry across the whole of Europe. They have created a networking forum for the exchange of ideas between statistical practitioners. He has spread this passion by speaking in many cities across Europe from Tel Aviv, through Turin, Budapest, Ljubljana, Copenhagen, Brussels, Sheffield, Newcastle and London.

List of Contributors

Claire Abson Sheffield Hallam University, Sheffield, UK

Alastair Allan University of Sheffield, Sheffield, UK

Douglas G. Altman University of Oxford, Oxford, UK

Patrick Andrews Hawkshaw Product Design Ltd, Crieff, Scotland, UK

Andrea Benn University of Brighton, Brighton, UK

Tom Bourner Professor Emeritus, Brighton Business School, University of Brighton, UK

Roland Caulcutt Caulcutt Associates, Salisbury, UK

Shirley Coleman ISRU, School of Maths and Stats, Newcastle University, UK

David de Vaus Emeritus Professor, FASSA, Institute for Social Science Research, University of Queensland, Brisbane, Australia

Keith Dugmore Demographic Decisions Ltd, London, UK

Aiden Fisher University of Adelaide, SA, Australia

Catherine Fraser-Martin Independent researcher, UK

Suzanne Fraser-Martin Independent researcher, UK

Felix Grant Lecturer and consultant, UK

David Green School of Mathematical Sciences, University of Adelaide, SA, Australia

Sue Greener Brighton Business School, University of Brighton, UK

Tony Greenfield Greenfield Research, UK (retired)

David J. Hand Imperial College, London, UK

Linda Heath Brighton Business School, University of Brighton, UK

Mark Hughes Brighton Business School, University of Brighton, UK

Garth R. Johnson Newcastle University, Newcastle, UK

Clifford E. Lunneborg R Foundation, Boston, MA, USA

Peter Lynn Institute for Social and Economic Research, University of Essex, Colchester, UK

Lluis Marco-Almagro Universitat Politecnica de Catalunya, BarcelonaTech, Barcelona, Spain

Vivien Martin Brighton Business School, University of Brighton, UK

Lowry McComb Durham University, UK

Andrew Metcalfe School of Mathematical Sciences, University of Adelaide, SA, Australia

Juliet Millican CUPP, University of Brighton, UK

Irena Ograjenšek Faculty of Economics, University of Ljubljana, Slovenia

Anand D. Pandyan Keele University, Newcastle under Lyme, UK

Roger Payne VSN International & Department of Computational and Systems Biology, Rothamsted Research, Harpenden, UK

Silvia Salini University of Milan, Italy

Sara Shinton Shinton Consulting Ltd, Galashiels, UK

Iveta Simera Centre for Statistics in Medicine, Nuffield Department of Orthopaedics, Rheumatology and Musculoskeletal Sciences University of Oxford

Stan Taylor School of Education, Durham University, Durham, UK

Frederike van Wijck Glasgow Caledonian University, Glasgow, Scotland, UK

Preface to the Third Edition

"Would you like to produce a third edition?" asked Heather. "Wiley have bought the rights from Hodder. I read the second edition and looked for competition. There is no other book about research methods as good as yours".

She assured me that I could apply my own style and that she and others at Wiley would help me as much as they could.

Years have passed since the first edition, and I have grown old, so this is my last work for the scientific and technical literature. I have depended entirely on all authors of chapters, for whose patience and understanding I am immensely grateful. They are all erudite and enthusiastic about their own subjects and eager to inspire you, our students, to do first-class research. I hope my own story will also inspire you.

This is a personal story. Perhaps this is the wrong place for a personal story but I want to tell it, as my attempt to inspire you.

"Is statistics a science?" is a hackneyed old question. It discomforts me. The question is needless. It is needless because it is predicated by the assumption that there are many sciences.

We have split science into several separate sciences, but the splits are artificial.

What is my science? I am a scientist. (No splits.)

We do split science into subject areas for pedagogical convenience in schools and universities. I do remember most of the chemistry, physics and biology I learned at school 70 years ago. I could not claim to be a chemist, physicist or biologist. But I would not say, as I was once shocked to hear a statistician say, "I cannot discuss the design of an aerofoil because I am not an aeronautical engineer; I am a statistician".

Statistics is a part of science, but it is not 'a science'; it is a subject area within science just as is chemistry. And it has no discrete boundary, as neither does chemistry.

Statistics provides method to science:

> Do you ever notice something; describe it; ask yourself, "What is it? Why is it? Where it it? Is it useful to me or to anybody else? Does it have any relationship to anything else?" Then you have the makings of a scientist.

But, and this is where the usefulness of statistics arrives, do you then invent a working assumption, called a *hypothesis*, that is consistent with what you have observed? If you do, can you then use the hypothesis to make predictions?

Now, you must see clearly that statistics is an essential tool of science. You can test your predictions by experiments or further observations and modify the hypothesis in the light of your results. The scientific method insists that you keep revising your hypothesis and experimenting until you can detect no discrepancies between your hypothesis and your observations. You may then, correctly in the scientific realm, tell the world that you have a theory that may explain a class of phenomena.

A *theory*, by my description and by dictionary definition, *is a framework within which observations are explained and predictions are made.*

I once proposed a curriculum approach to representation of statistics as the cement for binding science's subject areas. This was in a paper, *The polymath consultant,* at the first meeting of ICOTS (International Conference on Teaching Statistics). *The Times* newspaper published a short version of it. The UK secretary of state for education, Keith Joseph, was interested enough to invite me to discuss it, and he encouraged me to promote the idea in universities. Nobody else took any notice. Yet I still believe that there was an idea that could be developed as part of our search for the future of data analysis. We must teach that statistical methods are just as part of, and just as applicable, in social studies as they are in physics and chemistry; and that they are as useful in linguistics, history and geography as much as they are in engineering and marketing.

Collections of worked practical cases, such as those by Cox and Snell (1981), must help and we need more of them. A recent book (Greenfield and Metcalfe 2007) aims at this with more than 50 worked cases about school absence, metro noise levels, water fluoridation, diamond prospecting, wine tasting, compulsive gambling, prosthetic heart valves and many more.

Evidence is the life-blood of science and scepticism is its spark of life. Data analysis is the flux of evidence. We should continue to ensure that all scientists, in all subject areas, and these include *you*, perceive it as such. Always you must be sceptical about any assertion that has no evidential support. *Nullius in verba.*

Frances Ashcroft, a truly great scientist of this, the twenty-first century, tells us in a recent book how her own research excited her.

> I discovered that the KATP channel sits in the membrane that envelops the beta-cell and regulates its electrical activity and thereby insulin release.... The breakthrough came late at night when I was working alone.... I was ecstatic. I was dancing in the air, shot high into the sky on the rocket of excitement with the stars exploding in vivid colours all around me. Even recalling that moment sends excitement fizzing through my veins, and puts a smile on my face.
>
> There is nothing — nothing at all — that compares to the exhilaration of discovery, of being the first person on the planet to see something new and understand what it means. It comes all too rarely to a scientist, perhaps just once in a lifetime, and usually requires years of hard grind to get there. But the delight of discovery is truly magical, a life-transforming event that keeps you at the bench even when times are tough. It makes science an addictive pursuit.
>
> That night I felt like stout Cortez, silent upon his peak in Darien, gazing out across not the Pacific Ocean, but a landscape of the mind. It was crystal clear where my mental journey must take me, what experiments were needed and what the implications were.
>
> Next morning, all certainty swept away, I felt sure my beautiful result was merely a mistake. There was only one way to find out. Repeat the experiment — again and again and again. That is the daily drudgery of a scientific life: it is very far from the ecstasy of discovery.

The Spark of Life Electricity in the Human Body
Frances Ashcroft

Such reporting inspired me to read the rest of the book even though, in her last paragraph, she warns that all of us, including you, cannot expect winning without drudgery. Thomas Edison expressed this well:

Genius is one percent inspiration, ninety-nine percent perspiration.
  Spoken statement (c. 1903); published in Harper's Monthly (September 1932)

Many writers in the past have felt the same elation as Frances Ashcroft. John Keats, for example, recorded that feeling:

> *Then felt I like some watcher of the skies*
> *When a new planet swims into his ken;*
> *Or like stout Cortez when with eagle eyes*
> *He star'd at the Pacific — and all his men*
> *Look'd at each other with a wild surmise —*
> *Silent, upon a peak in Darien.*
> John Keats

Mary Shelley told us how Doctor Frankenstein's feeling went further from the beauty of scientific achievement to disgust at what he had done.

The different accidents of life are not so changeable as the feelings of human nature. I had worked hard for nearly two years, for the sole purpose of infusing life into an inanimate body. For this I had deprived myself of rest and health. I had desired it with an ardour that far exceeded moderation; but now that I had finished, the beauty of the dream vanished, and breathless horror and disgust filled my heart.
                                           Frankenstein (chapter 5) Mary Wolstencroft Shelley

If, as a scientist, you can keep powering the bellows that inflame your spark of inspiration into a bright light of scientific achievement, scientists will acknowledge that you are one of them. But, first, you must be sure that you believe you are a scientist. You must have started somewhere, sometime. Here is how and when and where I started.

"Tell Father that lunch is ready," said Mother. "He's in the garage".

I loved Sunday lunch when I was six, especially when it was roast beef, Yorkshire pudding, dark green cabbage and rich gravy. I went to the garage to summon Father to the table where he would display his knife sharpening and carving skills.

He was on the floor, asleep, and his face had a bluish greenish tinge.

I ran to Mother. She quickly opened the doors and windows and called an ambulance. She dragged him onto the lawn and pumped his chest. He breathed and his face turned grey. An ambulance arrived. The men put a mask over his face. It was connected by a rubber tube to a cylinder of oxygen. His face turned pink. The ambulance drove away and we had lunch, a little late.

Carbon black is an amorphous carbon with a high surface-area-to-volume ratio. It is used as a pigment and reinforcement in rubber and plastic products. It also helps to conduct heat away from the tread and belt area of the tyre, reducing thermal damage and increasing tyre life. It is very expensive. It was even more expensive when I was six and Father thought he could make a lot of money by producing it from cheap by-products, usually discarded, from coal distillation or coke making. One of these was naphtha and, as a chemical engineer, he knew where he could get as much as he wanted very cheap. In those days, the Americans made most of the world's carbon black from natural gas and it cost about £5000 a ton (imperial spelling): a lot, especially if you convert that into today's money. Nowadays, with many more sources of materials and more efficient production, it is worth about £500 a tonne (note the SI spelling).

Father explained as much of this to me as I could understand and he showed me his experimental machine. So far as I can remember, nearly 60 years later, it comprised a rotating drum with cooling water circulating through it. There was a row of tiny jets through which he pumped naphtha that burned, with only partial combustion, so that carbon black deposited on the cold drum surface. Naphtha is a crude mix of oils that drained out of the bottoms of coke ovens where it was used by burning to heat the ovens.

I learned, when I was six, that Father was a scientist, an engineer and an experimentalist. But his research had its perils, including the possibility of carbon monoxide poisoning.

Father explained many things to me over the years.

He drove to a coke oven plant in Scunthorpe when I was 11. I went too but, in case I was bored, I carried a *Just William* book (by Richmal Compton). At the ovens, we sat all night measuring things as they happened. Father told me, "This is called 'dynamic measuring' ". I watched, fascinated, as ink flowed onto rotating graphs. In the morning, Father analysed the data and advised the works manager on how to improve his benzole production. Benzole (a mix of benzene and toluene) had been seen, by coke-makers, as a waste by-product. In 1942, it was an essential fuel for the Spitfire.

This was science in the raw and I was excited, never bored.

I was 15 when I bought an ancient motorbike that wouldn't go. Father commented that I would be a competent mechanic and understand internal combustion engines by the time I was allowed to drive it. A year later I had fixed it but petrol was rationed. I decided to make my own. I had fitted the bike with acetylene lights instead of electric lights. A local garage gave me a drum of (calcium) carbide that they no longer needed. Water drips onto carbide to produce acetylene which, at one time, was used for lighting. I also knew that acetylene (C_2H_2) could be polymerised to benzene (C_6H_6) by contact with red-hot iron or most alloys in which iron is the dominant component, at about 700 °C. I wound an electric fire element round the gap between two silicon tubes which I sealed into a large silicon jar connected by a rubber tube to an acetylene generator more than a metre from the jar. I intended to send acetylene into the jar; heat from the fire element would draw the gas up and then down again until benzene appeared. When, eventually, I had made a litre of benzene, I would experiment with mixed proportions of paraffin (not rationed) to discover the best mix to drive my motorbike.

Grandpa's brick garage was integral to the house. Father's was wooden and that is where I had my benzene plant. The power switch was by the door. I started the acetylene generator and let it run for about 10 minutes to expunge all air from the jar. Then I switched on the power; and watched.

Two minutes later, I saw: drip…drip…drip…from the silicon tubes.

Frances Ashcroft expressed my feelings later: "I was ecstatic. I was dancing in the air, shot high into the sky on the rocket of excitement with the stars exploding in vivid colours all around me."

I watched, enslaved by the sight of my success, but for only a few seconds.

A crack of thunder, a great white light, and the apparatus went through the roof and fell in the garden.

I studied the hole in the roof and caught the next bus to Worthing. I arrived home late at night. Father was still up. I said nothing. "Are you afraid of me?" he asked. "Yes". "No need", he said. "I am proud of you".

Weeks later, my physics master said Father had told him the story and he, too, was proud of me. "You will be a good experimentalist", he predicted, "but you will always be the servant of others unless you learn about patents".

When that teacher demonstrated the Michelson–Morley experiment, and said it proved that ether did not exist, I said perhaps it did but it may have properties that were hidden from the experiment. "You have a curious mind, Greenfield", he said. Father said I had a hypothesis as good as any, and he encouraged me to design an experiment to test it.

Red shift is generally accepted as evidence of universal expansion. Father again encouraged me to design an experiment to test an alternative hypothesis. Although we borrowed a quarry to set up an experiment, the apparatus we designed and built was not good enough.

At 18, I had my first provisional patent for a photographic colour method using the five oxides of vanadium. Kodak were interested but couldn't improve on my colours, the worst of which gave brown instead of green.

Thirty years later:

"I like your style", said the visiting professor.

His compliment came towards the end of the first course I gave on research methods to the medical faculty of Queen's University, Belfast.

The course arose from my experiences in steel research, in Sheffield University and in Queen's University. In all of these I had found a shocking inability among scientific researchers to write and to speak clearly about their research. Scientific books and papers are so mysterious, so arcane, so bewildering, that scientists can understand only those of their own speciality. They are obscure to others.

I had also been shocked in all these places by the lack of appreciation of statistical and other research methods. A short, sharp course was needed. The faculty dean agreed and encouraged me to run such a course, which I did every year for five years. Students ranged from new medical graduates to senior consultants and professors. Teachers included the dean, a librarian, computer staff, statisticians, professors of clinical psychology, epidemiology and chemical pathology, and the chairman of the research ethics committee.

"I didn't know I had a style", I replied. "What is it?"
"You always look as if you don't know what you will say next", the visiting professor told me.

I knew that was true. I always watch students' faces and look into their eyes to be sure that they understand what I am saying. That is an essence of teaching, but it is hard in a lecture hall with 150 students; it is easy in a classroom with no more than 30, which I had.

I can't see the eyes of students when they read a text book, but I can try to write in a style that will grab and keep their attention. Contributors to earlier editions of this book agreed. Unfortunately, the publisher's editors disagreed and, in my view, ruined the style of the second edition. They made changes to the text that were far beyond acceptable editing. They were changes with which I did not agree: changes that affected my style and the styles of other authors. They refused to repair the damage, and eventually I surrendered.

Felix Grant, author of one chapter, wrote to me: "Watching the progress of this whole spectacle has been an education. I shall never look at Hodder or their imprints in the same way again, not only as a writer but as a professional and as an educational book buyer. I admire your tenacity and integrity; I hope I have the same level of commitment to what matters, if it should come to that. I am, after a break of two decades, currently starting on the long process of steering a book of my own through 'another publisher'. If your experience with Hodder turns out to be typical of changes in publishing's attitudes to quality and verity over that time, I shall be very disappointed".

I apologise to Felix for my surrender, and to other authors and readers who feel the same. Felix wrote again for this edition. I wish he could see it. Sadly, he died a few months before we went to press.

Wiley have promised no such desecration. They like and enjoy the style that I have encouraged all authors to adopt. Their editors (Debbie Cox and Heather Kay) agree with John Gribbin who wrote, in his *New Scientist* review of the first edition, "The most useful book any new postgraduate could ever buy".

Debbie, Heather and Richard Davies have supported this project wonderfully. For myself, and for all the authors and readers, I thank them.

But some authors know that during the course of this work I have developed Parkinson's disease. This delayed production for four years. Heather recruited two angels to help. One is Sue Greener (see chapters 2, 12, 15, 16, 37, and 38) a delightful and positive writer and editor. The other is Liz, my wife, who eagerly follows Heather's instructions to keep me going and keeps in touch with Sue.

Now we have the best edition with revised chapters, new chapters and new writers. You will enjoy reading this book so much that you won't want to put it down. You will start with a journey through the general research scene.... This is where the hard work begins: collect data, analyse and interpret data, and write and publish articles, news items, technical reports and a thesis that you must present to your examiners.

You, the researcher, the problem solver, are responsible to a manager: in a company, a university or a government department. You must report results so that the manager can understand them enough to make decisions. Research does not end with design and analysis. You must interpret and communicate the results. Unless you can describe and explain your results to people who do not share your analytic skills, your results will be worthless. Read the book.

I have no data yet. It is a capital mistake to theorise before one has data. Insensibly one begins to twist fact to suit theories, instead of theories to suit facts.

Sherlock Holmes
A Scandal in Bohemia

Tony Greenfield

Preface to the Second Edition

'This just might be the most useful book any new post-graduate contemplating research could ever buy', wrote John Gribbin in his *New Scientist* review of the first edition of *Research Methods*. Agreement with that view came from post-graduate researchers. Supervisors and teachers welcomed the book as a prop, even *the* main course book, for post-graduate courses in research methods.

Comments and advice flowed in and technology advanced. The time arrived for a second edition. Liz Gooster replaced Nicki Dennis as the publisher's commissioning editor and we worked well together. Fortunately, most of the original contributors were willing to revise their chapters but we needed authors for new chapters and some for a few replacements. The Internet and the World Wide Web are here and they have had a profound influence on the ways of post-graduates. This needed to be reflected in many chapters: library, literature reviews, search for funds, information technology and computers, sources of population statistics. References to further reading via the WWW can be offered for almost every chapter.

Reviewers' comments on the first edition were almost all favourable. Here are a few:

- Good coverage of major topics relevant to our students (PG training course).
- Useful reference material for students with their dissertations and analysing results (MSc Oncology).
- Good introduction to many of the skills required by research students.
- I have used it a great deal myself. I needed to know about surveys and sampling.
- Most useful section is on presentation, particularly *Writing the thesis*.
- Aims and approach are sound, even if broad and ambitious. Text well written, concise and convincing. Well-structured. Argument easy to read and digest. Every reader should learn something. New PG students will learn a great deal.
- Great strength is the scope and interdisciplinary appeal.
- No comparable book.

Some chapters, I was told, may be of interest to some students but had little relevance to others. Well I believe that there is something in every chapter for almost all research students so I asked the authors to refer to many more illustrations of the diverse relevance of their advice. You will find the response in, for example, chapter 28 *Instrumentation for research*, and in chapter 33 *The value of mathematical models*.

Some comments could not be reconciled. For example:

- Students from engineering valued this book.
- Not specific enough for engineers.

Suggestions included:

- More emphasis on the Internet.
- Some exciting new developments in using the Internet for teaching and research are com-
  ing from chemistry departments around the world. Students can send their experimental
  data across the world and receive it back from computer, spectroscopic or other forms of
  processing. Similar services for other subjects.
- Have a chapter on WAP (Wireless Application Protocol) technology, knowledge based
  searches and alerts over the Internet and digital television and other multi-media systems.
- Chapter needed on navigating the WWW.
- Reflect diversity of software packages.
- More on creativity.

There is certainly much more emphasis on the Internet and on the WWW. The diversity of
software packages, particularly for statistical analysis, is discussed. We have four chapters on
creativity. I tried very hard to recruit an author to describe what is happening in chemistry,
looking in university departments and in industry, but I failed. 'He didn't ask me,' some reader
will say. Well please, dear reader, write to me soon, so that we can start to plan the third
edition. Nor could I find an author to tell us about WAP. For these, and other topics, I need
suggestions and volunteers.

I should also like your comments about how the book is used. I believe it is a good reference
text for any post-graduate student. I also believe it is a good framework for any postgraduate
research course. Do you agree?

A further suggestion was to construct a FAQ page on the publisher's website. I know this
is being considered. Perhaps, by the time this new edition reaches your desk, the page will be
there, ready for you to use.

Finally, thanks again to all contributors: those from the first edition for their continuing
support; the new recruits for this edition for putting their faith and effort into such a valuable
publication. And thanks to Liz Gooster for encouraging and helping me in my role as editor
of 'the most useful book any new post-graduate contemplating research could ever buy'.

Tony Greenfield

Preface to the First Edition

The government proposed in 1994 in their White Paper *Realising our Potential* that all graduates who wish to study for doctorates should first take a one-year master's course in research methods. Several universities have since introduced such courses and more are planned. This book is a response to that development. It is not intended to be a deeply detailed textbook, rather a set of notes for guidance, to nudge the student's mind into useful avenues, to tell him or her what help is available and to show how he or she can help themselves. This guidance includes many references for further study. As a set of notes it should be useful to all researchers, those studying for doctorates as well as for masters' degrees, for their lecturers too and, indeed, for anybody in any field of research even if a higher qualification is not expected.

The breadth of the subject rules out a single author: none but the most arrogant would pretend to such ability. The publishers and I therefore decided that we should seek contributions from many authors. This posed difficulties of recruitment, of meeting deadlines, of agreeing a common philosophy and adhering to it, and of imposing an editorial style without causing offence to the authors. These difficulties were resolved because there was one clear bond between all the authors: an enthusiasm to help young people to plan, manage, analyse and report their research better than they may otherwise. All of them are busy and successful as researchers and as teachers. I believe that all readers of this book will appreciate how much time and effort, as well as knowledge and experience, the contributors have devoted to its production.

Unusually, this preface is titled *Preface to the first edition*. This is because I have no doubt that there will be subsequent editions. The situation will change with the introduction of more courses on research methods and experience will accumulate. I invite all readers to tell me how it can be improved: what should be added, what should be omitted, and what should be rewritten. But if you like any of it, please write and tell me. I shall forward your comments on to the authors. They deserve your praise.

Tony Greenfield

Preface to the First Edition

Part I

First Steps

1

A View of Research

Tony Greenfield

Introduction

Research, depending on your viewpoint, is:

- a quest for knowledge and understanding;
- an interesting, and perhaps useful, experience;
- a course for qualification;
- a career;
- a style of life;
- an essential process for commercial success;
- a way to improve human quality of life;
- an ego boost for you; and/or
- a justification for funds for your department and its continued existence.

To me, research is an art aided by skills of inquiry, experimental design, data collection, measurement and analysis, by interpretation, and by presentation. A further skill, which can be acquired and developed, is creativity or invention.

This book is mainly about the former set of skills, inquiry to presentation. Further useful topics are described, such as: how to find funds, how to protect your intellectual property and how to share and use the results when your research is concluded.

But first: a few words about the origin of this book. The first edition was inspired by the government's proposal in 1994 that all graduates who wish to study for doctorates should first take a one-year master's course in research methods. Whether or not you agree with this, you may agree that some notes for guidance of post-graduate research students would be useful. Many universities have followed the government's proposal and have created research methods courses. There is already a place for this book. Whether you are studying a master's course in research methods, or doing some research for a master's degree or a doctorate, you can be guided by this book.

Research Methods for Postgraduates, Third Edition. Edited by Tony Greenfield with Sue Greener.
© 2016 John Wiley & Sons, Ltd. Published 2016 by John Wiley & Sons, Ltd.

However, research is a big subject, and it would not be possible to write a single volume about it in any depth. This book is intended to be a general reference on all aspects of research methods and should be used as notes for guidance. Its content is intended to be fairly simple and easily intelligible by most readers. There are references to more substantive texts.

The many viewpoints and components of research methods persuaded me that several contributors would be needed. Fortunately, there are enough qualified people in universities, consultancy and industry who volunteered eagerly to write one or more chapters each. I asked them to write in a light style that could be read easily with a view to the reader picking up the general themes. I believe that between us we have achieved this but leave it now to you, the reader, to judge.

If there are parts that you don't understand, or that could be expressed more clearly, or if there are important omissions, please write to me or the publishers. Everything can be improved, especially a book, and your opinions will help us.

Contents

These notes for your guidance have been divided into seven sections, with several chapters in each. Look through the contents list and see how the topics have been grouped. You may feel that some of the chapters are not for you. For example: do you know how to use the library? Of course you do! But do you? I suspect that many people who believe that they know how to find the right text at the right time will be happily surprised to discover how much easier the task becomes when qualified guidance is given. Surely you will want to know how to find funds for your research, but 'Ethics? Ethics has nothing to do with my research', you might say. It has. It has something to do with all research. Read the chapter and learn.

You will run into difficulties. You will find problems of management, of resources, of people. There's a chapter telling you who can help. I suggest you read it before you meet those problems. There are chapters too on planning your work, about keeping documents, about examining your research process and keeping it on course.

There are several types of research and we have classified them as: clinical trials, laboratory and industrial experiments, agricultural experiments, and surveys. These may seem to be distinct but there is a general philosophy running through all of them, expressed in different ways by the different writers. You may think that because your research fits into one class of research, you can ignore the other three chapters in that section. Please make the effort to read those other three chapters. You will be stimulated to discover a new slant on your research.

Glance quickly at the section on data analysis and you may think "I can leave that until much later, when I have some data to analyse." Scientific method is about observing the world and collecting information so that you can understand the world better. The way in which you do this must surely depend on how you will process the information when you have collected it. The data you collect will depend on how you will analyse the data. Analysis is an essential feature of research, and you will make easier progress with your research the more you understand analysis. To some people it is hard and daunting. They would prefer to ignore it. To other people it is a challenge. Whichever is your viewpoint, make it a challenge and face it now. Honestly, the more you understand how you will analyse and interpret data, the better will be your planning and management of the way you collect it. The design of a good experiment depends on how the data from the experiment will be analysed.

Mathematical modelling and simulation may seem to be remote from the reality that you want to investigate in your research. They are powerful tools in many situations. Social, medical, economic and political systems, as well as physical, chemical and biological ones, can be described as mathematical models which can then be used in computers to predict the behaviour of those systems under various conditions. This is a useful approach to many types of research. While you read through the examples included in Part V, keep asking yourself how each example may relate to your research project.

Whatever research you perform, you must present your results: in a thesis or dissertation, in reports and published papers and in stand-up talks or synchronous webinars to live audiences. There are many books about presentation, and some are recommended. Three chapters summarise the most useful points.

Other chapters offer good advice about how to buy and use computers and instrumentation, how to sample from populations and interview people, how to protect your intellectual property and how to progress in your career.

Creativity

Four chapters about creativity were added in the second edition. As a brief stimulant, I have preserved the following paragraphs.

Liam Hudson, in *Contrary Imaginations: A Psychological Study of the English Schoolboy* (Penguin, 1972), presents evidence that intelligence and creativity, as features of the human mind, are negatively correlated but that there are some fortunate people who are both intelligent and creative. The rare combination must be desirable in research, where we need both logic and imagination, where we need vision as well as the ability to plan and manage.

But what is creativity?

You are planning your research. You believe that every step on the way must be taken rationally. Indeed, that is the essence of most of this book: to guide you rationally through your work. But if you look at the most outstanding creative leaps in the history of science, you will see that they were all founded on an irrationality of thought. Well-known examples are: Watt's invention of the separate condenser for the steam engine as he strolled in the country, Poincare's theory of Fuchsian functions as he boarded a bus and Kekule's discovery of the benzene ring as he dozed by the fireside. So, be prepared to note any odd thought you might have at an unexpected time in an unexpected place. And don't discard unexpected results.

Just because something doesn't do what you planned it to do doesn't mean it's useless.
                                                                                          Thomas Alva Edison

Nevertheless, you can bring to bear some methods of intellectual discovery:

- **Analogy**: look for similarity between your problem and one for which the solution is known. Electrical circuits are envisioned as water flowing through tanks, pipes, pumps and valves; brain function is studied by comparison with computers. The more remote your analogy is from your problem, the more creative will be your solution.
- **By parts**: break the problem into a series of sub-problems which you hope will be more amenable to solution.

- By **random guesses**: Edison used it extensively, and brainstorming is a modern version of it.
- **Generalise**: if a specific problem is baffling, write a general version of it; an algebraic model leads to simplified solutions compared with tackling complicated arithmetic head on.
- **Add**: a difficult problem may be resolved by adding an auxiliary sub-problem.
- **Subtract**: drop some of the complicating features of the original problem; this is a trick used in simulation to make it more tractable.
- **Particularise**: look for a special case with a narrower set of conditions, such as tackling a two-dimensional example of a three-dimensional problem.
- **Stretch or contract**: some problems are more tractable if their scale or the range of variables is altered.
- **Invert**: look at the problem from the opposite viewpoint; instead of "When will this train arrive at Oxford?" ask, "When will Oxford arrive at this train?"
- **Restructure**: in clinical studies, we do not ask if a treatment will cure a disease, but whether an inert treatment will fail to cure the disease.
- The method of **Pappus**: assume the problem is solved and calculate backwards.
- The method of **Tertullus**: assume a solution is impossible and try to prove why.

Check each of these approaches, asking yourself how you might bring it to bear on your problem. Then, if you need any more stimulation, read the following:

> The *Art of Scientific Investigation* is a book by W.I.B. Beveridge published in 1950 but still, over half a century later, stimulating to read;
>
> G. Polya's *How to Solve It* offers practical recipes; and
>
> Arthur Koestler's *The Act of Creation* has a discussion of the working of the mind.

2

The Research Journey: Four Steps to Success

Tom Bourner and Sue Greener

Research can seem daunting to those who are new to it. This chapter has two aims:

- to provide an overview of the research journey from start to finish; and
- to demystify the business of research.

Most research projects take quite a long time to complete. Research degrees, for example, usually take at least three years of full-time research. Completing a part-time research degree usually takes correspondingly longer. At the start, research can seem like an ill-defined mish-mash of activities littered with hidden pitfalls. When you are in the middle of your research, it is sometimes difficult to see the woods for the trees. In this chapter, I suggest a map for keeping in perspective your research project as a whole. The map is designed to give you an overview of the whole research process as though you were in a helicopter looking down on it so that you can keep it in view from beginning to end.

Research can also seem like a mysterious process: an arcane art practised by the cleverest people, employing obscure jargon and demanding an awesome depth of knowledge. This chapter offers an antidote to this intimidating view of research.

Let's start by considering a problem with which you will probably be familiar. Suppose that you are a young person wanting to get a flat; how would you go about it? Well, if you're like most people, you would probably start online with a search for some relevant maps and local property prices, review what is currently advertised online and then get the local paper and visit some local estate agents and letting agencies to get an overview of what is available, perhaps printing details of flats you think could be possible. You would compare the features (such as price, size, location and amenities) of the different flats and make a short list of ones which are most likely to meet your needs. You'd probably then personally look over the ones that you'd selected for your short list. Afterwards, when you'd made your decision and you're sitting in your new home, you'd probably reflect on the process: the extent to which your

Research Methods for Postgraduates, Third Edition. Edited by Tony Greenfield with Sue Greener.
© 2016 John Wiley & Sons, Ltd. Published 2016 by John Wiley & Sons, Ltd.

flat meets your original aspirations, what your first-hand experience has told you about the housing market, what you've learned from the experience and so on.

Now, if you take off your flat-hunter's spectacles and put on instead a pair of researcher's spectacles, you will observe some similarities between that process of flat hunting and the process of research. Firstly, you did a literature review (local newspapers, online property sites and estate agents' blurbs) to get an overview of the field. Secondly, you developed a theory of which of the available flats would be to your requirements (your short list). Thirdly, you tested the theory by inspecting those on your short list. Finally, you reflected on the experience and your results. Stated formally, the process contains four parts:

- Part 1: reviewing the field
- Part 2: theory building
- Part 3: theory testing
- Part 4: reflecting and integrating.

Perhaps this sequence seems familiar. Perhaps you recognise it from other significant decisions you have made in your life: choosing a college, buying a mobile phone or choosing a job.

With some decisions, it's not possible to go through all the stages. For example, when you choose a job, the final test of your theory that you have chosen the right job is by doing the job. Unfortunately, this is possible only after you've committed yourself to the job. Perhaps that's why so many unsatisfactory job decisions are made.[1]

Once you recognise that you are already familiar with each of the major parts of the research process through your experience of making the larger decisions of your life, you will have a valuable resource to draw on. Reflection on those experiences will also give you an indication of the possible pitfalls.

That four-part process can help you to put what you are doing into a broader picture when you start to get bogged down in the detail of research. It can also be useful in designing your research project.

Let us examine the parts of the process in more detail.

Part 1: Reviewing the Field

Many research projects arise from a study of current thinking in a field. The research project follows from identifying a gap in the literature. Most other research projects arise from awareness of a problem that is worth solving. In either case, a good start is an overview of current thinking in the field.

In case you are impatient with this part of the process and want to start immediately with fieldwork, here are some reasons for spending time and effort on a review of the field:

- to identify gaps in current knowledge;
- to avoid reinventing the wheel (at the very least this will save time, and it can stop you from making the same mistakes as others);

---

[1] The literature on labour turnover often refers to the period immediately following recruitment as the *induction crisis*, when job expectations are tested by the job realities.

- to carry on from where others have already reached (reviewing the field allows you to build on the platform of existing knowledge and ideas);
- to identify other people working in the same and related fields (they provide you with a researcher network, which is a valuable resource indeed);
- to increase your *breadth* of knowledge of the area in which your subject is located;
- to identify the seminal works in your area;
- to provide the intellectual *context* for your own work (this will enable you to position your project in terms of related work);
- to identify opposing views;
- to put your own work in perspective;
- to provide evidence that you can access the previous significant work in an area;
- to discover transferable information and ideas (information and insights that may be relevant to your own project); and
- to discover transferable *research* methods (research methods that could be relevant to your own project).

Part 2: Theory Building

In some ways, theory building is the most personal and creative part of the research process. Some people find it the most exciting and challenging part of the whole business.

In some cases, data collection precedes theory building and, in other cases, it follows it. Have you ever bought a used car? If so, you may have identified some possibles before narrowing down to a few probables. You collected data and then formed a theory about which of the cars would best meet your needs. In that situation, theory building followed data collection. The process of developing a theory by inspecting individual cases has a special name: *induction*.

Our flat-hunting example is another illustration of induction. If, each time you are sent the details of a flat in a certain area of town, you notice that it is more expensive than you can afford, you may form the theory that all the flats in that area are too expensive for you. Acting on that theory, you may ask the estate agents to stop sending details of flats in that area. That is the process of induction at work again: forming a theory from information about specific instances. Induction is a type of generalisation.

The other side of the coin from induction is *deduction* which involves reaching conclusions about specific instances from general principles. Here is an example of deduction: 'I can't afford to live in Mayfair so don't bother to send me the details of any flats in that part of town'. In this example 'I can't afford to live in Mayfair' is the generalisation, and deduction leads me to the conclusion about any specific flat in Mayfair that I can't afford it.

Induction is a thought process that takes you from the specific to the general. Deduction is a thought process that takes you from the general to the specific.

We have seen how a theory can emerge from the data. However, theory can also emerge from armchair theorising, introspection, deduction following a review of the literature, personal experience, a fortuitous remark, a brainstorm, an apt metaphor or pure inspiration. Creativity has a role to play in all aspects of the research process, but especially in the theory-building part.

I said earlier that data collection can precede theory building and that it can *follow* it. In the case of induction, data collection comes first. When data collection *follows* theory building, then it is usually for the purpose of testing the theory. That is the part of the research process that we turn to next.

Part 3: Theory Testing

> Experience has shown each one of us it is very easy to deceive ourselves, to believe something
> which later experience shows us is not so.
>
>   Carl Rogers (1955)

When flat hunting, we wanted to check whether those attractive-sounding apartments, reported by the estate agent, would really meet our needs. Likewise, when we are doing research, we will want to check if the theory (or theories) that we have formulated fulfil our hopes and expectations.

The sort of theory testing we do will depend on our ambitions and claims for our theory. If we want to claim that our theory applies *generally*,[2] then we may want to use statistical methods (known as *inferential statistics*) which have been developed to enable us to make claims about whole populations from information about a sample from a population.

If, however, your claims are only about the accuracy of your theory in the context of a particular situation,[3] then theory testing may involve checking your conclusions (theory) from other perspectives. You may have looked at estate agents' brochures, and now you want to look at the flats themselves, talk to the neighbours and so on. In research in the social sciences, the term *triangulation* is used to describe the process of checking if different data sources and different methods allow you to reach the same conclusions.

Testing theory can take many forms. At one extreme, you may simply invite the reader of a research report to test the conclusions against his own experiences. The test is: does the reader say, 'Aha! I can now make sense of my own experience in a new and convincing way'? But if the reader is unlikely to have first-hand experience for testing the researcher's theory, or if the claims being made involve a high level of generality, then the theory-testing stage will be more formal and elaborate. At some level, however, theory testing is likely to be part of any research process.

Part 4: Reflection and Integration

> Knowledge doesn't exist in a vacuum, and your knowledge only has value in relation to other
> people's.
>
>   A.D. Jankowitz (1991)

Reflection and integration comprise the last stage of the research journey. There may be many things on which you want to reflect: what you have learned about the process of research, what you could have done differently and what you have learned about yourself. But there is one matter for reflection that is a crucial part of the research process itself. It will affect how your research is judged and the impact of your research. You must reflect on *how your research findings relate to current thinking* in the field of your research topic.

Your reflection on how your research results relate to current thinking will include your assessment of where your research *fits into* the field of knowledge. It will contain your assessment of *your contribution* to the field. In this part of the research process, you are likely

---

[2] For example, 'All two bedroom flats in Mayfair are more expensive than all two bedroom flats in Leytonstone'.
[3] For example, 'The flat that suits me best among those whose details have been sent is Number 10 Railway Cuttings'.

to return to your review of current thinking that you made at the outset and reassess it in the light of your results. It's as if the current thinking in your field of study is a partially complete jigsaw puzzle and you are detecting where your own new piece of the jigsaw fits in.

Relating the outcomes of your research to current thinking in the field may simply involve showing how it *adds* to what is already known in the field. This would be the case when you have filled a gap in the literature or found a solution to a particular problem in the field. It may involve seeking connections with current thinking. It may involve challenging some parts of the map of the current thinking in the field, so that you will be proposing some reconstruction of that map. It may involve testing the consistency of your research findings with current thinking. It may involve asking 'What if?' questions of your research findings.

Any of these ways of relating your research findings to current thinking in the field may present further questions and new avenues to explore. Successful research usually answers some questions but also raises new ones. It enables researchers to ask questions that would not have been asked before the research. New questions can be an important outcome of research. It is small wonder, therefore, that the final chapter of most research reports has a section containing suggestions for further research.

A good practical question to ask yourself is 'What are the implications of my research results for our understanding in this area?' The implications can take many forms. Here are a few:

- You may have filled a gap in the literature.
- You may have produced a possible solution to an identified problem in the field.
- Your results may challenge accepted ideas in the field (some earlier statements in the literature may seem less plausible in the light of your findings).
- Some earlier statements in the literature may seem *more* plausible in the light of your findings.
- Your work may help to clarify and specify the precise areas in which existing ideas apply and where they do not apply (it may help you to identify domains of application of those ideas).
- Your results may suggest a synthesis of existing ideas.
- You may have provided a new perspective on existing ideas in the field.
- Your work may suggest new *methods* for researching your topic.
- Your results may suggest new ideas, perhaps some new lines of investigation in the field.
- You may have generated new *questions* in the field.
- There may be implications for further research.

Most of all, this last stage in the research process is about seeking to *integrate* the fruits of your own research with current thinking in the field.

Summary and conclusions

It is sometimes difficult to keep in mind the whole research journey when all of your attention is focussed on crossing some particularly difficult ground. Our purpose in this chapter is to help you to keep the whole research process in perspective when you are engaged in a particular research activity. We have done this by giving you an overview map on which the whole journey is plotted in outline. We hope this will help you to *plan* your research journey.

We have related the process of research to the way that you find information needed for the larger decisions in your life. You already have much experience to draw upon in planning and performing your research.

We have suggested a four-part research process: (1) reviewing the field, (2) building theory, (3) testing theory and (4) reflecting and integrating.

There is a considerable diversity of approaches to research in different fields, but this four-part framework is sufficiently broad to encompass most research in the sciences, the social sciences and the humanities. Much of the literature on research focusses on different parts of the process. For example, in the social sciences it usually focusses on theory building, whereas in other sciences it may focus on theory testing.

Your four parts may not follow this sequence strictly. For example, after you have reviewed the literature, you may want to monitor developments in current thinking *while* you are collecting and analysing data. You may engage in some parts of the research process more than once. For example, you may find that data you collect for theory building enables you to test statements found in the literature. Or data collected to test a theory may suggest a new theory so that it becomes an element of theory building. When you finally present your research, it will probably be presented as a structured linear process through the four activities discussed, but that certainly doesn't mean you have to do them in just this order. In particular, published literature has a habit of accelerating in your field just when you are getting on with data analysis, so you need to keep it under review until your conclusions are clear.

You may not want to spend the same amount of time and energy on each of the four parts of the process. For example, theory building may be only a token part of your research project if your main contribution lies in testing a theory that you found in the literature. On the other hand, you may direct most of your effort towards theory *building*, so that theory *testing* may be little more than establishing the plausibility of your theory in the light of the data you've collected.

The four parts will be present in almost all research projects, at least conceptually. If one of the four parts seems to be missing from your own research project, you should discuss it with other researchers and, if you are registered for a research degree, with your supervisor. If you intend to omit one of the parts from your own research project, you must be able to state clearly why it has no role.

References

Jankowitz, A.D. (1991). *Business Research for Students*. London: Chapman and Hall.
Rogers, C. (1955). Persons or science: a philosophical question. *American Psychologist* 10(7): 267–278.

3

Managing Your Doctorate

Stan Taylor

Introduction

While the knowledge and skills that you gained as an undergraduate and/or in studying for
a master's degree will have given you some experience of and insight into the process of
research, they may not necessarily have equipped you to undertake a doctorate. To do that,
you need to undertake a research project which makes an original contribution to knowledge
and understanding in your field.

In striving to make that contribution, you will of course have the support of your supervisory
team, academic colleagues in the field, and fellow doctoral candidates. But ultimately the
responsibility is yours. You may have to create the project; you will certainly have to do
the research and make an original contribution; you will have to present it in an appropriate
form for examination; you will have to defend it at a viva; and, if you do all of these things
successfully, you will be awarded the degree and gain the title of *Dr*.

Doing all of these things is difficult enough, but you also have to do them within what,
by historical standards, is a relatively short time period. Until a few years ago, the prevailing
view was that, because doctorates involved making an original contribution to the field, they
could not be constrained by time and took as long as they took (see e.g. Taylor, 2014). But
candidates are now under intense pressures from research sponsors, academic departments,
supervisors, families and in many cases bank managers to complete within three or four years
in the cases of full-time candidates (or six or seven years part-time).

The purpose or aim of this section of the book is to help you to think about how you are
going to go about managing your doctorate. With that in mind, the objectives are to consider:

- how to approach and start your doctorate;
- how to plan it;
- how to organise it;
- how to manage your research project;
- how to manage your relationship with your supervisory team;

Research Methods for Postgraduates, Third Edition. Edited by Tony Greenfield with Sue Greener.
© 2016 John Wiley & Sons, Ltd. Published 2016 by John Wiley & Sons, Ltd.

- how to manage yourself;
- how to complete your submission;
- how to prepare for your viva; and
- how to disseminate your findings.

Hopefully, thinking about these topics will help you to improve your chances of finishing on time and gaining the degree.

Approaching and Starting Your Doctorate

If you are thinking about doing a doctorate, it makes sense to firstly enquire about precisely what you are letting yourself in for. In other words, what is a doctorate?

The answer depends crucially upon two things; the sort of doctorate you are undertaking and how the standards are defined.

Doctorates can be divided into four main types: professional doctorates which usually involve completing a substantial taught component and producing a thesis or portfolio which contributes to practice in the relevant field (see e.g. Kott and Hendel, 2011); practice-based doctorates which involve the production of an artefact (e.g. sculptures, paintings or musical pieces) accompanied by an exegesis of the process and the contribution to the artistic form (see e.g. Paltridge *et al.*, 2011); doctorates by publication which include papers published in journals with a linking commentary (see e.g. Sharmini *et al.*, 2014); and the traditional PhD which involves submitting a thesis. The requirements for each are different, and it is important that you establish the appropriate ones for your degree.

The second factor is the definition of the standards for the degree. In many Western European countries, doctorates are defined in law and there are common standards. But in other countries, including Australia, the United Kingdom and the United States, there is not: it is up to individual institutions to define the standards (see e.g. Tinkler and Jackson, 2004; Pearce, 2005). So you need to consult your institution's definition of the standards for doctorates, which will usually be found in its regulations and may well be included in your postgraduate handbook.

While there will be variations, there should be the common elements that the research part of the doctorate:

- should be on a specified and approved topic;
- should constitute a substantial piece of research;
- should make an original contribution in the relevant domains of scholarship and/or practice; and
- should have been disseminated or should be worthy in principle of dissemination to the relevant research and, where appropriate, practice communities.

These common elements would be fine if there were standard definitions of *substantial*, *original contribution* or *worthy of dissemination*. In practice, they can be interpreted in different ways. So, to take the PhD for example, at one extreme a doctorate can be seen as a multi-volume exhaustive account of a subject, opening up new vistas in the discipline, and ready to go into proof as a major book; at the other, it may be seen as the application of an intensive research training to produce a narrow account of a detailed aspect of a phenomenon and leading to two or three papers in specialised journals.

If, then, the nature of the beast is contested, what can you do? By far, the best strategy is to look up a few successful doctoral submissions recently undertaken in a cognate area to that which you are considering researching, and ask yourself:

- Were they on 'big' topics or relatively narrow ones?
- Did they involve a massive amount of detailed research, or was this limited and narrowly focussed?
- Did they aim to open up broad new fields of knowledge or push the barriers out slightly in a tightly defined area?
- What made them, in whole or part, publishable?

Once you have done this, it can be very useful indeed to discuss examples with your supervisory team both to clarify matters in your own mind and to ensure that there is a shared understanding of what exactly is involved in your doctorate.

With that established, you can move on to your research topic. Many doctoral students in the sciences and engineering are recruited to undertake a specific piece of research which has been planned by their supervisors (i.e. they can step into a ready-made research project). But others, particularly in the arts, humanities and social sciences, start with a broad proposal which has to be refined into a detailed one (i.e. create a research project).

This can be a frustrating, difficult and demoralising experience, as your initial ideas and expectations often turn out to be too ambitious and you find yourself thrashing around in an intellectual vacuum and moving from the macro level of your subject through the meta and so to the micro level in the search for a topic (see e.g. Kiley, 2009).

You should, however, receive strong support at this stage from your supervisory team, who should act as a sounding board for your ideas and give you the benefit of their experience. With their assistance, you should eventually come up with a topic which is:

- *viable* – can be undertaken with available or specified obtainable resources, including the human one of a supervisory team with relevant experience and expertise in the area, as well as material ones such as library, computing facilities, labs and so on;
- *ethical* – meets relevant ethical guidelines set by institutions, disciplines and research sponsors;
- *doable* – can be done by someone with your knowledge and skills or with specified additions to these and within the time available;
- *sustainable* – will hold your interest and maintain your commitment over three or four years (or the part-time equivalent);
- *original* – has the potential to make an original contribution to knowledge and/or practice and/or culture;
- *publishable* – leads to publication or other forms of dissemination among the relevant research, practice or cultural communities; and
- *acceptable* – conforms otherwise to the requirements and standards likely to be applied by examiners in the discipline.

While it is never possible to guarantee the outcome of the research project – if the latter was known, the research would by definition not be original – it cannot be stressed enough that both you and your supervisory team must be reasonably satisfied on all of these counts before you proceed.

Planning Your Research Project

One of the key themes of the literature on managing research projects (Gattrell, 2006; Watts, 2008; Phillips and Pugh, 2010) is the need to plan. This is because planning:

Enables you to allocate time to your studies – Completing a doctorate full-time within three or four years is challenging; completing one part-time is even more so because candidates often have jobs and family responsibilities as well, so that doctoral study is likely to be their second or third priority (see McCulloch and Stokes, 2008; Gardner and Gopaul, 2012). Whether full- or part-time, it is vital that you try to allocate time over the course of your studies to work on the doctorate.

Enables you to allocate the right amount of time to each stage – As will be seen, all doctorates involve undertaking a number of tasks within the time that is available, and clearly it is important that you allocate an appropriate time length to each task. If you don't do this, then you can spend considerably more time than you have available on particular parts of the thesis and not enough on others, which can be to the detriment of completion and of the quality of the final product.

Helps to maintain your direction – One of the benefits of planning is that you know where you are, where you have been and where you are going. This gives you a more business-like approach to your work and can also, as Cryer (2006: 130) puts it, give you a sense of security.

Supplies evidence of progress – A plan supplies evidence of progress to others, for example to your supervisory team or to the panel which is considering your 'stepping up' from initial to full candidacy, usually towards the end of the first year of your studies.

Enables concerns to be externalised – A plan enables you to externalise concerns about progress by discussing them with your supervisory team and peers rather than internalising them and chasing them round endlessly in your own mind.

Provides a basis for reflection – Finally, a plan provides a basis for reflection on your progress and, if necessary, for rescheduling.

If, then, there are significant benefits to planning your doctorate, the next question concerns how to go about planning it. As in planning any project, this involves trying to answer the two questions of: what am I going to do (tasks)? And when am I going to do it (time)?

What Am I Going to Do?

While doctorates vary hugely in their structures and format and in how they are undertaken, most involve the 10 components of:

- reviewing and evaluating general and specific literary and other sources in the area of the topic;

- identifying the gap, problem, issue, conflict or other trigger for the research, and formulating appropriate investigations, potential solutions, questions, theories and alternative interpretations;
- deciding upon an appropriate paradigm, conceptual framework or methodology within which to conduct the research; operationalising concepts, theories and hypotheses; determining statistical procedures and significance levels before undertaking survey or experimental work; and so on;
- where appropriate, obtaining appropriate ethical clearances for the research and determining intellectual property rights (see Chapter 5);
- undertaking the substantive research, involving one or a combination of:
  - reading original documents,
  - subject interviewing,
  - elite interviewing,
  - mass interviewing,
  - scientific fieldwork,
  - scientific experimentation,
  - problem solving, and
  - computer modelling;
- sifting, checking and analysing material;
- interpreting findings and results, and assessing their implications for the initial focus of the inquiry;
- reformulating the field of knowledge in the light of your findings;
- summarising what your research has achieved, how it has been achieved, what conclusions have been reached and, if appropriate, directions for future research; and
- writing the drafts of your thesis.

These, then, are the core components – and there may be more, depending upon the topic – which will define your tasks over the next few years.

When Am I Going to Do It?

While it is difficult at the start of your research to assign precise time values to each of these components and some, of course, may overlap, you should, in conjunction with your supervisory team, try to plan a schedule for completing the stages of the research within a reasonable time limit.

So you may, for example, plan in outline to spend the first six months refining the topic, reviewing and evaluating the literature, dealing with methodological questions and obtaining ethical clearances; 18 months undertaking the substantive research; three months analysing, interpreting and reformulating; and three months gluing the whole together, writing your summary and conclusions and finally polishing the thesis for submission. Then, with these broad boundaries in place, you may be able to break up the components into sub-components and try to set targets for their completion.

It is absolutely vital that, in specifying time for each of the components of the research, you make a clear allocation for writing up that part of your work and linking it to what has gone before. This, as a number of studies (Kamler and Thompson, 2006; Badley, 2009; Wolff, 2010) have shown, is critical for five reasons:

- it gets you straight into the habit of writing rather than leaving this until the end, when it can be difficult to acquire;
- it encourages you to reflect upon what you have done before you go on and may highlight problems/avenues of exploration;
- it provides a continuing record of your achievements so far;
- it enables your supervisory team and university progress panels to see exactly what you have done and advise you how to proceed; and
- if you write throughout, you build up a portfolio of your work, which you can then fashion into the shape of the final thesis rather than face the beginning from scratch to assemble your submission.

Planning, then, can have considerable benefits. But it is important to note that plans should be frameworks and not cages. Given that a research project is ultimately concerned with making an original contribution to knowledge and/or practice, it obviously involves a creative process. As the research unfolds and new discoveries are made which lead off in different directions, plans have to be modified and changed, often in major ways. It is for this reason that, in many universities, the precise title of the thesis is left until just before final submission.

Organising Your Research

With an outline of a realistic plan in place, you now have to decide how you are going to try to meet it in an organised way. This involves organising your time, your working conditions and your materials.

Time

At the start of a research degree, three, four (or, in the case of part-time students, six) years can seem a very long time away, and you may think that there is little or no urgency in progressing your studies. But, as Kearns *et al.* (2008) have pointed out, it is all too easy to procrastinate and waste time, leading to delays in the research project.

In order to avoid this, you need to actively manage your time. This involves:

- setting realistic tasks for the week ahead;
- prioritising them;
- timetabling them;
- reviewing whether you have achieved them; and
- where appropriate, planning how to catch up.

Such active time management is vital if you are to complete your research on time.

Working Conditions

Research can be very demanding, and you need an appropriate working environment where you can read, reflect, think, evaluate and write. You need:

- office space, preferably within the department where you are based, with designated maximum numbers;
- necessary equipment, with a minimum of a desk, chair, lamp, bookcase and lockable filing cabinet;
- where appropriate, a carrel in the library or bench space in the laboratory;
- access to computing facilities, preferably ones dedicated to postgraduate use;
- private common space for you to study when necessary on your own; and
- access to the departmental staff common room and to the postgraduate common room for social purposes.

Preferably before signing up for a doctorate, you should ensure that these facilities are available to you. If there are deficiencies, you should bring these to the attention of your department as factors which are likely to delay the completion of your research.

Materials

In the course of your doctorate, you can expect to accumulate a vast amount of materials relating to your research, and it is important that you organise these properly. There is nothing worse than writing up your work, finding that you cannot trace the precise source of that key point or that your reference is inadequate and having to interrupt the flow while you try to find the original and sort it out. You should:

- decide with your supervisory team which referencing system you will use in your thesis and note the exact requirements for footnotes or endnotes and for your bibliography;
- assume anything that you read may eventually be cited and take full details of the reference. A master list of these should be stored in an electronic database, for example Endnote, which can then later be manipulated and sorted to form your bibliography;
- establish a filing system and file your materials under appropriate headings (this may be manual or computerised, but in the latter case copies must be backed up regularly); and
- index your filing system so that it is easy to find any document.

Organising your material in this way should mean the minimum of delay while you are completing your work, and hence the least interruption to your creative flow.

Managing Your Research

Many students starting off on their research projects assume that, once they have settled upon a topic, the rest will flow seamlessly until they complete their research. This is entirely understandable because so many of us presume that the way that research is undertaken is as it is set out in journal articles and in books. However, published work, as Martin (1992: 2) has put it in the science context:

> present[s] a mythical reconstruction of what actually happened. All of what are in retrospect mistaken ideas, badly designed experiments, and incorrect calculations are omitted. The paper presents the research as if it is carefully thought out, planned and executed according to a neat rigorous process.

So what is finally published is usually only the tip of the iceberg; the nine-tenths of sometimes barren blood, sweat, toil and tears is hidden. Except in very rare cases, research is inherently a messy business, and many things can and do go wrong. Challenges can include:

- **Maintaining the focus of the research** – As the research progresses, highways and byways of new exploration open up which just have to be investigated because they could be vital to the outcome. So every avenue is investigated until you become lost in the maze of possibilities and unable to find your way back to where you should be at that stage of the project.
- **Coming to grips with theoretical and methodological challenges** – Research can require you to grapple with a whole range of new and difficult philosophical, theoretical, empirical and experimental challenges. As Kiley (2015) has shown, it can require a major effort to try to identify, tackle and resolve these, particularly when you are really itching to undertake the substantive research.
- **Frustration with the outcomes** – You can expect a range of challenges as you undertake the substantive research. As Newsome (2008: 17)[1] has put it for the case of chemistry:

> Scientific research often entails repeating the same experiment many times. This can be a frustrating and stressful process. The desirable results could take weeks or months to produce as the success rate is often low, but the researcher is necessarily working to a deadline.
>
> Serendipity as well as skill and experience reduce the time it takes. But even with both on his or her side, the researcher cannot predict how long it will take to get a desirable result.
>
> This makes planning difficult. Relying on equipment and machinery that may not always be in service exacerbates these issues….
>
> [Students] expressed that they had not anticipated the reality at the start of their PhD (most PhD students will not have undertaken primary research before).

- **Inconsistencies in results** – With the substantive research accomplished, you experience difficulties in analysing and interpreting it – evidence which is contradictory, experiments which yield unexpected results, cast-iron assumptions which are apparently falsified, equations which should be stable but turn out to be unstable, simulation results which defy predictions, variables which behave badly and so on.

---

[1] From *The Chemistry PhD: the impact on women's retention*, reproduced by permission of the Royal Society of Chemistry and the UK Resource Centre for Women in SET.

But, if you can't avoid one or more of these challenges, part of becoming an independent researcher is learning how to manage them. You can try to do this by:

- **Acknowledging the existence of a problem** – As an undergraduate or master's student, you may have sailed through with effortless brilliance, and it can be an immense shock to encounter challenges of the kinds outlined here; acknowledging them can be seen as weakness or failure (see Lovitts, 2005). So you may not feel like telling anybody that you are experiencing problems, perhaps least of all your supervisory team. But such problems are experienced throughout virtually the entire research community, and admission should not be conceived of as a weakness, but as a strength. So, if you feel that you are falling behind, acknowledge that there is a problem, reflect upon the reasons why and try to identify a solution.
- **Seeking help** – If you can't quite define the problem or can't find a solution, seek help. You might initially explain the problem to a fellow student, preferably one more advanced in his or her studies, or if appropriate a post-doc – often they have experienced similar problems and will be able to help you out. If you are still blocked, go to see your supervisors, tell them that you are stuck, tell them why and ask them for suggestions. They may not have answers – after all, you are the emerging expert in your field – but they should support you to find a way forward.

Managing Your Relationship with Your Supervisory Team

Normally, you will be supervised by a team consisting of a principal or main supervisor or director of studies and at least one second supervisor. They will be your guides and mentors over the course of your doctorate, and it is important that you establish and maintain a good relationship with them.

One thing that you need to do right at the start is to sort out what you can expect from your supervisory team and what they can expect from you. Studies (see e.g. Meyer *et al.*, 2007; Engebretson *et al.*, 2008; Stubb *et al.*, 2014) suggest that a mismatch of expectations is one of the principal causes of frustration on the part of both candidates and supervisors, and it can lead to delays in progressing the project. Your supervisory team may well ask you to complete one of the standard questionnaires on expectations (see e.g. Hair, 2006; Lee, 2011) and discuss it with them, and/or take you through the relevant section of your institution's postgraduate handbook concerned with the responsibilities of the parties and/or negotiate a learning agreement. The important thing is that you are clear about what you can expect from each other in the relationship.

But that relationship is not static, and it should change over time. At the beginning of the project, you may well need high levels of both academic and personal support from your supervisory team to help you to get started on the research project and to enable you to undertake it. As the project proceeds, however, you should take hold of the reins more and more as you develop into an independent researcher. Towards the end, you may need high levels of support again as you create drafts of your final submission and rely heavily upon your supervisory team for feedback on whether it is acceptable and will meet the standard for the award.

However, these adjustments may not happen automatically, with supervisors sometimes being unwilling to cut candidates the slack to do things on their own or candidates sometimes being reluctant to embrace opportunities for independent work. Neither of these states of affairs is healthy – in the former the candidate can feel stifled, and in the latter the supervisory team can despair of the candidate ever flying the nest – and it is important that issues are addressed. For that reason, it can be helpful for you and your supervisory team to have regular discussions not just about the progress of your research, but also about the progress of the relationship, perhaps using the tool developed by Gurr (2001) for that purpose.

As surveys (see e.g. Bennett and Turner, 2013) have consistently shown, the vast majority of research students are very satisfied with the overall quality of supervision they receive, and the chances are statistically very high that you will have a good experience. But, in some cases, problems may arise including lack of interest in your research project, lack of time to give you effective supervision and disagreement within the supervisory team.

Research students are a source of prestige and income to academic departments, and staff can be put under considerable pressure to take on students even if the topic of their proposed research is marginal to their interests and expertise. A supervisor without adequate knowledge of the topic can delay completion (see e.g. Jacobsonn, 2009; Wadesango and Machingambi, 2011) or, in extremis, even cost the candidate their degree. If you feel that your supervisory team does not have the expertise to offer you effective supervision, you should ask that alternative arrangements should be put in place – it is the responsibility of the institution which accepted you to ensure that your supervisory team is appropriately qualified in the field.

Academic staff are under severe pressure to perform in research, teaching, service and administration, and time is at a premium (see Hammond *et al.*, 2010). That said, many institutions now have supervisor workload policies which allow time for supervision, and there is no excuse for the occasional horror stories about students having meetings with supervisors in railway stations or airports or receiving feedback on their draft chapters months after they have been handed in. If a team agrees to supervise you, then they take on a commitment and should be prepared to fulfil it. So, if you are having problems, try to discuss them with your supervisory team. If this does not work, your institution will have procedures for raising issues or complaints with Directors of Postgraduate Studies, heads of departments, heads of graduate schools, or independent counsellors or ombudspersons (see Pearson, 2012), which should be set out in the graduate student handbook. In extreme cases, a new supervisory team may be the only way forward.

While supervisory teams can have considerable advantages over single supervisors in terms of expertise, access and maintaining continuity (see e.g. Manathunga, 2012), they can also introduce an added complication in terms of potential disagreement between supervisors. Supervisors can be at odds in terms of intellectual traditions (particularly if they are from different disciplines or different strands within a discipline; see Guerin and Green, 2013), conceptions of the research project (particularly if drawn from academic and industrial backgrounds, respectively; see Malfoy, 2011) and supervisory styles (particularly if both are determined to direct the research project themselves; see Bui, 2014). As Gunnarsson *et al.* (2013) have pointed out, such disagreements can leave students trying to satisfy everybody but actually satisfying no one and delaying the research.

So you need to be alert to signs of explicit or implicit disagreement between your supervisors and, as far as possible, ensure that they are discussed and resolved as early as possible in the research project. If differences persist, and you continue to find yourself the 'piggy in the middle', then it may be worth taking this up with the appropriate person in your department or institution.

Managing Yourself

In addition to managing your thesis and your relationship with your supervisor over the course of your studies, you also have to manage yourself (i.e. cope with the slings and arrows of life as a researcher). These can include initial adjustment, isolation and loneliness, and mid-thesis crisis.

Initial Adjustment

In the taught programmes you took as an undergraduate or taught postgraduate, knowledge was pedagogically processed, packaged and controlled so that it 'works'. But, as Delamont *et al.* (2000) have pointed out, there is a world of difference between following intellectual paths which have been well trodden by others and blazing a new trail across the landscape. Things can and do go wrong, it can be a frightening experience and it is not unknown for postgraduates in their first few months to feel as if they have been left up the creek without a paddle.

While a few years ago, this period was regarded as one you had to survive to progress, institutions now offer help in the form of comprehensive induction programmes which are designed to support students to adjust. Also supervisors are now placing greater emphasis upon the need to support candidates during this critical period, and some institutions have established 'buddying' schemes whereby candidates further on in their research support new ones. So, if you do feel lost and adrift, you are by no means alone, and you can seek and expect support to help you to adjust to postgraduate life.

Isolation and Loneliness

One of the most consistent findings of the literature on research students over the past three decades or more (see e.g. Rudd, 1985; Becher *et al.*, 1994; Gardener, 2009; Janta *et al.*, 2012; McAlpine, 2013) is the extent to which they can suffer from isolation, both intellectually and/or socially. An element of isolation may be inherent in studying for a research degree; you are required to make an independent and original contribution to knowledge and understanding, which implies that the work must be undertaken on your own. That said, there is a difference between working on your own but with support from your supervisors and possibly other members of the research team and fellow graduate students and working in an almost complete intellectual and social vacuum.

Perhaps the candidates most vulnerable to the extreme of isolation are the 'lone scholars' in the arts, humanities and social sciences ploughing their own furrows often at a remove from the interests of their supervisors or fellow doctoral candidates. But even students working in research groups in the laboratory sciences can feel isolated (see UK Resource Centre for SET and Royal Society of Chemistry and Biochemical Society, 2009), while part-time students in all disciplines may also suffer because of their geographical remoteness from the university and fellow students (see Gattrell, 2006; McCulloch and Stokes, 2008; Gardner and Gopaul, 2012). Such isolation can lead to or reinforce mental health issues which can beset doctoral students (Hopwood *et al.*, 2011; Else, 2015).

Here Cryer's (2006) advice is pertinent, that you actively need to take steps to ward off isolation. Examples of ways to do this include:

- using social media to keep in active touch with supervisors, colleagues and your department;
- participating in mentoring schemes whereby a graduate or a more advanced doctoral student is teamed with a less advanced one;
- joining journal clubs or writing groups with fellow students;
- playing a part in departmental and/or graduate school academic and social activities; and
- participating in the postgraduate sections of subject associations.

Mid-thesis Crisis

You commence your project all bright-eyed and bushy-tailed at the apparently glamourous prospect of undertaking original research. But, perhaps a year to 18 months into the research project, the glamour can begin to wear off as you begin to lose your sense of direction, feel overwhelmed by what little you seem to have done and by how much is left to be done, suffer loss of confidence or are bored out of your mind by the endless grind. In consequence, as a number of commentators (see e.g. Delamont *et al.*, 2004; Cryer, 2006; Phillips and Pugh, 2010; Roed, 2012; McAlpine, 2013) have noted, candidates may undergo a mid-thesis crisis and suffer a consequent loss of motivation.

Mid-thesis crisis is well illustrated by one of Rudd's (1985: 44) respondents, who said:

> One starts … to question why you're there, because the bandwagon effect of undergraduate work disappears about a year after you've got into the PhD, because you are no longer involved in that corporate body, you don't have to attend lectures, you don't have official things you must do, you don't have to attend a specific tutorial. In other words, there are no longer any clear stepping-stones. The only stepping stone is to submit your thesis, and you ask "Why?" I think people are disillusioned after about eighteen months. I think that's a normal phenomenon you know – 'I'm pissed off with the whole thing. I'm not going to do it.'

While there is no simple solution to this problem – if you want to complete the research, you have to continue it – it might be an idea to try to do something else relating to your thesis (write or rewrite an earlier chapter) or even take a short break and then come back to it with a fresher mind. But, if you contemplate a break, do stick to a defined time – there are many ex–research students who took a breather from their studies and then procrastinated about returning until it was far too late.

Completing Your Submission

After spending two and often nearer three years (or *pro rata* part-time) of your life doing the research project, you are then faced with what is the last major task of producing the final submission. This task is absolutely crucial because it is primarily upon the basis of your submission that you will be examined at the end of the day. So it is important to consider what your case is, what you are going to write, how much you are going to write, the style in which you will write it and how you will go about the writing process.

What Is My Case or Argument?

If you have written up as you have gone along with your research, you will have a record of what you have done and achieved. You have not, however, necessarily met the requirements for a doctorate; a submission needs to be far more than a record of what you have done but entails presenting a case supported by appropriate argument and evidence based upon your research.

This can be difficult to articulate, and if it is, then it is well worth spending some time with your supervisory team or research colleagues discussing how you might best translate what you have written into the form of a case or argument.

What Am I Going to Write?

Once you have a clear idea of your case, you then need a framework within which you can write it. You may have already derived such a framework in the course of your research but, if the focus has changed substantially, you may need to create one. This, as Kiley (2009, 2015) and Trafford and Lesham (2009) have shown, can prove a real sticking point and cause significant delays in completing the thesis.

One strategy which might be helpful is to think of yourself as an explorer who has followed a journey across unknown territory and who is producing a guidebook for others to follow (see e.g. Taylor and Beasley, 2005; Neville, 2008). As the author of the guidebook, you need to explain:

- your starting point and why you decided to embark upon the journey (the literature and the aims and objectives of the research);
- how you decided to undertake the journey (the methodology);
- the route you followed and the discoveries you made on the way (the substantive research chapters);
- how, in the light of the above, you redrew the route (analysis and interpretation); and
- where you arrived at the end of your journey, how it differed from your starting point and where you would go from here (conclusions, original contribution, and directions for future research in the subject).

You could start by – literally – drawing an outline map of the above and ensuring that the various stages link together and are reasonably consistent with each other so that, in general, the route is clear and can be easily followed. Then, within each part of the route (i.e. each chapter), you need to decide what needs to be said to take your reader through that stage and lead him or her onto the next one, bearing in mind that the deviations up highways and byways which were so fascinating to you might be irrelevant to others (i.e. concentrating upon the essentials and leaving extraneous materials to footnotes, endnotes or appendices).

If you do this, you should have a master-map of the route as a whole and detailed guides to each of the sections of it (i.e. a template for your submission). You should now try this out on your supervisory team and ask for comments before proceeding; there is nothing more dispiriting than completing a draft and then finding that there are fatal flaws in the structure which necessitate a major re-write.

How Much Am I Going to Write?

You need to decide not only what you are going to write about, but also how much you will write in total and for each part of the submission.

With regard to the total, most institutions have word limits on submissions, and it is important that you find out what these are and what they include (e.g. whether they relate just to the text or include the bibliography and any appendices as well).

Perhaps more importantly at this stage, you need to allocate at least rough targets for each part of the submission. So, for example, you should not normally aim for half to be taken up by the literature review, a further quarter by the methodology and only a quarter for the original scholarship; you need to arrive at a relevant weighting for the components. You need to determine an appropriate balance with your supervisory team and arrive at a set of rough allocations under each of the headings of your submission.

What Style Am I Going to Write It In?

A doctoral submission is, of course, a work of scholarship, and it will have to be written in the style known as *academic writing* (see e.g. Murray and Moore, 2006). While there is no objective definition, academic writing can be said to be characterised by:

- **Explicitness of intention** – At the start of the chapter or section, you need to make explicit what you are trying to do.
- **Clarity and coherence of argument and analysis** – Arguments need to be made and analyses presented in ways which are clear, can be followed easily and can be linked to other relevant parts of the thesis.
- **Respect for the conventions of writing in the subject or discipline** – You need to respect whatever conventions pertain in your subject or discipline. For example, it may be permissible in some to argue in the first person or the second person, but this would be frowned upon in others.
- **Substantiation of points** – Points, arguments, contentions and so on should be substantiated by appropriate references to the literature and/or other parts of the thesis. In the case of a research thesis, it is preferable to err on the side of over- rather than under-referencing.
- **Linkages** – There should be clear linkages between what you have said, what has gone before and what will come after in the thesis.

It can be very useful to ask your supervisory team to identify a particular example which is a 'model' of academic writing in your discipline and to discuss it with them before starting your own writing.

How Will I Go about the Writing Process?

One of the features of the research degree experience is that it almost always takes longer to write the submission than was allowed for in the planning. The reason for this is that, as Kamler and Thompson (2006) have argued, you now have the benefit of hindsight over the research project as a whole, and the writing process should yield new insights and connections. But, while incorporating the benefits of hindsight is an essential part of the research, it can mean that you end up writing against the clock, and to do this you need a high degree of discipline and hence self-management

By definition, self-management is an individual thing (i.e. it depends on your preferred writing routines). Some work best with set times each day for writing and need no interruptions, some work best intermittently and need to have background music or the like and some work best with targets (e.g. 1000 words a day no matter how long it takes). It is important that, whatever it is, you identify your preferred writing routine and, as far as possible, stick to it. In doing so, it may be helpful to remember what Delamont *et al.* (2004) have identified as the golden rules of writing. These are:

- the more you write, the easier it gets;
- if you write every day, it becomes a habit;
- tiny bits of writing add up to a lot of writing; and
- the longer you don't write, the more difficult it is to get back in the habit.

Drafting and Submission

Once you have a first draft, you should show it to your supervisory team and ask whether it is likely to make the grade. If not, your supervisors should indicate the additional work required; if it looks likely to pass, you then begin the process of polishing what will initially be a rough diamond into the final gem.

When, usually after several iterations, you have produced the final draft, it is back to your supervisory team for a final review and, with luck, the green light to go ahead and submit the work for examination. In some institutions, it is a requirement that your supervisory team approves your thesis for submission, while in others it is your decision alone. In the latter case, if your supervisory team have reservations, you should pay heed and do any additional work to satisfy them before submitting.

Additionally, before deciding to submit, you should check your institution's regulations. Usually, you will have to conform to a range of stylistic conventions for the presentation of the thesis (e.g. concerning page margins and layout), as well as provide an abstract and a contents page, and meet any other requirements, such as a word count.

You also need to look at the presentation, for, as Carter (2008: 368) has put it:

> If the work is poorly presented, and some theories or material well-known to the examiner badly represented, then there is more likelihood that the unknown things will be investigated further.... The candidate's work of revision, proofing and editing, checking and rechecking is thus worth doing.

Only when you are satisfied on this score should you submit the work electronically and/or print off the usually multiple copies to be submitted to your institution.

Preparing for Your Viva

Once you have formally intimated that you will submit, the wheels are set in motion for the process of examination, beginning with the appointment of examiners.

Usually, the head of department will ask for nominations from your supervisory team. They should discuss the matter with you and come up with suggestions. If you have reservations, for example that Dr X in your department has consistently been sceptical of the value of your research or that Professor Y might take unkindly to the fact that your thesis has refuted his or her life's work, you can and should raise them. There is, of course, no guarantee that your preferences will be taken into account, and it may be that you end up with Dr X and/or Professor Y.

Once examiners are appointed, there is usually a hiatus of a few weeks while they read, digest and form an opinion about your work, following which the statutory oral examination, the viva, is arranged.

In the United States and right across Europe, vivas are public events and you can attend one prior to your own. But in the United Kingdom, they are normally held in private. In the absence of precedents, as Delamont *et al.* (2004) have pointed out, even the most able students can feel anxious and stressed by the prospect of the viva.

But the stress can be reduced by finding out about the process of the viva, planning for it and practising.

Finding Out about the Process of the Viva

Until a few years ago, the only way to find out about the process of the viva was to ask your supervisory team or post-docs in your research group about their own experiences as both research students and examiners. But, in recent years, many institutions have produced handbooks for the examiners of research degrees which set out processes for the viva and can be consulted by candidates, and some have provided specific training sessions (see Wellington, 2009). In addition, there is a literature on the examination of research degrees (see Tinkler and Jackson, 2004; Pearce, 2006; Trafford and Lesham, 2008) and how to survive the viva (Murray, 2009), and institutions may also have video resources (Angel Productions, 2009). You should try to find out as much as you can from these sources so that you have a clear idea of the process and what it entails.

Planning for the Viva

While, because every thesis is unique, every viva is unique, it is still possible to undertake some planning in advance for your viva. In particular, you should:

> **Re-read your submission** – Since you have written the document, there is an obvious temptation to think that you know it backwards and don't need to re-read it in the hiatus between sending it in and the viva. But, even within a short space of time it is possible to forget about some of the detail of what you have written and, if this translates to the viva, it can give examiners the impression that you do not know your own work.

> **Keep Up with the Literature** – In the interim between submission and examination, it is perfectly possible for new work to come out which has a direct or indirect bearing upon your topic. In order to demonstrate to examiners the currency of your knowledge and your command of the subject, you should check all likely sources and think about the implications of your work for any new papers and vice versa.

> **Prepare for Questions** – You can try to anticipate and prepare for the sorts of questions you are likely to be asked. Again, there is no standard format, but it

might be helpful to think about the kinds of things that you might be asked. Helpful advice about how to go about this can be found in Trafford and Lesham (2008), but it is worth remembering the point made by Murray (2009) that you cannot prepare for every eventuality and you have be prepared to think on your feet.

Practising

Once you have planned and prepared, it can be very useful to ask your supervisory team to arrange a mock viva in which departmental colleagues ask questions on a key part or parts of your submission and then give you some feedback upon your performance. This can be invaluable in anticipating lines of inquiry and in improving your presentation skills (see Hartley and Fox, 2004).

On the Day

On the day itself, you may well feel nervous, but remember that – within limits – this is a physiological sharpening of the senses that will help you respond more effectively.

You should go to the viva as well-rested as possible, wearing smart but comfortable clothes – remember, you will be sitting for a couple of hours and possibly longer.

You should take with you:

- a copy of your thesis (if appropriate, annotated with Post-Its to highlight 'killer' facts or quotes so that you can find them quickly);
- a pen and paper if you need to jot down questions or possibly draw diagrams; and
- a list of corrections.

Following Cryer (2006: 251–252), you should:

- wait for questions to be asked of you by the examiners;
- ask for clarification if questions are unclear;
- take whatever time you need to answer them;
- defend your thesis without becoming wholly defensive – remember that, as Mullins and Kiley (2002) have put it, 'It's a PhD, not a Nobel Prize', and be prepared if necessary to concede points;

At the end of the viva, the chair should ask you to leave while the examiners deliberate, and afterwards you will normally be called back in to hear their verdict.

Outcomes

The examiners' verdict takes the form of a recommendation to the relevant body in your institution. Recommendations vary between institutions (see Pearce 2004) but can include:

- the immediate award of the degree;
- the award of the degree subject to minor corrections;
- revision of the submission, and re-submission and re-examination within a specified period;

- the award of a lower research degree; and
- that no award be made.

The chances are low that the recommendation will be the immediate award of the degree. Because of the minor errors that creep into even the best presented theses, in most cases it will be to award the degree subject to making minor corrections.

In some cases, it will be referral for re-writing, re-submission and re-examination. While this is disappointing, it is important to remember that examiners should only recommend this if they consider that the work submitted has the potential for the award, and that they are required to specify in detail what you need to do to make the grade. So it is effectively a re-sit, a further opportunity to gain the award.

Other outcomes (i.e. the award of a lower degree or no award at all) are mercifully rare. But, if they do occur to you, and you have reason to believe that this relates to the prejudices of examiners or to irregularities in the examination procedure, you may wish to consider appealing. Where appeals are successful, then usually the submission is re-examined by another examiner external to the institution whose decision is final.

In the majority of cases, however, you only need to do one thing after the viva – celebrate.

Disseminating Your Findings

If you are thinking about an academic career, you should be aware that competition is fierce and, as well as completing your doctorate, you will need to have publications in high-impact journals or prestigious conference proceedings to stand a chance of being short-listed for a faculty position (see Thein and Beach, 2010; Jepsen *et al.*, 2012). But even if you are not thinking about an academic career (see Chapter 41), you should recognise that, unless there are pressing reasons for withholding results (e.g. pending patent applications), you have an obligation to disseminate your findings; there is no point in making discoveries if no one knows about them.

However, as Kamler (2008) has shown, many early career researchers (particularly in the arts, humanities and social sciences) deposit their theses or dissertations in libraries (physical or electronic) and they never again see the light of day. Studies of this surprising reluctance to disseminate have been investigated in a number of studies (see e.g. Cuthbert and Spark, 2008; Kamler, 2008; Belcher, 2009) which have concluded that the root cause is a lack of knowledge and understanding of how to go about it.

So one thing that you need to do is to learn how to go about disseminating your research. You can ask your supervisory team for help in terms of identifying suitable journals and/or conferences and tailoring your work towards their requirements. Where appropriate, your supervisor(s) may suggest that you write a paper jointly, which offers an opportunity to find out the 'dos and don'ts'. Other useful activities might include joining 'writing for publication' groups where you present your research to peers acting as referees, submitting papers to online departmental journals, attending workshops on 'how to publish' or 'writing press releases', or reading guides to publication (see e.g. Wilkinson, 2015). In addition, social media, including Twitter, Facebook, Linked-in and Research Gate, are becoming increasingly important for disseminating research and raising profiles (see e.g. Sheldon and Doran, 2015), and you should make full use of them as well.

Conclusions

Doctorates are awarded for making an original contribution to knowledge and understanding and/or practice. To do this within the relatively short space of time available, you have to actively manage your doctorate. While effective management will not turn an inadequate research project into a successful submission, its absence can mean that completion is delayed or even that a promising topic comes to nothing because the project is abandoned.

So, in undertaking your doctorate, remember that while you are being examined explicitly on your contribution, you are also being examined implicitly upon your ability to manage the process of producing it, and you need both to ultimately gain the award.

References

Badley, G. (2009). Academic writing as shaping and re-shaping. *Teaching in Higher Education* 14(2): 209–229.

Becher, T., Henkel, M., and Kogan, M. (1994). *Graduate Education in Britain*. London: Jessica Kingsley.

Belcher, W.L. (2009). Reflections on ten years of teaching writing for publication to graduate students and junior faculty. *Journal of Scholarly Publishing* 40(2): 184–199.

Bennett, P., and Turner, G. (2013). PRES 2013: results from the Research Students' Post-Experience Survey. Higher Education Academy. https://www.heacademy.ac.uk/sites/default/files/resources/PRES_2013_UK_report.pdf

Bui, T. (2014). Student-supervisor expectations in the doctoral supervision process for business and management students. *Business and Management Education in Higher Education* 1: 12–27.

Carter, S. (2008). Examining the doctoral thesis: a discussion. *Innovations in Education and Teaching International* 45(4): 365–374.

Cryer, P. (2006). *The Research Student's Guide to Success*, 3rd ed. Buckingham: Open University Press.

Cuthbert, D., and Spark, C. (2008). Getting a GriP: examining the outcomes of a pilot programme to support graduate research students in writing for publication. *Studies in Higher Education* 33(1): 77–86.

Delamont, S., Atkinson, P., and Parry, O. (2000). *The Doctoral Experience: Success and Failure in Graduate School*. London: Falmer.

Delamont, S., Atkinson, P., and Parry, O. (2004). *Supervising the PhD: A Guide to Success*, 2nd ed. Buckingham: Open University Press and Society for Research into Higher Education.

Else, H. (2015). Four in 10 say PhD affects mental and physical health. *Times Higher Education*, 9 April, p. 8.

Engebretson, K., Smith, K., McLaughlin, D., Seibold, C., and Ryan, E. (2008). The changing reality of research education in Australia and implications for supervision: a review of the literature. *Teaching in Higher Education* 13(1): 1–15.

Gardner, S.K. (2009). *The Development of Doctoral Students: Phases of Challenge and Support*. San Francisco: Jossey-Bass.

Gardner, S.K., and Gopaul, B. (2012). The part-time doctoral student experience. *International Journal of Doctoral Studies* 7: 63–78.

Gattrell, C. (2006). *Managing Part-Time Study*. Maidenhead: Open University Press.

Guerin, C., and Green, I. (2013). 'They're the bosses': feedback in team supervision. *Journal of Further and Higher Education*. doi:10.1080/0309877X.2013.83109

Gunnarsson, R, Jonasson, G. and Billhult, A. (2013). The experience of disagreement between students and supervisors in PhD education: a qualitative study. *BMC Medical Education* 13: 134.

Gurr, G. (2001). Negotiating the 'rackety bridge' – a dynamic model for aligning supervisory style with research student development. *Higher Education Research and Development* 20(1): 81–92.

Hair, M. (2006). Superqual: A tool to explore the initial expectations of PhD students and supervisors. *Active Learning in Higher Education* 7(1): 9–23.

Hammond, J., Ryland, K., Tennant, M., and Boud, D. (2010). *Building Research Supervision and Training across Australian Universities*. Australian Learning and Teaching Council. http://www.academia.edu/2314934/Building_Research_Supervision_and_Training_across_Australian_Universities

Hartley, J., and Fox, C. (2004). Assessing the mock viva; the experiences of British doctoral students. *Studies in Higher Education* 29(6): 727–738.

Hopwood, N., Alexander, P., Harris-Huemmert, S., McAlpine, L., and Wagstaff, S. (2011). The hidden realities of life as a doctoral student. In V. Kumar and A. Lee (eds.), *Doctoral Education in International Context: Connecting Local, Regional and Global Perspectives*. Serdang, Malaysia: Universiti Putra Malaysia Press.

Janta, H., Lugosi, P., and Brown, L. (2014). Coping with loneliness: a netographic study of doctoral students. *Journal of Further and Higher Education* 38(4): 553–571.

Jepsen, D., Verhagyi, M., and Edwards, D. (2012). Academics attitudes towards PhD students' teaching: preparing research higher degree students for an academic career. *Journal of Higher Education Policy and Management* 34(6): 629–645.

Kamler, B. (2008). Rethinking doctoral publication practices: writing from and beyond the thesis. *Studies in Higher Education* 33(3): 283–294.

Kamler, B., and Thompson, P. (2006). *Helping Doctoral Students Write: Pedagogies for Supervision*. London: Routledge.

Kearns, H., Gardiner, M., and Marshall, K. (2008). Innovation in PhD completion: the hardy shall succeed and be happy. *Higher Education Research and Development* 27(1): 77–89.

Kiley, M. (2009). Identifying threshold concepts and proposing strategies to support doctoral candidates. *Innovations in Education and Teaching International* 46(3): 293–304.

Kiley, M. (2015). 'I didn't have a clue what they were talking about': PhD candidates and theory. *Innovations in Education and Teaching International* 52(1): 52–63.

Kot, F., and Hendel, D. (2011). Emergence and growth of professional doctorates in the United States, United Kingdom, Canada and Australia: a comparative analysis. *Studies in Higher Education* 37(3): 345–364.

Lovitts, B.E. (2005). Being a good course taker is not enough: a theoretical perspective on the transition to independent research. *Studies in Higher Education* 30(2): 137–154.

Malfoy, J. (2011). The impact of university-industry research on doctoral programmes and practices. *Studies in Higher Education* 36(5): 571–584.

Manathunga, C. (2012). Supervisors watching supervisors: the deconstructive possibilities and tensions of team supervision. *Australian Universities Review* 54(1): 29–37.

Martin, B. (1992). Scientific fraud and the power structure of science. *Prometheus* 10(1): 83–98.

McAlpine, L. (2013). Doctoral supervision: not an individual but a collective institutional responsibility. *Journal for the Study of Education and Development* 36(3): 259–280.

McCulloch, A., and Stokes, P. (2008). *The Silent Majority: Meeting the Needs of Part-time Research Students*. London: Society for Research into Higher Education.

Meyer, J., Shanahan, M., and Laugksch, R. (2007). Students' conceptions of research: an exploration of contrasting patterns of variation. *Scandinavian Journal of Educational Research* 51(4): 415–433.

Mullins, G., and Kiley, M. (2002). It's a PhD, not a Nobel Prize: how experienced examiners assess research degrees. *Studies in Higher Education* 27(4): 369–386.

Murray, R. (2009). *How to Survive Your Viva*, 2nd ed. Maidenhead: Open University Press.

Murray, R. (2011). *How to Write a Thesis*, 3rd ed. Buckingham: Open University Press.

Murray, R., and Moore, S. (2006). *The Handbook of Academic Writing*. Maidenhead: Open University Press.

Neville, B. (2008). Creating a research community. *Qualitative Research Journal* 8(1): 37–46.

Newsome, J. (2008). *The Chemistry PhD: The Impact on Women's Retention*. London: Resource Centre for SET and Royal Society of Chemistry.

Paltridge, B, Starfield, S., Ravelli, L., and Nicholson, S. (2011). Doctoral writing II: the visual and performing arts: two ends of a continuum. *Studies in Higher Education* 37(8): 989–1003.

Pearce, L. (2004). *How to Examine a Thesis*. Buckingham: Open University Press and Society for Research into Higher Education.

Pearson, M. (2012). Building bridges: higher degree student retention and counselling support. *Journal of Higher Education Policy and Management* 34(2): 187–199.

Phillips, E., and Pugh, D. (2010). *How to Get a PhD*, 5th ed. Buckingham: Open University Press.

Roed, J. (2012). *Labour of love: Emotions and Identities in Doctoral Supervision*. EdD thesis, University of Sussex. http://sro.sussex.ac.uk/42949/

Rudd, E. (1975). *The Highest Education: A Study of Graduate Education in Britain*. London: Routledge and Keegan Paul.

Rudd, E. (1985) *A New Look at Postgraduate Failure*. London: Society for Research into Higher Education and National Foundation for Educational Research.

Sharmini, S., Spronken-Smith, R., Golding, C., and Harland, T. (2014). Assessing the doctoral thesis when it includes published work. *Assessment and Evaluation in Higher Education* 40(1): 89–102.

Sheldon, K., and Doran, H. (2015). Are doctoral candidates switched on to the impact of social media? In *2nd International Conference on Developments in Doctoral Education and Training*, University of Oxford, 30 March.

Stubb, J., Pyhalto, K., and Lonka, K. (2014). Conceptions of research: the doctoral student experience in three domains. *Studies in Higher Education* 39(2): 251–264.

Taylor, S. (2012). Changes in doctoral education: implications for supervisors in developing early career researchers. *International Journal of Researcher Development* 3(2): 118–138.

Taylor, S., and Beasley, N. (2005). *A Handbook for Doctoral Supervisors*. London: RoutledgeFalmer.

Tinkler, P., and Jackson, C. (2004). *The Doctoral Examination Process: A Handbook for Students, Examiners and Supervisors*. Buckingham: Open University Press and Society for Research into Higher Education.

Trafford, V., and Lesham, S. (2008). *Stepping Stones to Achieving Your Doctorate*. Maidenhead: Open University Press.

Trafford, V., and Lesham, S. (2009). Doctorateness as a threshold concept. *Innovations in Education and Teaching International* 46(3): 305–316.

UK Resource Centre for SET and Royal Society of Chemistry and Biochemical Society. (2009). *The Molecular Science PhD and Women's Retention: A Survey and Comparison with Chemistry*. London: UK Resource Centre for SET and Royal Society of Chemistry and Biochemical Society.

Wadesango, N., and Machingambi, S. (2011). Post graduate students' experiences with research supervisors. *Journal of Sociology and Social Anthropology* 2(1): 31–37.

Watts, J. (2008). Challenges of supervising part-time PhD students: towards student-centred practice. *Teaching in Higher Education* 13(3): 369–373.

Wellington, J. (2009). Supporting students' preparation for the viva; their pre-conceptions and implications for practice. *Teaching in Higher Education* 15(1): 71–84.

Wilkinson, A. (2015). The rules of the game: a short guide for PhD students and new academics on publishing in academic journals. *Innovation in Education and Teaching International* 52(1): 99–107.

Wolff, L. (2010). Learning through writing: reconceptualising the research supervision process. *International Journal of Teaching and Learning in Higher Education* 22(3): 229–237.

4

Documenting Your Work

Vivien Martin

Whatever type of research you do, you will need to keep records of what you do, how, when, where and why. You may not think that this is important, may even think that you will easily remember everything and can write it up later. You won't and you can't!

Why Document?

Keeping clear and accurate records is important for you as the researcher. Your perception of what you do and why you are doing it will change in subtle ways as your research progresses. As you become clearer about some aspects, you forget earlier doubts. As your findings accumulate you form firmer ideas, and if some findings do not confirm hesitant proposals you are likely to reform them or forget them and may concentrate on those that seem to offer interesting results. Much of the richness of your original thinking and planning is lost, and hesitant directions which do not look immediately rewarding may be prematurely closed. Without good records, you will forget earlier ideas.

Research is often talked about more for its results than for its processes. In developing your research, you will have studied research methods and taken time to make a plan of your proposed work. You may think in terms of the broad question you plan to address and the ways in which you will explore the question. You will consider the research methods and try to choose appropriate ones for your intended study. You will probably plan the process carefully to ensure that your data collection is suitably rigorous. You will expect to write your "Methods" chapter explaining how you have planned. You may not, however, have thought about how you will demonstrate that you worked in a methodical way.

Many people declare their intentions in the 'Methods' section and then jump straight to describing results. You would expect to have to substantiate your results, to offer supporting evidence for everything you claim to have found. So why not expect to do the same for your methods, for the process you have used to collect and analyse your data? Experienced

Research Methods for Postgraduates, Third Edition. Edited by Tony Greenfield with Sue Greener.
© 2016 John Wiley & Sons, Ltd. Published 2016 by John Wiley & Sons, Ltd.

researchers know that the plan is only the starting point and that many illuminating and annoying hitches will occur between plan and report. Much of the learning for you and your reader will be in the detail of the process planned and the discovery of what really happened: why changes were made, what could have been anticipated and what could not, what caused time to slip and which expectations were unrealistic.

One advantage of having records of your process is in being able to supply and use the detail of the 'whole story' whenever you may need it. If all goes reasonably well, you can use the detail to substantiate your discussion of the planned and the actual methods used. If anything goes wrong, you can use the detail to explore and explain. Some researchers despair if their original idea or hypothesis either seems to be unprovable or is even disproven – but with good records of the process, there is still much to say about why this might have happened and perhaps evidence to recommend a different approach another time.

One more good reason for methodical record keeping is that at least you are writing something and you won't have to face the intimidating blank sheet when you start to write up the full report. You will have a lot to start with, and it will contribute to many sections including:

the introduction of why you are doing what you are doing;

the background for the study;

the context;

the choice of methods;

the report of findings; and, possibly,

a discussion of the strengths and weaknesses in your study.

How can you ignore such potential value from what is simply a little self-discipline?

Keeping clear and accurate records is also very important to ensure that your research can be taken seriously by your audience. The way in which you organize and manage your data is part of the way in which you are able to demonstrate that your research has achieved a noteworthy result. In many areas of work, it is essential to use recording methods (including journals, laboratory notes and software files) as everyday documenting procedures. These records have to be kept in a way that fulfils the purpose of recording, and the resulting records have to be available to those with authority to use them, consult them and inspect them.

There may be issues of confidentiality, and if the records contain information about individuals who could be identified the records may be subject to the legal conditions regarding data protection. These conditions are no different for a researcher, and you should ensure that your proposals for record keeping are legal, conform to any regulations within the organization in which you are researching and are approved by the relevant ethics committee, if there is one for your area of research.

Documenting your work includes several aspects of the work. Perhaps the most obvious is that you must document the data that are central to the idea that research is a structured study intended to produce useful results. The process of researching is also therefore important, because you must be able to defend the structure of your study and explain why the process you use is appropriate for the study you undertake. Another aspect of the research is the confidence that your audience, your readers, can place in you as the researcher. You can increase confidence in you, the researcher, by demonstrating that you have taken a careful and consistent approach to documentation of your work from the very early stages to completion.

You might think of your documenting approach as consisting of two levels: the overall aspect of how you prepare, plan, structure and carry out the processes of your research, and the specific aspect of how you document your data.

Documenting Data

Data is a plural word – the singular form is *datum*, which is rarely used because your data are all the quantitative and qualitative features of the variables that become part of your research. Consider this statement from guidance given by the Massachusetts Institute of Technology (MIT, 2016): "Data management includes all aspects of handling, organizing, documenting and enhancing data to enable sustainability and sharing".

You need to devise a way of documenting how you handle (select, collect and store), organize (how you label, code, describe, summarise and group) and enhance your data (collect additional data, improve the quality of the data you collect). The purpose of this is to ensure that the data are recorded well enough to continue to exist in some form (this is why recording is essential as a means of capturing data from events that happen at one moment in time but that you want to study or compare with others at a later date). This sustainable record then enables you to write about, discuss and share the results of your research with others.

Consider at an early stage in your research how you can make your data 'user-friendly' to ensure that any reader of your work will be able to understand exactly what you collected and how you went about the selection and collection. You can find some useful guidelines from the UK Data Archive (www.data-archive.ac.uk). With some types of research, the long-term storage is very important, and there are now ways of storing digital data on servers so that they can be accessed by other researchers over long periods. In the early stages of your research, you should think about where the data will be published, who might then have access to it and what the implications of this are for the usefulness of your research and for preservation of confidentiality.

How to Document?

The methods that you choose to keep records of the process of your research will reflect the type of research and the conditions in which you work as a researcher. Your record-keeping system can be a great help if you plan it carefully and acknowledge the possibility of your own weaknesses from the start. The system you plan can be designed to compensate, in part, for your personal weaknesses. If you recognise that you do not always make careful notes or keep all the details that might be needed later, design a system for yourself that will always jog your memory whilst you are able to remember the detail. For example, if your work will include interviewing people, use a record form that requires you to fill in the essential information. If you will conduct a number of telephone interviews, you could record these on a form that prompted you to keep a full record. You can design a form as the basis of your recording system by thinking of the essential information that you would need if you had to return to ask more, perhaps as much as a year later. As a starting point, I suggest that you need to record:

date of interview;

name (of person being interviewed);

name of interviewer, if you are not conducting all of the interviews yourself;

how the interview was conducted (in person or by phone);

contact details of the interviewee (address, phone number and email);

title of interviewee if nature of work or affiliations are an issue, and possibly address and contact details of any relevant organization;

location of interview (this might influence the nature of the responses);

purpose of interview (your purpose will shape the ways in which you question);

content of interview (this may be verbatim recording, if necessary);

any agreed action (this may include giving the interviewee an opportunity to review the record kept of the interview); and

anything that might have caused the interviewee to modify their responses (e.g. environmental disturbances, interruptions or the presence of colleagues).

You may want to include other headings relevant to your research. Keeping notes like these is essentially a way of cataloguing your data and will help you to take a consistent approach to your analysis and interpretation. It doesn't matter in what format or media these forms are kept; this will depend on what you can use and access regularly. Software such as Evernote (www.evernote.com) is free and can be available on all devices through the web, but by all means use paper or notebooks if that suits your method of working. There is much more on how you could do this in this chapter.

When you begin a research project, it often seems that it is so fresh in your mind that you think that you would always remember the people and organizations involved. However, it is amazing how quickly such details slip your mind and how quickly your interests and concerns can move to have a different focus.

If your records are to be of use to you, there are some characteristics that are essential to try to achieve. Records must be accurate, or they will be misleading. They must be relevant to the research that you are conducting. The record must be written clearly – an illegible record is useless. The record must be as complete as possible; this often means that it should include, as a minimum, details of the source of the data, the full or summarised content and the date that the record was made. This information in a record will help you to decide how reliable it is for potential use in your research.

Consider using different methods of recording for different stages or different aspects of the study so that the method used fits as naturally as possible with the way in which you are working. There are a number of methods that you might consider using.

Diary

Keep a diary or journal of the research from first idea to completion of write-up. The diary might be handwritten in a hardback book or in several small notebooks that you date and number as you fill them. You might prefer to use file paper and ring files so that you can rearrange the pages as themes emerge (but if you do this, number or date the pages so that you remember where they originated). You may prefer to keep the diary in a computer programme, but this would restrict you to making entries only when you have access to the computer.

You may also consider using the diary as a major part of your research and analyse its contents occasionally. There are software programmes with search facilities that can help you to do this. Other ways of pulling out themes and recurring interests once there is enough material in your diary include:

using highlighter pens;

applying different coloured Post-Its to significant pages;

colour coding entries with stick-on dots;

using coloured pens; and

photocopying and cutting and pasting to make up reconstructed pages.

A useful source of further information about this approach is in Judith Bell's book, *Doing Your Research Project: A Guide for First-Time Researchers* (2014).

In some research, you might ask your respondents to keep diaries. You might do this to elicit recurring issues, maybe to identify critical incidents in their work or lives, perhaps to identify their problem areas and responses to problems. People are often prepared to collaborate to assist in research and will often agree to keep a diary for a researcher. However, if you have a particular focus in your research, it will be necessary to give clear direction about the focal area, the types of issues that you would like them to address and the length of time that you would like them to keep the diary. If your research is not so closely focussed and you want them to join you in conducting a broad inquiry into the issues that they encounter over a period of time, make that clear to them. This approach to research brings some other issues into consideration. If you ask people to keep personal records with the expectation that they will reveal the contents to you, you are asking for access to very personal matters. Before you set off into research of this nature, consider the implications of forming such a relationship with your prospective respondents. In *Collecting and Interpreting Qualitative Materials* (1998), Denzin and Lincoln offer very helpful thoughts about the responsibilities of a researcher who engages in an enquiry into the experience of others.

Recording Observations

In a study involving observation, you will need a way to record the issues relating to your research rather than everything going on. Are you an outsider observing a setting, or are you a participant observer? What are the implications of each position?

Your note making will need a mixture of writing, diagramming, mapping and drawing. The setting will need to be recorded in terms of anything which may affect your data collection – this might include a map of the setting (showing physical features like doors, windows and furniture, if it is a room). If the research concerns movements of people or animals, this will need to be shown, probably using diagrams to show paths taken and timing, perhaps coded to indicate who moved or how the movement was made. You may find it helpful to devise grids to tick as things occur or checklists of things to look for and mark off. Observations which include listening to speech or sounds will need to be recorded in some way, perhaps with tape recording or perhaps with diagrams showing frequency of speech and types of interaction. Video recording might be sensible in some studies, but similar issues arise as for photography.

Judith Bell (2014) introduces methods suitable for recording interaction between people. In any study of this sort, you will need pilot studies to derive a good method for recording observations and to be sure that you record what you intend. You will also need to check that you do not change the setting too much by use of your recording process. For example, if you choose a method of recording that is not unobtrusive, you might accidentally change how your subjects behave by attracting them to look at the interesting diagrams you are drawing!

Could you benefit from structured observation, like activity sampling with a strict framework for when and how observations will be made and recorded? More information about how to do this can be found in *Management Research: An Introduction* (2002) by Mark Easterby-Smith and his colleagues.

To some extent, the method you choose for recording observations will relate to your approach as a researcher, and it would be wise to be aware of the approaches used traditionally by your discipline or related disciplines. Anthropology field notes record observations, as do the interaction diagrams used by social scientists and the activity samples of organisational behaviour practitioners. Your choice of method should relate to the traditions and expectations of your disciplinary area.

Laboratory Notes

Laboratory notes apply more to experimental research. If you set up an experiment to prove your hunch, you will need to keep very accurate and methodical records to defend your findings. You should be careful to establish a method of note keeping from the beginning of your studies so that your records are consistent. You must even record mistakes and omissions, like missed entries or lost information. The planning is very important, so consult literature from your discipline. It can be very helpful to look at recent work in your area of study to see how other people have kept records, particularly if there are examples of similar research. Remember that the records that you keep might have to be inspected by others if your claims are to be verified, so ensure that notes are clearly written and could be understood by someone else. You may find it useful to write up laboratory notes, either in a word-processed format or by using e-portfolio software.

Recording Voices

If your research involves interviews or focus groups, how will you record what people say? People usually speak more quickly than a researcher can write, so although it is often possible to make notes, catching 'verbatim' quotes is more difficult. You may think your notes are sufficient, but they will be your precis of chat you have heard, a brief analysis of what was said – not what was actually said. One way to record what was actually said is to explain to the interviewee that you need to write down every word and that pauses will be needed.

This is often acceptable if the interview raises questions that require thoughtful answers. It is sometimes more acceptable than tape recording.

Audio recording is the obvious way to record sound, but it does have some hazards. The first consideration is the effect of having an obvious voice recorder between you and your interviewee or group. It can inhibit conversation, and it can present difficulties of confidentiality if respondents might be recognised and would rather be anonymous. If you are tempted to record but not to tell respondents, consider the ethical issues and the constraints your covert approach might bring in terms of the use of your data. You might also consider what the consequences might be for you or the subjects of your recording if you were found out. *Focus Groups: Theory and Practice* (2007) by Stewart, Shamdasani and Rook outlines the issues that you will need to consider if you are planning to hold a focus group to obtain data.

If you think that it would not present particular problems for your interviewees, it is best to explain why you want a recording and how you will use it, and then to record openly. Plan to have an appropriate charge or power source for the voice recorder so that you don't have to disrupt the interview. Choose the recording device with care and ensure that it can record effectively in the setting, particularly in group work. Learn how to use it confidently before you are in the interview. Consider the advantages and disadvantages of using voice-activated equipment as the pauses between voices might be significant and important to record. You are likely to find a pause facility on the device useful when you play back to make notes. Some researchers recommend using variable-speed playback to help at the analysis stage. If you have produced a digital recording file, this will help for review and storage.

If you are not a good note maker, consider using a personal voice recorder to record your research process, the thoughts and additional ideas that occur and what actually happens as you go along. This is essentially the diary approach, but keeping an oral diary instead of a written one. You will need to write something eventually if you are making a research report, but you could then refer to your recordings and quote passages as appropriate, or you could have a typed transcript made as you complete tapes. A digital voice recorder is inexpensive and may be more helpful for this purpose than using your mobile to keep track of ideas and holding a large amount of data.

If you use tapes in a tape recorder, do be sure to label them as you use them with date and contents. If you use a large number of tapes, consider how you will file them so that it is easy to find any particular one when you want it later. If the tapes are crucial to your research, consider making backup copies. If you use digital recordings, the key is to label them clearly with date, time and something which will help you find the right ones quickly.

If your research will require you to document long oral reports, a number of other concerns arise relating to the personal nature of biographical accounts. You may need to keep a very detailed record containing personal information about the individual's life and experience. Yow's 'Recording Oral History' (1994) is a helpful text and offers examples of forms that you might use.

Card Index

For some record keeping, the card index is ideal. It may seem a bit old-fashioned or associated too much with office files, but there is a very practical advantage in its portability and the portability of packs of file cards. Researchers can carry a few cards around with them and use them wherever they are to make an instant record. The cards can be filed in a system once you are back in your study area with the file box or boxes. An example of this is to use cards to record the sources in a literature search, perhaps to note key quotes. The cards can be arranged in a helpful sequence at the point when you write the literature review, then rearranged when you compile your bibliography.

Computer Database

The main advantage of a card index over a computer one used to be in its portability, but now laptop computers have filled this gap. If you have access to a laptop, check out the software now available to help researchers at many stages of research. There are some very useful ones that will link stages of the literature review in a similar way to a manual system with card files. Some of the popular bibliographic packages include Pro-Cite, Reference Manager, EndNote Plus, Papyrus and Bib/Search. You should record the full reference for every book or journal article that you might want to list in your final bibliography: the publication date and details, and the page numbers that are particularly relevant.

Your searches will probably include web pages. For bibliographic purposes, you should keep similar records; the full URL, the correct title of the resource and the author's name if possible. Also note the date(s) when you used the site. It is useful to keep records of your online searches so that you can save time if you need to revisit sites. Make sure that you keep a copy of the copyright statement on a page that you might want to refer to, and note any citation instructions. You will also need to note the last date on which you visited a site that you mention in your references. If you enter into any email correspondence, remember to save the documents as these too can be useful references. Alternatively, using a database such as Endnote or Mendeley may be a great help to keep a searchable record of your web sources.

If your research involves collection of a considerable amount of data from which you will make a content analysis, consider putting the whole data collection into a package that will subsequently help with the analysis using key words and phrases. You may be able to do this directly with a laptop if you can key in your data as you collect it. With a desktop model it may be less convenient, and you should consider the advantages and disadvantages carefully if it will mean transcribing written notes which might be used effectively without computerising.

Again, a mobile device can be used very effectively for diary keeping and for keeping a personal log of progress through your research process.

Mapping and Making Diagrams

Some sorts of research will involve complex ideas that can be described and recorded visually as maps. This can be useful if there is not yet an obvious sequence or priority, so making lists may be less appropriate. It may also allow ideas to be linked in ways that sequential writing does not.

Mind mapping is one approach which works very well for some people. It can be used to develop detailed thoughts around a central theme. You write the theme in a central bubble and then draw branches out from it as related thoughts arise. Main thoughts are the lines out from the bubble, and branches from these represent aspects of the main idea. More information about this method can be found in *Use Your Head* (2010) and other books by Tony Buzan and also some downloads from his website, www.thinkbuzan.com. The method is useful at the early stages of generating ideas and connecting them. Some people like to use this method to explore their first thoughts about an idea as you might if you were to carry out a brainstorming session with a group of people. It is also very useful in planning chapters of writing and mapping out how to report various aspects of research studies. Another use of the mind map is to make a quick record of a day or an event when it is more important to catch the elements and associated thoughts and feelings than to make a linear time-related record.

More conventional mapping may be useful in research records to record where something is located or to record differences over time when the research relates to physical changes in an environment or the use that people or animals make of an environment. More comment on this method is made under recording observations.

Another recording system for group work is cognitive mapping, which is familiar in strategy development workshops in business and management research. The idea is based on mapping perceptions of the setting and has connections with the Repertory Grid technique. Groups can work with coloured cards and walls of flip-chart paper to map out issues and relationships, and the group map can be the basis of subsequent planning. More information about how to use this can be found in *Management Research: An Introduction* (Easterby-Smith *et al.*, 2002).

Remember that diagrams can save the use of a lot of words and explanations. Consider use of flow diagrams, Venn diagrams, multiple-cause diagrams, fishbone diagrams and force-field diagrams.

Drawing

This is similar in some ways to mapping but is less formal. You might make a drawing of something as a record, which is a traditional approach of many archaeologists, palaeontologists and other researchers whose work involves recording objects.

One way of using drawing is as a projective technique. For example, if I asked you to make a drawing of your research as you see it at the moment, you might draw yourself struggling to climb a mountain or disappearing down a black hole, or perhaps more cheerfully relaxing on a beach in the sun. The drawing could be the basis of a discussion about why you chose the images and what this means for you in terms of the progress of your research. The drawing would record the initial stimulus for the discussion and could be part of your record keeping of the process of your research. This has been described in words, and clearly words could be used for this sort of process instead of drawing, for example in using metaphor to liken your feelings as a researcher to something else: "As a researcher, I feel like a fish out of water". Once you, or a colleague, has said this, it is possible to explore the feelings that have prompted the remark and to look for ways in which this person can be supported to feel more at home as a researcher.

A use of drawing with individuals or groups is the compiling of a Rich Picture. The idea of this is to capture a situation in as much of its complexity as possible, showing all its component groups and individuals, sites and connections, communications, conflicts, inputs and outputs, messy areas and so on. It is drawn with pictures, symbols and connecting lines, in any way that makes sense to the individual or the group. It is drawn with discussion if it is a group work, and the drawing captures the discussion and as many aspects of the situation as members of the group can come up with. For an individual, it captures the thought process and the personal perception. These can be used only as a record, but they are more usually the first stage in making an analysis of problem areas and muddled systems. In its most formal form, this is part of a process called Soft Systems Analysis as described in the book by Peter Checkland, *Systems Thinking and Systems Practice* (1981).

Photographs in Record Keeping

If your research concerns something that can be communicated effectively in visual records, then it may be appropriate to use photographs. Consider all the usual research issues in deciding how and when to use them because the subjective choice of viewpoint, span of view and selectivity are all choices made by the recording photographer. If you take photographs yourself to support your research, make sure you record:

> when you took it (date and time, which might be recorded digitally);
>
> where you were;
>
> what you were looking at; and
>
> the reason why you took it.

The last point is important because of the possibility that you will look at the prints and wonder what one or two were supposed to be about. When you look at the subject you photograph, you know exactly what you are focussing on, but the camera will record everything evenly unless you are sufficiently skilled to ensure that you focus on your subject and reduce the importance of everything else. If you are not a skilled photographer, you will need your notes to make good use of the prints. If you use the prints in writing up your research, you will need to reference them and to link them to your text. An example of use of photographs in research is their use in marketing research. The researcher was evaluating potential shop sites in different high streets, looking at which areas had the most people shopping at particular times of the day. Photographs taken regularly from the same spots were used with maps of streets with shops marked on them and numbers of people present at each time noted. The photographs added visual information which was richer than what the other methods could produce without them.

You might find that you want to use descriptions of photographs, films, videos or other imagery so that you can identify characteristics and make comments and comparisons. There is some helpful advice about this from the Joint Information Systems Committee (2016).

Memoranda and Correspondence

As soon as you start to research, you will produce and receive all sorts of related notes, phone calls, emails, letters, comments and so on. Consider right from the beginning how you might keep these in an accessible form in case they are useful later. It is easy to lose the more trivial things, including scraps of paper with notes and phone numbers, handwritten memos and letters that you write yourself requesting information. A personal example of why keeping these somewhere is important was in some of my own research when I wrote off for college prospectuses, put them all in a box and pondered about how to analyse them but did not use them for several years. The original idea changed and developed, and I went to use them in a comparison with current ones. I found that they were not consistently dated and that some prospectuses were for one academic year and some for another, so I needed my original letter to fix the point at which the prospectuses had been requested. It was sheer luck that I found a copy of that letter. Now I file all correspondence relating to research even if it does not seem immediately relevant.

It is easy to disregard email messages as part of the correspondence that contributes to your research records. If you anticipate using email for anything related to your research, think about how you will file this correspondence. It is easy to form the habit of copying all relevant email messages, both those you send and those you receive, into your permanent records. You can then file them according to their respective topics, as you would file paper records.

Memoranda of all sorts can influence and shape research and might subsequently be seen as crucial to the formation of your research idea. Try to record how you became interested in doing a particular piece of research. If it relates in any way to photographs you discovered in the attic or an article in a newspaper or a set of receipts or bus tickets, keep all the evidence. For this type of material, I suggest just putting things in a cardboard box labelled 'Research' rather than trying to develop a filing system. You will soon find out when you want to retrieve something whether you need a more sophisticated system. If you find that this type of material is becoming important in your research, consider scanning the images into digital files so that you can incorporate them into your text easily when appropriate.

Storage and Retrieval

Remember that records are no use at all unless you can retrieve them when you need them, preferably quickly and accurately. If all else fails, throwing everything relevant into a cardboard box will at least preserve the material, but the task of retrieval may be so daunting that you never actually get round to searching for the item that you want. Some thought given to storage of records and retrieval of data will pay off later! Consider storing in date order, in topics, in themes, in labelled envelopes, in transparent envelopes in ring files, in labelled boxes, in card files, in computer files … much depends on your material and your own preferred methods. There is little in the research process that is more annoying than having to retrace your steps to find the exact reference for a quotation or the exact source of a piece of information.

In Conclusion

However you decide to document, the best advice I was ever given is to MAKE COPIES! Now that most of our documentation is digital, it is essential to save carefully and to make a habit of regularly copying your work to ensure that even if you lose or damage work, a recent version will still be available. People do lose work often – don't be one of them! Keep three copies regularly: one locally at your own workstation, one on an external hard drive and one in a different physical location (think fires, earthquakes and other disasters). I always keep a copy in my workplace and at home. Also remember to label each version – it is surprisingly easy to muddle them when you have several copies.

When you know that you can demonstrate that you have used good practice in organizing and managing all your record keeping, you can be assured that readers will have confidence in that aspect of your research.

References

Bell, J. (2014). *Doing Your Research Project*. Maidenhead: Open University Press, Milton Keynes/McGraw-Hill.

Buzan, T. (2010). *Use Your Head*. London: BBC Books.

Buzan, T. (2016). Think Buzan. http://www.thinkbuzan.com

Checkland, P. (1981). *Systems Thinking, Systems Practice*. London: John Wiley & Sons Ltd.

Denzin, N.K., and Lincoln, Y.S., eds. (1998). *Collecting and Interpreting Qualitative Materials*. London: Sage Publications Ltd.

Easterby-Smith, M., Thorpe, R., and Lowe, A. (2002). *Management Research: An Introduction*. London: Sage Publications Ltd.

Joint Information Systems Committee (JISC). (2016). Approaches to describing images. http://www.jiscdigitalmedia.acuk/stillimages/advice/approaches-to-describing-images/

Massachusetts Institute of Technology (MIT). (2016). Data management. http://libraries.mit.edu/guides/subjects/data-management/index.html

Plutchik, R. (1974) *Foundations of Experimental Research*. London: Harper and Row.

Stewart, D.W., Shamdasani, P.N., and Rook, D.W. (2007). *Focus Groups: Theory and Practice*. London: Sage Publications Ltd.

UK Data Archive. (2016). Home page. http://www.data-archive.ac.uk

Yow, V.R. (1994) *Recording Oral History*. London: Sage Publications Ltd.

5

Ethics of Research

Tony Greenfield

Kant's Wonder

Two things fill my mind with ever-increasing wonder and awe the more often and the more intensely the reflection dwells on them:

> *the starry heavens above me*
> *and*
> *the moral law within me*

Immanuel Kant (1724–1804), *Critique of Pure Reason*

Introduction

Honesty must be central to every aspect of your research.

Even if you have designed and run the perfect experiment, you must be honest in gathering, recording and reporting all data from the experiment, in analysing the data and in reporting any inferences.

Honesty must be the pillar of your code of ethics.

This code is a set of rules, not necessarily written, which is understood and agreed by everyone involved in scientific research.

Morality is different: it is personal, subjective and often opinionated.

So ethics of research is *not* about moral questions relating to projects such as experiments using aborted foetuses or the release of genetically altered viruses for the control of crop pests or the development of weapons of war.

The ethics of medical research have rightly demanded attention which has led to legislation, international agreements and declarations, regulatory authorities and local committees empowered to approve and monitor research projects. This emphasis, which arises from human concern, has extended to research about animals and the environment. There has been a lot of publicity and debate in all media about the ethics of research in these areas.

Research Methods for Postgraduates, Third Edition. Edited by Tony Greenfield with Sue Greener.
© 2016 John Wiley & Sons, Ltd. Published 2016 by John Wiley & Sons, Ltd.

If you are researching in some area of medicine, either human or animal, or into some aspect of the environment you will almost certainly have thought about ethical aspects of your intentions. But if your research is in some other area such as sociology, education, physics, chemistry, materials, electronics, computing, mechanics or industrial manufacturing, you may think that there are no ethical questions for you to consider.

You would be wrong to think that.

Fraud is an obvious ethical matter but, surprisingly, so are experimental design, planning, management and execution; and so is publication.

If you know yourself to be thoroughly honest, you must be confident that you will never be deliberately unethical. Unfortunately, no matter how good a person you are and how well intentioned, there is the possibility, indeed it is very likely, that you will be inadvertently unethical, insomuch as you infringe the accepted code of research behaviour. Anybody who embarks on research is at risk of such inadvertent unethical behaviour. Avoidance demands good advice at all stages. Where will you find that advice? Start here and follow the leads.

Background

We start with some definitions and, in the rest of the chapter, look at some good principles and bad behaviour.

Ethics, in its widest sense, as the principles of good human behaviour, is one of the issues for which philosophers have striven to provide guidance. Plato, in about 400 BC, proposed that there were *forms* of all things, including a *form of the good*. We could never experience true forms, but could at least approach them through knowledge.

For Plato, bad behaviour was the result of ignorance. Despite the enormous influence of his ideas, especially on religious belief, few people today would accept them in their original guise. For example, philosophers of the post-modern school hold the view that there are no absolute standards, and that morality can only be culturally determined.

There have been many philosophers, and theories, in the intervening years. Kant emphasised the *will* and the importance of intention. His categorical *imperative* is stated by Russell (1946) as: 'Act as if the maxim of your action were to become through your will a general natural law'. The utilitarians, Jeremy Bentham and others, concentrated on consequences and the 'greatest-happiness to the greatest-number' principle. Their ideas had a great, and generally highly beneficial, influence on British government during the middle of the 19th century, and probably still exert their influence today. Nietzsche's intense individualism was in stark contrast to this. He argued that such paradigms would stifle creativity.

The works of the major philosophers are not usually easy reading and, given the other demands on your time, you may think a more appealing understanding of ethics was given by Charles Kingsley in his children's adventure *The Water Babies*:

> She is the loveliest fairy in the world and her name is Mrs Doasyouwouldbedoneby.

If philosophers cannot agree on the basic principles of ethics, and commentators cannot always agree about the correct interpretation of their work, it is hardly surprising that there is even more diversity of opinion about the practical application of those principles. Some philosophers, such as Nietzsche, had their ideas grotesquely misrepresented, and then reinterpreted in a more generous light. Hollingdale's translation of Nietzsche's *Thus Spake Zarathustra* (1969) sold well in the late 1960s. Despite all the controversy, there is enough common ground to establish codes of conduct which are generally accepted.

Codes of Conduct

Most professional organisations have their own codes of conduct that are largely about the ethical standards that are expected of members. One of the best known of these codes of conduct is embodied in the Declaration of Helsinki. This was first adopted by the World Medical Association (WMA) at Helsinki in 1964. There have been several amendments since then. The latest revised declaration, its seventh, was issued by the WMA in 2013.

Even if your research may be far removed from 'biomedical research involving human objects', which is what the Helsinki Declaration is about, you should read it. Many of the points can be interpreted more widely. One, from an early edition of the declaration, that clearly applies to all research without exception, and that includes yours, is: 'It is unethical to conduct research which is badly planned or poorly executed'.

This has been restated in the latest edition as: 'The design and performance of each experimental procedure involving human subjects should be clearly formulated in an experimental protocol'.

My view is that this principle should apply to all research, whether or not involving human subjects.

Here is a further selection of points:

- … research … must conform to generally accepted scientific principles … based on adequately performed … experimentation and on a thorough knowledge of the scientific literature.
- Every … research project … should be preceded by careful assessment of predictable risks in comparison with foreseeable benefits.
- In publication of the results of … research … preserve the accuracy of the results. Report of experimentation not in accordance with the principles … should not be accepted for publication.
- The research protocol should always contain a statement of the ethical considerations involved.
- Special caution must be exercised in the conduct of research which may affect the environment.

Since all research involves the collection, analysis, interpretation and presentation of data, some points from the codes of conduct of statisticians are worthy of mention.

The Royal Statistical Society declares:

- Professional membership of the Society is an assurance of ability and integrity.
- … within their chosen fields … have an appropriate knowledge and understanding of relevant legislation, regulations and standards and … comply with such requirements.
- … have regard to basic human rights and … avoid any actions that adversely affect such rights.
- … identities of subjects should be kept confidential unless consent for disclosure is explicitly obtained.

- … not disclose or authorise to be disclosed, or use for personal gain or to benefit a third party, confidential information … except with prior written consent.
- … seek to avoid being put in a position where they might become privy to or party to activities or information concerning activities which would conflict with their responsibilities.
- Whilst free to engage in controversy, no fellow shall cast doubt on the professional competence or another without good cause.
- … shall not lay claim to any level of competence which they do not possess.
- … any professional opinion … shall be objective and reliable.

Some points from the code of conduct of the Institute of Statisticians (now merged with the Royal Statistical Society) were:

> The primary concern … the public interest and the preservation of professional standards.
> Fellows should not allow any misleading summary of data to be issued in their name.
> A statistical analysis may need to be amplified by a description of the way the data were selected and the way any apparently erroneous data were corrected or rejected. Explicit statements may also be needed about the assumptions made when selecting a method of analysis. Views or opinions based on general knowledge or belief should be clearly distinguished from views or opinions derived from the statistical analysis being reported.
> Standards of integrity required of a professional statistician should not normally conflict with the interests of a client or employer. If such a conflict does occur, the public interest and professional standards shall be paramount.

None of these points needs elaboration. You can judge which apply to your research. However, thinking in terms of medical research, the ethical implications of statistically substandard research may be summarised as:

- misuse of patients – put at risk or inconvenience for no benefit; subsequently given inferior treatment;
- misuse of resources – diverted from more worthwhile use; and/or
- misleading published results – future research misdirected.

It is worth remembering that:

> *precise conclusions cannot be drawn from inadequate data.*
> Pearson and Hartley, *Biometrika Tables for Statisticians* (1962)

Politics

Facts are sometimes distorted for political advantage. The ways in which this is done may be applied also in scientific research, so some discussion of them is appropriate with a warning to be on your guard.

There is no official code of conduct about 'official statistics': those tables and graphs that are published by government departments and reach the public through newspapers, radio and television. But there is wide concern in Britain, and in most countries in the world, about the way that governments handle the figures. In Britain, for example, we are told that unemployment figures are expected to fall again, that the economy is recovering, that the poor are better off and that more is being spent on the National Health Service. Can we believe such statements?

Some tricks of official statistics:

- burying unfavourable statistics in a mass of detail;
- changing definitions (what constitutes a major hospital project; items included and method of calculation of the retail prices index; who is unemployed); and
- discrediting authors of unfavourable reports.

Cutting Corners

Some of these guidelines are illustrated in the following anecdote from my own experience.

Pharmaceutical companies are naturally eager to conclude clinical trials quickly and favourably. This eagerness constitutes a commercial pressure on clinical research departments or agencies, and all others involved. This is fair enough provided no corners are cut and the highest ethical standards are maintained. Generally this is so, but sometimes statistical analysis reveals that it is far from the case.

What should I, the statistician, do then?

I believe that I must state my opinion firmly, without fear of loss of business or even of a libel action. I should do this just as if the trial had been conducted properly and the results had been entirely favourable with the expectation that the company would respect and honour my work and opinion.

The following example is of a trial that was designed and conducted by the pharmaceutical company. The data, already coded and entered into a computer file, came my way for analysis because the company was in difficulties, some of which will be revealed.

The trial was an open, randomised, phase 3, multicentre study (see Chapter 19). The protocol specified that 150 eligible patients would be recruited by 12 investigators.

In fact 32 patients were recruited by five investigators, one of whom recruited only one patient.

Of those 32, only 21 patients were clinically evaluable and only seven were microbiologically evaluable.

The general conclusion was that nothing useful emerged from this study except for a strong message to the pharmaceutical company that they must pay closer attention to the design, planning, management and execution of trials than had been exhibited in this case.

Scientific integrity demanded comments on these aspects of this study. Recall that the Declaration of Helsinki (1975, sect. 4.2) states:

> It is unethical to conduct research which is badly planned or poorly executed.

It is generally accepted, by medical research ethics committees, that if the number of patients is too small to obtain a useful and significant result, then patients will have been submitted fruitlessly to inconvenience, discomfort, doubt of outcome, and risk. Such a trial would be unethical.

The writers of the protocol assumed that if there were a 75% evaluability rate, approximately 112 of the 150 patients would be eligible for efficacy analysis. In fact, 21 of the recruited patients were clinically evaluable (65%).

The assumptions of a cure rate of 85% for the better of the two treatments and a 23% difference between the two with a significance level of 0.05 and a power of 0.8 indicated that 112 patients would be sufficient to detect that difference. In fact, the total cure rate was 43%, and there was far from enough information to test for any differences of outcomes between the two treatment groups. No differences were indicated.

It was improper to embark on this trial without confidently expecting 150 eligible patients to present. There was nothing in the protocol to show that the necessary number would present in a specified time. In fact, a time was not specified.

Having embarked on the study, it was not ethical to stop it, without clear evidence that one treatment was inferior, before the specified number of patients had been recruited.

The data collection form provided for the collection of 1488 items of information on each patient. Much of this information, particularly relating to return visits, was returned as blanks.

Catch-all data forms may have a semblance of thoroughness to the uninitiated, but they demonstrate a lack of forethought and an absence of scientific planning.

Apart from demographic data collected to demonstrate the success of random allocation of patients to treatments and general homogeneity of the sample, all other data collected in a clinical trial should be related in some way to clearly stateable hypotheses.

The only hypotheses implicit in the protocol, although not explicitly stated, were:

1. There is no difference in the clinical improvement rates between the treatments.
2. There is no difference in the microbiological responses between the two treatments.
3. There is no difference in the incidences of adverse events between the two treatments.

These are straightforward hypotheses which may have been tested if 150 patients had been recruited. If any of the many variables of haematology, blood chemistry, medical history, age, sex, race, height and weight, surgical procedures and other medications may have influenced the outcomes, then the relevant hypotheses should have been stated. It could be left to the statistician to decide how to use these extra variables and what multivariate techniques to apply. However, the expected relationships should be stated in advance so that they could be taken into account in sample size determination.

A medical research ethics committee may have been misled by the protocol into believing that the study was well designed and would be well executed. The section on statistical methods contained an 'outline of statistical analysis plan' which appeared to be thorough. However, a responsible and careful committee would also look at the data collection form and question its potential, not only for collecting the necessary data but also for facilitating data processing and statistical analysis. They may question first the desirability of collecting so much information and how it was proposed to use it all in testing hypotheses. The plan suggested tabulations and complete listings but without any indication of how these would be interpreted. While clinical judgement may be needed to assess the effect of a treatment, it is not an appropriate tool to use when data recorded from clinical trials are interpreted. Clinical judgement is not necessarily reproducible between investigators, whereas formal statistical analysis is reproducible.

The most striking feature of the data collection form was that there was no indication as to how the data were to be coded and entered into computer files. The consequence of this failing was that the data had been entered into the computer files in formats which are very difficult to manage.

Pharmaceutical companies should understand that it is usual to have a pilot study for testing data collection forms for their suitability for:

1. use by investigators;
2. coding for data entry; and
3. statistical analysis.

The investigators had not generally completed the forms properly. This may be because the forms were badly designed.

Haematology and clinical chemistry data called for individual assessment of 'significant abnormality' which was not defined. If it had been intended to be 'outside normal ranges', this could be left to calculation provided the laboratory normal ranges were given, although it is well known that these are contentious. The normal ranges were provided for the various centres, but the data collection forms had not been designed to include this information.

Because the data collection forms were poorly designed, they could not be expected to encourage co-operation by the investigators, who responded by leaving many questions unanswered or improperly answered.

If the regulatory authorities were aware of the nature of this study, it is likely that they would admonish the company for conducting trials unethically.

The results of this trial were not satisfactory. This is because the trial was poorly designed and inefficiently and incompletely executed, and because the data forms were badly designed and unsuitable for data coding, computer entry and statistical analysis.

The unavoidable conclusion was that the conduct of this trial was not ethical.

As the statistician responsible for analysing the data and reporting the results, that is what I told the pharmaceutical company.

Fraud

While much unethical science is inadvertent, caused mainly by poor management, there is a long history of scientific fraud reaching back several centuries. Charles Babbage, who was Lucasian professor of mathematics at Cambridge University (a chair held by many great scientists including Isaac Newton and Stephen Hawking), published a book in 1830 entitled *The Decline of Science in England*.

Read that again. The date was *eighteen* thirty.

One chapter in his book was about scientific fraud under which he described four methods of fraud: *hoaxing*, *forging*, *trimming* and *cooking*. To these, I would add *obfuscation*. For the first four, I cannot do better than quote him directly.

Hoaxing

In the year 1788, M Gioeni, a knight of Malta, published an account of a new family of Testacea of which he described, with great minuteness, one species. It consisted of two rounded triangular valves, united by the body of the animal to a smaller valve in front. He gave figures of the animal, and of its parts; described its structure, its mode of advancing along the sand, the figure of the tract it left, and estimated the velocity of its course at about two-thirds of an inch per minute ... no such animal exists.

There have been many more hoaxes since Babbage's day, including the saga of the Piltdown Man.

Forging

Forging differs from hoaxing, inasmuch as in the latter the deceit is intended to last for a time, and then be discovered, to the ridicule of those who have credited it; whereas the forger is one who, wishing to acquire a reputation for science, records observations which he has never made.... The observations of the second comet of 1784, which was only seen by the Chevalier D'Angos, were long suspected to be a forgery and were at length proved to be so by the calculations and reasoning of Encke. The pretended observations did not accord amongst each other in giving any possible orbit.

Statistical methods now exist to discover forged data. Examples may be found in industrial research and in clinical trials. If you are tempted to forge your data, be warned. A good examiner will detect your forgery and you will be humiliated.

There can be great pressure on a student to complete a research project within the time specified by the university rules or before his grant expires. Under such pressure, the student may be tempted to forge data which he has never observed. Or, if he has made some measurements but they don't properly meet his expectation, he may be tempted to cook the results. Cooking is also described in this chapter by Babbage.

Trimming

Trimming consists in clipping off little bits here and there from those observations which differ most in excess from the mean, and in sticking them on to those which are too small … the average given by the observations of the trimmer is the same, whether they are trimmed or untrimmed. His object is to gain a reputation for extreme accuracy in making observations.... He has more sense or less adventure than the cook.

Cooking

This is an art of various forms, the object of which is to give to ordinary observations the appearance and character of those of the highest degree of accuracy.

One of its numerous processes is to make multitudes of observations, and out of these to select those only which agree, or very nearly agree. If a hundred observations are made, the cook must be very unlucky if he cannot pick out fifteen or twenty which will do for serving up.

Another approved receipt, when the observations to be used will not come within the limit of accuracy, is to calculate them by two different formulae. The difference in the constants, employed in those formulae has sometimes a most happy effect in promoting unanimity amongst discordant measures. If still greater accuracy is required, three or more formulae can be used.

It sometimes happens that the constant quantities in formulae given by the highest authorities, although they differ amongst themselves, yet they will not suit the materials. This is precisely the point in which the skill of the artist is shown; and an accomplished cook will carry himself triumphantly through it, provided happily some mean value of such constants will fit his observations. He will discuss the relative merits of formulae … and with admirable candour assigning their proper share of applause to Bessel, to Gauss, and to Laplace, he will take that mean value of the constant used by three such philosophers which will make his own observations accord to a miracle.

Obfuscation

Obfuscation means 'to make something obscure'. It is a deliberate act which is intended to convey the impression of erudition, of being learned, of great scholarship. Hence it is fraudulent. There is a style of academic writing, increasingly common in recent years, that is long-winded with long paragraphs, long sentences, long words, passive statements and tortuous structures (see Chapter 38). It is intended to deceive, and it does so easily because the reader, even an examiner, is tempted to skim such verbosity and subsequently fears to confess he has not understood every word.

It is a trick that is apparent today in many academic papers and theses, but it was not uncommon a hundred years ago.

> The researches of many commentators have already thrown much darkness on this subject, and it is probable that, if they continue, we shall soon know nothing at all about it.
>
> Mark Twain

Perhaps some people can't help writing obscurely, but if a post-graduate research student does so we should be suspicious.

> People who write obscurely are either unskilled in writing or up to mischief.
>
> Peter Medawar

Unnecessarily esoteric mathematics should be avoided. For example, it is not necessary to preface straightforward calculus, as applied to an engineering problem, with references to Hilbert spaces and sigma field. Simple numerical examples can be a great help to your readers.

Obscure writing is not the only ethical problem in presenting your results. The other big problem is plagiarism. This is dealt with in a separate chapter (Chapter 6).

Advice

How can you, an inexperienced student, know how to avoid any of the problems, to be sure that your research is ethical? Only by seeking advice. The librarian is there to help you (Chapter 10); your supervisor is there to help you (Chapter 9); and there are statisticians.

Why Statisticians?

A statistician is objective. Although he may know little about your special subject, he can advise you about how to do things fairly, how to achieve balance, how to measure and record information (which is what research is about), how to analyse data (Chapters 29 and 30), how to design your experiments (Chapters 19 to 24), how to avoid making too many measurements, how to be sure you are making enough, how to avoid bias, how to achieve high precision and how to present your results clearly and succinctly.

In measurement and analysis, 'Hard science is easy. Soft science is hard'. Anything qualitative is so difficult to measure, analyse and interpret that special care must be taken to avoid subjective judgement and misinterpretation, either inadvertent or deliberate. There is no excuse for qualitative research which lacks rigour.

> When you can measure what you are speaking about and express it in numbers, you know something about it: when you cannot measure it, cannot express it in numbers, your knowledge is of a meagre and unsatisfactory kind. It may be the beginning of knowledge, but you have scarcely in your thought advanced to the stage of science.
>
> Lord Kelvin

While in the social sciences it is acceptable to study very small numbers of cases in order to develop the richness and detail of the study (in some situations, there may be only one case to study), you should take extra care when drawing conclusions. The case or cases must be understood against the background of wider published research so that we know how different it may be and in what ways. Then you need to decide what such a difference means. To suggest that small qualitative studies can offer generalizable conclusions is at best dangerous, at worst unethical.

The Researcher's Prayer

> *Grant, oh God, thy benedictions*
> *On my theory's predictions*
> *Lest the facts, when verified,*
> *Show thy servant to have lied.*
> *May they make me BSc,*
> *A PhD and then*
> *A DSc and FRS,*
> *A Times Obit. Amen.*
> *Oh, Lord, I pray, forgive me please,*
> *My unsuccessful syntheses,*
> *Thou know'st, of course - in thy position -*
> *I'm up against such competition.*
> *Let not the hardened editor,*
> *With referee to quote,*
> *Cut all my explanation out*
> *And print it as a note.*

> Proceedings of the Chemical Society January 1963: 8–10
> (quoted in A Random Walk in Science, an anthology
> published by the Institute of Physics in 1973; Weber, 1973)

References

Babbage, C. (1830). *The Decline of Science in England*. London: Fellowes.

Nietzsche, F.W. (1969). *Thus Spake Zarathustra: A Book for Everyone and No One* (trans. and intro. R.J. Hollingdale). Harmondsworth: Penguin Books.

Pearson, E.S., and Hartley, H.O. (1962). *Biometrika Tables for Statisticians*, 2 vols. Cambridge: Cambridge University Press.

Russell, B. (1946). *History of Western Philosophy*. London: George Allen & Unwin.

Weber, R.L. (1973). *A Random Walk in Science: An Anthology*. London: Institute of Physics.

6

Plagiarism

Tony Greenfield

> *Plagiarise,*
> Let no one else's work evade your eyes,
> Remember why the good Lord made your eyes,
> So don't shade your eyes,
> But Plagiarise, Plagiarise, Plagiarise …
> Only be sure always to call it please, *"research."*
>
> Tom Lehrer, American satirist

Plagiarism is the theft of ideas and text from other people's work.

You will be a plagiarist if you steal ideas or text from other people's work and present it as if it is your own.

You will also be a plagiarist if you present the work of other people as your own if they give or sell the work to you.

Research demands that you use all relevant knowledge that you can find. But much knowledge belongs to somebody else. There is universal understanding that you may use whatever information has been published, so long as you attribute the ownership to the original author.

Writers of many books and articles published in the 18th and 19th centuries ignored that understanding but, through the 20th century, proper referencing became normal and expected practice. But there is still bad practice, and it is getting worse.

Some Examples

Stealing from a Colleague

Sometimes two or more students will share a project. One will be more assiduous than the other and finish her report on time, while the other plays games. The second, in a panic, will

Research Methods for Postgraduates, Third Edition. Edited by Tony Greenfield with Sue Greener.
© 2016 John Wiley & Sons, Ltd. Published 2016 by John Wiley & Sons, Ltd.

borrow his colleague's report and copy chunks of text from it, handing it in to his examiner with a sigh of relief. But a good examiner will recognise this and may even detect, from the writing style, which student is the originator and which is the plagiarist. The plagiarist will at least fail but, more justly, the college will expel him and his career will be finished.

If you start to write early, you will avoid the panic of late submission and the temptation to plagiarise.

Collusion

Collusion can be as bad as copying. If you and another student work together, but you then present the work as if it is only yours, then you have stolen credit. The lecturers who set the work for students to share should clearly state what is expected of each individual student. If they don't, ask them.

Stealing from Published Papers and Books

It is so easy now, working at your computer, to copy from existing publications and paste it into your own text. You alter a few words here and there, and pass off the descriptions of research methods, results and discussions as your own. 'Look at this', you tell your friends and your family. 'Listen to me', you say to the scientific conference. 'Publish this', you ask the academic journal. You'll be caught. Even if you are not caught, you will live the lie until you die.

Quoting without Citation

Some students may claim that it is in their culture to copy original text from a well-known publication, without citation. They may even assert that it would be discourteous to change the quotation in any way. But, without citation and without indicating what is the copied text and what is the student's own work, it is still a theft of ideas and writing that will mislead readers into belief that the work is entirely the student's own.

Beware too of copying from websites. Much information cannot be trusted. Many universities discourage references from Wikipedia because much text is not referenced and, even if it is, anybody can change it.

Buying It

A new form of plagiarism arrived with the personal computer and the internet.

There are many essay-writing, and even thesis-writing, services on the web. Don't kid yourself that you can pay and get away with it. Your university is well aware of these services and will catch you if you have used them.

Google is now aware that plagiarism is threatening the integrity of university degrees and will therefore ban any advertisements for essay-writing services.

Data Recycling

A postgraduate student collects some data over several years, and writes and publishes a thesis and some associated papers. The student moves away, but the data stay in the department. Later researchers join the department, find the data, squeeze out more analysis and publish more papers. It takes little skill to make the data look like new results.

If you choose a topic and follow it through several years and several scientific and medical journals, you will soon find papers recycling data and, in each case, presenting the data as if they had been collected only for that paper.

Self-plagiarism

Some researchers are so eager to build long lists of publications that they repeatedly publish the same data and results in several journals with different titles, sometimes with a different selection and order of co-authors. If you are tempted to do this, you must come clean: tell the journal editors that you are doing it but that you are adding new information or interpretation. If they detect dishonesty, they should bar your later work.

Editors need to be sure that they always publish research papers with new results not published elsewhere. They usually ask at least two referees to review each submitted paper. The referees must tell the editor if they suspect any self-, or other, plagiarism.

Stopping Plagiarism

There are two steps in stopping plagiarism by students: deterrence and detection.

Deterrence

Department heads should tell all new students about plagiarism, defining it clearly, with examples. They should tell all students about the penalties that detected plagiarism will incur.

You can protect your own work by deterring others. Let it be known that you are publishing your work early. This doesn't mean going through all the stages of writing formally for a journal, submitting your paper, responding to referees and resubmitting. You can publish early and simply by giving a draft to your supervisor or by storing it on your university's research archive. If these services don't exist, ask for them to be provided. These safeguards will ensure that any attempt to steal your work will be detected.

You will ensure that nobody can accuse you of plagiarism if you include, with your submitted work, a statement that all unreferenced work is yours.

You will also ensure that nobody can accuse you of plagiarism if you reference all work, including quotations, and be totally clear in the text which material is from the referenced work and which is from you. Give credit where it is due, but be sure that you are credited for your own work. Refer to chapter 4 for a guide to documenting.

A difficulty with references is that too many destroy the flow of writing and will bore the reader. So, there is no need to reference the origins of common knowledge, such as process flow charts, differential calculus or Student's *t*-test.

Detection

An experienced supervisor for a specific subject should be able to detect plagiarism in any work submitted by her students. You can enter short phrases, enclosed in quotes, into Google and other browsers to check for matches. There are now programmes that can help with this. They can also help you, the student. These programmes will search a library of documents for any text that matches any of your text.

The most widely used programme is Turnitin, and it is recommended by JISC.

The JISC is the Joint Information Systems Committee that was created by the higher and further education councils of the four countries of the United Kingdom. Its mission is to provide world-class leadership in the innovative use of information and communications technology to support education and research.

You can access *Turnitin* through the JISC website: www.jisc.ac.uk. If you visit www.jiscpas.ac.uk/turnitinuk.php, you will find the full services including some training videos, for example www.turnitin.com/static/training_support/student_training.html.

Turnitin checks submitted documents against a vast database of billions of pages of both current and archived material, which includes previously submitted student papers. Archived material includes internet sources, books, newspapers and journals from both academic and professional sources. It will return a submitted paper as a customised originality report.

Provided your department subscribes to Turnitin, you, the student, may use the programme to check that your own work is free from any accusation of plagiarism. You can also check that earlier work you have referenced has not, itself, been plagiarised.

Conclusion

Plagiarism has been with us since writing began, but it is increasing. Universities are concerned, but so is the honest student who wants to protect his own work. He must be wary, especially when he lends his work to other students. He must understand how other people plagiarise so that he can be vigilant. Modern technology has brought new ways to plagiarise, but it has also brought new ways to protect against plagiarism and new ways to detect it.

7

Critically Reviewing Your Own Research

Tom Bourner and Juliet Millican

Introduction

This chapter is about reading, reviewing and refereeing your own research. Its main aim is to help you critically assess your research so that you can be satisfied that it's the best you can do.

This is an important issue for at least four reasons: (1) when you've finished your work, it is likely to be assessed by someone else, so it's good to be able to assess it yourself first; (2) being able to review your own work enhances the likelihood of a successful outcome to your research project; (3) being able to assess the quality of research is something that a good researcher should be able to do; and (4) to learn from your own experience.

Whether your research is for a post-graduate research degree or a funded research project or aimed at a publication in an academic journal, it will be assessed by someone else when it is complete. If it is for a post-graduate research degree, it will be assessed by examiners. If it is a funded research project, then your funder will assess it or have it assessed. And if it is aimed at publication in an academic journal, then it will be subject to the peer-review system, which means it will be reviewed by other academics in the field to decide if it is worthy of publication.

If you can review your own research, then you will be in a position to make the changes needed to bring it up to a standard that will satisfy your examiners, the funding body or a journal's reviewers. Failure to meet the standard in any of these situations is likely to be a painful experience.

As a competent researcher, you will be expected to be able to assess the quality of research, at least within your fields of expertise. There will come a time when you are likely to be asked to referee a proposal for a research project, a research report or an article that has been submitted to a journal in your field. At that time, it will be assumed that you have developed

Research Methods for Postgraduates, Third Edition. Edited by Tony Greenfield with Sue Greener.
© 2016 John Wiley & Sons, Ltd. Published 2016 by John Wiley & Sons, Ltd.

the skill of reviewing research. So you might as well start to develop that skill by reviewing your own work as soon as you can.

The more effectively you can assess your own research, the better position you are in to learn from your own experience of doing research. This will enable you to benefit from the process of continuous improvement from research project to research project.

This chapter identifies criteria used in assessing a research project, and this will enable you to assess your own research before it is assessed by someone else.

A Checklist from the Experts in Reviewing Research

How can you critically review your own research? Let's start with an easier question. How can you review your own research?[1] To answer this question, we could ask the experts. Who are the experts on reviewing research? They are the editors of peer-reviewed academic journals. They are experts because they regularly have to assess the research articles submitted by researchers for publication in journals. They have to judge the quality of the papers submitted to decide whether they are worthy of publication.

When the editor of an academic journal receives an article submitted for publication, it will normally be sent to two reviewers for a judgement on whether it is worthy of publication in the journal. There is a danger, of course, that different reviewers will apply different criteria in making their assessments and will therefore reach different conclusions. In order to guard against this, the reviewers normally receive a checklist of questions which embody the criteria the journal regards as most important.

If a post-graduate researcher had access to these lists of questions, then they would be in a position to review their own research and thus make sure their own research 'ticks all the boxes'. That would clearly be worthwhile, whether they were in fact submitting an article to an academic journal or assessing whether their research-based dissertation is up to standard or whether a funded research project will be well received.

So what *are* these questions that appear on the checklists for reviewers of peer-reviewed academic journals? There is no single checklist that all the journals use, but there is considerable overlap in the criteria used by the different journals. Table 7.1 contains a list of questions compiled by the writers of this chapter on the basis of their experience of reviewing articles for academic journals.

With this list, you'll be able to review your own research or that of anyone else. But what is the rationale for each of these questions? Let's take a look.

Does the Paper State, at the Outset, Its Aim(s), Its Objective(s) or the Problem(s) It Is Seeking to Address?

There are two main reasons why it is important to set out clearly the aims, objectives or key issues at the beginning of your work – one is the writer, and one the reader. If you write these at the start of your work, a bit like re-articulating the question in an academic essay, you will fix them in your mind and clarify them for yourself. It gives you a much better chance of

---

[1] We'll address the issue of *criticality* a little later in this chapter.

Table 7.1 The kinds of questions used by reviewers of articles for academic journals.

- Does the paper state, at the outset, its aim(s), its objective(s) or the problem(s) it is seeking to address?
- Does the paper demonstrate an adequate understanding of the relevant literature in the field and cite an appropriate range of literature sources?
- Has the paper ignored any published work that is significant and relevant to the issue it addresses?
- Is the paper clear that the methods employed are appropriate for the aims of the research?
- Does the paper provide enough detail of the method(s) to enable a reader to repeat the study if they were in a position to do so?
- Are the findings analysed appropriately?
- Are the results presented clearly?
- Does the paper discuss its key results in the light of published literature on the subject?
- Does the paper identify how its results make an original contribution to new knowledge?
- Do the conclusions tie together other elements of the paper?
- Is it clear how the conclusions are supported by the research findings or results?
- Does the paper identify clearly any implications for practice and/or society?
- Does the paper identify any implications or questions for further research?
- Has attention been paid to clarity of expression, including length of sentences, use of jargon and use of acronyms?

staying on track and being focussed in your writing. It also means your reader, whether that be an examiner, publisher journal audience or research funder, will be clear what it is you are trying to address and whether your writing is of interest to them and their areas of concern.

Does the Paper Demonstrate an Adequate Understanding of the Relevant Literature in the Field and Cite an Appropriate Range of Literature Sources?

As you begin to write and publish at the post-graduate level, you are dealing with areas of new knowledge. Your work needs to be starting from the perspective of what has been written and published already on this subject and how you have taken these ideas forward. Your work will not be taken seriously at an academic level unless it demonstrates how it has taken earlier findings into account, and it is of course difficult to answer the 'So what?' question if someone else has already reached similar conclusions and you appear not to know about them. So a thorough review of the literature will give a firm base from which to start writing.

Has the Paper Ignored Any Published Work That Is Significant and Relevant to the Issue It Addresses?

While it is not always possible to be sure you have not missed one or two published articles that relate to your research, the sophistication of current internet searches gives you a better chance of identifying the most important. By overlooking these, you may miss important findings that have a bearing on what your own research is saying. Those reading your research, either to decide whether to publish or for their own purposes, will be likely to know these, and if you ignore them it will reflect badly on the thoroughness of your own work.

Is the Paper Clear That the Methods Employed Are Appropriate for the Aims of the Research?

While there are a broad range of research methods you might choose from, matching method to research aims is an important process. You will need, for example, to take account of your resources, your question areas, your data sources and the sensitivity of the subject you are addressing. You also need to make these methods clear to your reader so they can understand how you arrived at your findings.

Does the Paper Provide Enough Detail of the Method(s) to Enable a Reader to Repeat the Study if They Were in a Position to Do So?

If a reader is sufficiently interested in what you did and what you found, they may want to replicate your study to confirm your findings … or refute them. If they can't replicate your study exactly, they many want to apply it to their own context to see if the situation is the same for them. While their context will invariably be different and they may not be able to repeat the study exactly, it could be valuable for them to get as close as possible, to try and test out how general your findings are. And if they are not in a position to repeat the study, then it is important to provide enough detail for the reader to assess the soundness of all the steps taken in the process of the research.

Are the Findings Analysed Appropriately?

Deciding on a process for analysing findings is important in order to ensure that the key issues emerging from the data are picked up. It is worth asking yourself these questions again when reviewing your work: 'Did I find what I expected?' 'Is there anything that surprises me?' and 'Could I have analysed from a different perspective?' Triangulating your data and looking at whether results can be verified from a different source will help to clarify whether or not you have sufficiently analysed your data or whether you are indeed reading too much into your results. Triangulation where possible, and leaving time and reflection between analysing and writing up your analysis, will enable you to take a fresh look at your conclusions from a different perspective.

Are the Results Presented Clearly?

When you are deep into a research project and immersed in your findings, it is often difficult to see whether or not a presentation of your results is clear to other people. But the impact of your research is undermined if your readers are unable to see clearly what you have found and how. Letting time pass between one reading and the next, sharing your report with other people, getting feedback from a range of peers and asking yourself 'Is there another way I could present or display my findings?' will help to ensure they are as clear as possible. Visual diagrams, tables and bullet points all might help to make your findings stand out in a way that words cannot.

Does the Paper Discuss Its Key Results in the Light of Published Literature on the Subject?

It is important to relate your findings back to the literature. The purpose of research is to identify new knowledge, and you need to show not only how your research builds on what has been done before but also what it adds to existing knowledge. Without this, your report is incomplete.

Does the Paper Identify How Results Make an Original Contribution to New Knowledge?

It is useful to reflect on what is different about the angle you have taken and the particular perspective you bring to a subject. This could be the methods you have used, the different concepts you have used to inform your starting point or the area in which you have based your research. It does not mean being a maverick for the sake of it or straying away from rigorous and recognised research approaches. However, you will need to identify where your work is located within the current map of knowledge on the subject and how your work adds to that of others.

Why Is It Important That the Article Makes a Contribution to Knowledge That Is Significant*?*

As a new researcher, or even as an experienced researcher, you need to answer the 'So what?' question of your research at an early stage. It is easy to get immersed in the process of research and to lose sight of the value or purpose of your findings. You may have studied this or researched that or answered the following questions, but so what? Why would anyone want to know? What is the impact of your research on the rest of the world? It is important to keep this in mind as you begin to write up your findings and to make this clear to your readers in the conclusions.

Do the Conclusions Tie Together Other Elements of the Paper?

Like any piece of academic work, it is important that your conclusions follow on from the different sections of your research report and bring these together in a coherent way. If you are unable to do this, it is perhaps an indication that an element or a series of ideas in the report might be better written up elsewhere and does not contribute to this particular piece of work. Beware of putting everything in just because you have data on it. Consider how the different bits of your work contribute to your original question and consequently to your conclusion.

If they don't, then your conclusions are either inadequate or the findings relate to a different question which you may want to pursue at a later stage.

Is It Clear How the Conclusions Are Supported by the Research Findings or Results?

You might find the concept of a 'dissertation audit trail' helpful for reviewing your research at the end. It involves taking each of your conclusions and then asking how they are supported by previous stages in the research process. In other words, having completed your research, you then ask each of the following questions in turn:

• How do your conclusions depend on your research findings?
• How do your research findings depend on your data?
• How do your data depend on your research method?
• How does your research method depend on what you were seeking to find out?

The aim is to show how each part of your research is supported by the previous stage. Occasionally, we have encountered a research report (e.g. a dissertation) where each part taken separately seems sound. However, when it is taken all together, there is a disconnect between some of the parts. The most common place to find such a disconnect is between the research findings and the conclusions. It is not unknown for a research student to present their research findings and then for their 'Conclusions' chapter to include one or more preconceived conclusions which were formed long before the research was started and which are unrelated to the actual research findings. If you keep in mind the idea of a 'research audit trail', you are unlikely to fall into that pitfall.

Does the Paper Identify Any Implications for Practice and/or Society?

Identifying your findings is not sufficient on their own. You need to relate these outward to recognise their broader impact on practice or on our understanding of the world. Increasingly, the impact of a piece of work is becoming part of how we judge its validity, and the universities' Research Excellence framework awards 25% of its marks in assessing quality to the impact of the findings. Making the implications clear in your article is part of answering the 'So what?' question we often ask about research.

Does the Paper Identify Clearly Any Implications or Questions for Further Research?

It is likely that any findings or new areas your research uncovers will point to new areas for investigation and, just as your work has built on that of others before you, so others will build on your work. Including implications or questions for future research is evidence of how you see your contribution within the context of a broader knowledge field and have considered its implications as well as its impact.

Has Attention Been Paid to Clarity of Expression, Including Length of Sentences, Use of Jargon and Use of Acronyms?

You can raise the likelihood that your article will be published if you make sure it is well written; if you want it to be useful, you will need to be sure it can be understood. Online journal articles are often accessed by people from other, but related, discipline areas, and if your article is steeped in jargon or acronyms, these may be familiar to colleagues who know your field but will make it difficult for others outside to make sense of it. Long sentences are inhibiting, particularly when a reader is trying to grasp new or difficult concepts. A clear, well-written article that can communicate to a layperson is of far more use than one that is dense and excluding.

The Issue of Criticality

So far, we've looked at reviewing your own research. In this section, we go a stage further and look at *critically* reviewing your own research.

What does it mean to *critically* review your own work? It means to *think* critically about what you have done and what you have written. But what does it mean to think critically about your work? It means to ask searching questions of your work. What kinds of searching questions? Table 7.2 contains a dozen questions that are used by critical thinkers to interrogate ideas, arguments, evidence and conclusions.

Table 7.2   Questions as tools for critical thinking.

1. What *explicit* assumptions are being made? Can they be challenged?
2. What *implicit* or *taken-for-granted* assumptions are being made? Can they be challenged?
3. How logical is the reasoning?
4. How sound is the evidence for the assertion(s)?
5. Whose interests and what interests are served by the assertions?
6. What values underpin the reasoning?
7. What meaning is conveyed by the terminology employed and the language used?
8. What are the implications of the conclusions?
9. What alternative conclusions can be drawn from the evidence?
10. What is being privileged and what is off-the-agenda in this discourse?
11. What is the context of this discourse? From what different perspectives can the discourse be viewed?
12. How generalisable are the conclusions?

When people talk about a critical review of the literature, they mean reviewing the literature in a questioning way, and Table 7.2 shows the sort of questions they have in mind. These same questions can be used to critically review your own work.

Conclusions

Reviewing your own work means more than just reading what you've written in a passive way. It means engaging with it actively, and this chapter has provided you with the tools to do so.

Table 7.1 offers a checklist of questions of the kind that academic journals use to peer-review articles submitted for publication. This will help you to be sure that you've included all the elements that are expected to appear in a completed piece of research. The appendix to this chapter provides a process for using such a checklist.

When you are sure you have all those elements in place, then Table 7.2 offers a set of questions to help you interrogate the validity of what you have written. Using these questions to *critically* review your work gives you your own quality control facility. They will enable you to test – and, if necessary, challenge – your own assertions and conclusions.

The use of the tools provided in this chapter by Tables 7.1 and 7.2 should ensure that your completed research project meets the most stringent assessment it might encounter.

Appendix

Self-refereeing an Article or Research Report

Checklist for Self-refereeing a Draft Article Prior to Its Submission to an Academic Journal

When an article is submitted to an academic journal for publication, it is normally sent out to at least one referee (and usually two) to be 'peer-reviewed' to assess whether it is suitable for publication. The purpose of this checklist is to enable the author to referee their own paper prior to submission to make it more likely that their paper will be accepted for publication. The checklist can also be used by the author of any form of research report, however, as it covers all the main criteria by which any research report is assessed.

		Mark (out of 5)[1]
Aims, objectives or problem	Does the paper state, at the outset, its aim(s), objective(s) or the problem(s) it is seeking to address?	
	Is the paper clear at the outset about the *significance* of the aims or problem it is addressing?	
Awareness of other work in the field	Does the paper demonstrate an adequate understanding of the relevant literature in the field and cite an appropriate range of literature sources?	
	Are you confident that the paper has ignored no published work that is significant and relevant to the issue it addresses?	
Method	Is the paper clear that the methods employed are appropriate for the aims of the research?	
	Does the paper provide enough detail of the method to enable a reader to repeat the study it they were in a position to do so?	
Findings/results	Are the findings analysed appropriately?	
	Are the results presented clearly?	

		Mark (out of 5)[1]
Discussion	Does the paper discuss its key results in the light of published literature on the subject?	
	Does the paper identify how results make an original contribution to new knowledge?	
Conclusions	Do the conclusions tie together other elements of the paper?	
	Is it clear how the conclusions are supported by the research findings or results?	
Implications	Does the paper identify clearly any implications for practice and/or society?	
	Does the paper identify clearly any implications or questions for further research?	
Readability	Has attention been paid to clarity of expression, including length of sentences, use of jargon and use of acronyms?	
	Has the paper been made as reader-friendly as possible?	

The Bottom Line[2]

Significant, original contribution to new knowledge	Does the paper make an *original* contribution to new knowledge?	
	Does the paper make a contribution to knowledge that is *significant*?	

[1] 5, Definitely; 4, probably; 3, possibly; 2, probably not; 1, not yet.
[2] The basic purpose of research is to make a significant, original contribution to new knowledge.

Part II
Support

8

Research Proposals for Funding

Lowry McComb

Introduction

In this chapter, we will describe the basic principles of applying for funding for a research project. Much of what is covered here will be applicable to other situations, such as applying for funding to attend a conference or for a fellowship after you have completed your research degree.

Find a Suitable Funder

In the current financial climate, most funders of research have great pressure on their budgets for research while, also, the numbers of applications for funding are increasing. In many instances, success rates for applications are as low as 10%. At the same time, the organisations which provide resources for research funders (which will be the government or other public bodies) are requiring them to show value-for-money and to demonstrate impact on society from the research that is funded.

All this means that many funders will now have a well-developed strategy for the types of research which they will support and will be looking for projects which produce impact beyond publication of the results in academic publications. It is thus vitally important that you make your application to a funder from which you have a realistic chance of support. First of all, preparing a good research proposal is a major undertaking on your part, and you will not want to waste your time preparing an application which is unlikely to be successful. Secondly, many funders are now introducing procedures to restrict the number of applications they have to process, and a failed application may affect either your or your institution's ability to submit future proposals to this funder.

The first part of this process should be to produce a short list of potential funders for the sort of research you want to do. You should seek as much advice as you can – your supervisory team is likely to be your first port of call for this information. Other possibilities are outlined in this volume.

Research Methods for Postgraduates, Third Edition. Edited by Tony Greenfield with Sue Greener.
© 2016 John Wiley & Sons, Ltd. Published 2016 by John Wiley & Sons, Ltd.

Having identified this short list, you should then spend some time analysing the strategy of each of these funders to really understand the sort of research projects that they are interested in supporting. Sources for this sort of information include:

- the funder's web site and other publications;
- talking to other researchers who have submitted proposals (successful or unsuccessful) to this funder;
- talking to other researchers who have been involved in evaluating proposals submitted to that funder (probably the most effective way of gaining real insight into a funder's strategy); and
- your institution's Research Office, or an equivalent organisation which facilitates the submission of research proposals.

If you are unsure of any details, then telephone or email the funder for clarification – most will be very happy to answer questions.

Hopefully, you will now have identified an organisation which, in principle, will be able to fund your research idea. It may be that you have realised that your research idea may need to be slightly modified to fit in with the funder's requirements – it is much easier to do this re-scoping now rather than when you are much further on with producing your proposal.

Plan Your Proposal

The next stage should be planning the production of the proposal – I suggest that you treat the writing and the submission of the proposal as mini-projects in themselves. The first point to emphasise is that proposals which are put together rapidly at the last minute to meet a submission deadline are rarely successful. A rushed proposal is normally quite obvious to evaluators and will immediately beg the question "If they cannot properly manage the writing of the proposal, how will they be able to manage the project itself?"

The first stage in planning your proposal writing is to be aware of the final date for submissions and work backwards from there. You will probably need the approval of your institution to submit a research proposal (they will, for instance, want to check all the details, including costings, to make sure the project is doable – legally, the contract for most research projects is between the funder and the host institution). This approval process will take some time to complete – for instance, in my university we need to submit final drafts of research-funding proposals to the Research Office two weeks before the funder's closing date.

So, the latest date for the institutional approval process is the effective submission date you will need to work to – your planning for the rest of the proposal writing should be geared to this.

You should also clarify the application process which the funder expects, such as:

- Must the application be in a particular format?
- Is there an application form?
- What should the application address?
- Do you need to nominate referees or the like?

Ethical Approval

At this stage, you also need to be aware of whether your project will need to go through an ethical approval process and what the timescale is. If you are working with human or animal subjects or using human tissue, then you will need ethical approval; however, many other areas of research may also require this clearance. You will need to check with both your research funder and your institution (probably via your departmental research ethics officer) as to their ethical approval procedures.

Your potential funder may require you to obtain ethical approval for the project via your institutional approval procedure before submission or may perform their own ethical assessment as part of the application process. Also, your institution may require your project to go through its ethical approval process before submission. If you will need approval before submission, then be sure to build in time for this into your planning of the writing of your application.

Write Your Proposal

In writing your proposal, you should follow precisely the guidelines given by the funder and take care to include everything that the funder is looking for in an application. Also, follow all rules and layout instructions – these may seem trivial, but ignoring them may mean that your proposal is automatically rejected.

You also need to write your proposal in a way that your readers – those who will be evaluating the proposal – can readily understand. In the evaluation process, your application will be read by a number of people (see the 'What Happens Next?' section), some of whom will have specialist knowledge of your research area while others will not be specialists. This means that you must write your proposal in a way that the non-specialist can understand, while at the same time providing enough detail for the specialist reviewer. Thus you should avoid using technical language as much as possible, especially in the parts of your application dealing with the wider objectives and impact of the work.

When writing your proposal, there are always a number of questions which you should answer (and these are the ones an evaluator will be judging you by).

What Are You Trying to Do?

You need to define carefully the scope of the project and identify what the aim of the project is and what the specific objectives are. Most funders will not be looking for open-ended projects – they will feel confident with projects with a definite end point, which could then lead into a new application for a follow-up project. Funders will judge the success of the project by whether you meet these aims and objectives.

Why Do This Project?

You need to convince the funders that it is important that this project is done and also why they should sponsor it

To answer the first question, you will need to describe the intellectual problem that your research hopes to solve and give the background to the proposal. You will need to describe other work that has been done that is relevant to the proposal, paying particular attention to studies which have previously been funded by the sponsor you are applying to. Do not forget to include relevant work which has been done by those who are likely to be referees or evaluators for your proposal!

As well as the academic justification for the proposal, you will also need to describe how your research fits in with your funder's strategic objectives. This may include considering the potential non-academic impact of your proposal and how this might be exploited. For example, if the work is likely to lead to an academic monograph, what are the opportunities for a parallel publication aimed at the non-academic market? Will your research generate any technological advances which could be patented and commercially exploited?

Why Are You Doing This Project?

You may have written a very strong research proposal, but you will still have to convince the funder that you are the best person to do this piece of work. Your proposal should show that:

- You have the necessary academic experience and ability to do research at this level – this could be evidenced through a strong publication record in your discipline and your track record to date.
- You have the necessary research skills to carry out this piece of work. You should establish that you can use the majority of the research techniques that you plan to use. Funders will not expect you to have all the necessary skills now, but if you are not experienced in using some of the necessary techniques, you should describe how you will be trained in these.
- You will have access to the necessary resources to carry out this work. This could cover, for instance, lab space, specialist computing facilities, access to archives, specialist library resources and so on.
- You have the ability to manage the research project – this could be evidenced through other projects that you have successfully managed.

How Will You Do It?

This will be a major part of your application and should describe in detail how you will go about doing the research. You will need to break the whole programme down into the constituent tasks and identify who will carry out each of these tasks, as well as the non-staff resources which each will require. You will need to identify and justify the methodology which you will employ for each of these tasks. You should also specify a number of milestones (which could be activities such as construction of a piece of equipment, completing the data collection in a survey, submission of a paper for publication, a collaboration meeting and so on) which can then be used to judge the progress of the project.

What Will You Need to Do It?

You will need to specify accurately all the resources needed to complete the project and cost these. As part of the evaluation process, reviewers will consider if you have all the resources needed to successfully finish your programme, so it is in your interests to be realistic. Few funders will be impressed if you come back during the project needing more for resources. Once you have identified what you need to carry out the project, your research office will need to be involved in the costing. They will also be able to advise you on which resources a particular funder will be prepared to include, and what will need to be covered by your institution.

All projects will have their own particular resource needs – however, as a starting point, you should consider the following categories.

Staff

You need to consider carefully the staff effort which will be required to carry out the project. This should include your own time as principal investigator, any other academic staff collaborating on the project, research assistants who will be employed to deliver the project, technician and secretary support and so on. You should quantify the effort needed in terms of their full-time equivalent (FTE) effort.

You should then calculate the direct cost of employing each of the staff who are contributing to the project – the direct cost is the salary (or fraction of a FTE salary) of the staff member plus the amount the institution will have to pay for the employer's National Insurance and pension contributions (or equivalent). Your research office will be able to provide you with detailed salary tables.

The precise categories of staff which you can charge to the project will depend on the funder's policies. In most cases, the costs of employing research assistants, technicians and secretaries will be an allowed cost. The position of principal and co-investigators is more complicated. Some funders will allow for their employment costs, and in this case you should charge your time to the project. In other cases, the funder will assume that your institution is responsible for your time and so will not cover your employment costs; in this case, you should make clear in your application that the institution is paying these as an institutional contribution to the total cost of the project. You should take advice from your Research Office on the whole area of staff costs.

Travel and Subsistence

You will need to include realistic estimates for travel and subsistence for essential travel associated with the project. Most funders will regard dissemination activities as essential, so you should include requests for funding to attend appropriate conferences to present the results of the project. However, in a project involving collaboration with colleagues in other institutions, most funders will expect a limited number of face-to-face meetings, with the majority of the interaction taking place by electronic means.

Consumables

Here, we include general running costs and materials that will be used in the project. Generally, you will not need to itemise these but, as with any other cost, you will need to provide careful justification.

Equipment

Here you should consider what large or specialist items of equipment will be needed for the project. Note that many funders will assume that the research will be taking place in a 'well-found laboratory' so they may not be prepared to fund small items of equipment (including personal computers etc.) that you will normally have access to. You will need to establish that this equipment is necessary to carry out the project and that you do not already have access to it. You should establish if it will be more economical to hire the equipment or purchase it outright. You will be required to provide full details of the proposed item, the supplier and the cost (possibly a quotation) and the time the equipment will be used for the project. You may also include maintenance costs for any major items.

It is becoming more normal for a university to fund expensive, common-user pieces of equipment (e.g. expensive analytical instruments) and recover the capital and running costs by requiring users to pay for their use of this equipment through their research grants. Costs for using this sort of shared equipment should be included here.

Other Items

Here you can include items like recruitment and advertising for staff. Publication costs (where appropriate) may also be allowed. Think carefully of any other financial implications that may be associated with carrying out the project.

Indirect Costs (Overheads)

UK universities are now expected to calculate the 'full economic cost' of externally funded research projects which they undertake. This includes both the costs that are directly attributable to the project – as described in this chapter – and the other costs to the institution hosting the project. These costs include:

- costs of accommodation (e.g. office and lab space, heating and lighting etc.);
- human resources costs;
- costs of running the finance and purchasing departments;
- central facilities, such as library and IT;
- staff training; and
- departmental services.

All these services will need to be provided by the institution to run the project. The process of calculating these costs and attributing them to specific projects would be too complex – the normal method which is used is to associate each category of staff with a notional indirect cost of employing them. There will thus be an indirect cost associated with each person contributing to the project, and this needs to be included in calculating the total cost. This whole area is complex, and you must consult with your research office (or equivalent) to obtain the correct figures for this. Note that in practice, the indirect costs per staff member will be roughly the same as the staff cost (salary plus institutional national insurance/pension contributions).

Although these indirect costs will increase the total amount applied for, you should not expect that this will increase the amount of money you have to spend. These are centrally incurred institutional costs, and the bulk of such funding will be retained by your institution. A fraction of this may filter down to your department, but you are unlikely to have this to spend – the precise situation here will vary from institution to institution.

How Will You Manage the Project?

The funder will need convincing that you, as the principal investigator, will be able to manage the project and to produce all the deliverables on time and to budget. In most applications, you will be expected to have a section describing the project management plan – and the amount of detail here will depend on the size and complexity of the project. As a minimum, the project should be broken down into a number of high-level tasks (normally referred to as *work-packages*). Your management plan should then include:

- details of who is in overall charge of the project;
- details of who is in charge of each of the work-packages;
- details of how the various work-packages are co-ordinated (normally via an overall project manager);
- timescales and interdependencies for completion of each of the work-packages; and
- appropriate milestones to enable measurement of the progress of the project.

Much of this can be best illustrated by including a Gantt chart as part of your project plan – this provides a visual representation of the timelines for the various tasks within the project, as well as their interdependencies. Various software packages are available to help produce these.

Of course, if you are successful in obtaining your funding, there will be many more practical aspects to managing the project – such as how your institution will expect you to manage the budget, the appointment process and management of staff associated with the project and so on. Normally, the funder will assume that these will be in place. However, many institutions now provide development courses for new principal investigators on research projects. If managing an award is new to you, then you should mention that you will undertake such training in your funding application.

How Much to Charge?

Your work on costing your research proposal will have resulted in a figure for the cost to your institution of carrying out this research. You then need to consider how much to charge the funder of the research. This may not be a straightforward decision, and you will need to involve your research office and, perhaps, your head of department in coming up with this figure.

Some funders of research, particularly those that are dispensing public money, will have strict sets of rules as to what can be funded. For instance, at present (in simple terms) the UK Research Councils will fund 80% of the full economic costs of a research project. Many charities (particularly those that fund medical research) will not contribute to the indirect costs of a project.

With other funders, particularly commercial or industrial sponsors, there may be more scope in pricing the project. Your institution may want to charge more than the full economic cost in order to generate some 'profit' from the project. If this is the case, then you might expect some of this will end up in your research account. However, it may be in the institution's interests to charge below full economic costs if, say, a pilot project could lead to much bigger projects in the future, or to establish a relationship with a new funder. Similarly, if the funder wants to retain the intellectual property rights (IPR) in the research, then the institution might want to charge a premium as compensation for the loss of opportunity costs associated with exploiting the IPR.

You should make it clear to the funder that the price you are quoting is exclusive of VAT. The situation here is complex, and your research office will be able to tell you if the project is liable for VAT.

Checks before Submission

It is important to realise that many funders will assess your application on other aspects beyond the quality of the proposed research and the quality of the individual (or team) which will carry out the project. Each funder will have its own set of criteria which will support its strategic aims. Some typical examples are:

- Will the completed project produce innovative economic, environmental, social and/or cultural benefit to the national and international community?
- Are there suitable plans for dissemination, commercial exploitation (if appropriate) and/or promotion of the research outcomes?

You should thus make yourself fully aware of the assessment criteria that the funder will use – these will normally be freely available in the documentation for the scheme you are applying to. In practice, it is likely that most proposals will score highly on the quality of the research proposal and of the research team. Often, it is these other criteria that will make the difference between a proposal being funded or not.

As you approach producing a final draft of your proposal, it is worth putting yourself in the place of a reviewer and to assess your proposal dispassionately using the published evaluation criteria. Here, you should check that you have covered all the areas that will be assessed – if you have not covered something in your proposal, you cannot expect to score well in that category!

You should also arrange to have your draft proposal read by both a subject expert and also by someone who is not an expert in your field. Any useful feedback from them should be incorporated into your final draft.

You need to make a final check of all the administrative details and check that your application meets these (e.g. length, typeface, required appendices etc.). The final draft should then be sent in good time to your research office (or equivalent) for the final approval process. Once this is done, the final submission can be made. In some cases, you will do this; in other cases, the funder will insist that the formal submission is made by your research office.

What Happens Next?

The normal procedure is that the funder will first do an administrative check that your application meets the scheme's requirement. This will include:

- Is the area of research within the scope of the funding scheme?
- Does the applicant meet the scheme's requirements?
- Is the application complete?

- Does the application exceed any length limits?
- Does the funding request come within the scheme's upper and lower limits?

In most cases, if the application fails any of these checks, it will be rejected without further review. It is thus vitally important that you check that your application complies with these administrative requirements.

The second stage in the process is that the application will be sent out to a number of expert reviewers who will be asked to grade your application in accordance with the scheme's evaluation criteria and also will back up their score with the reasons for their grading, considering both the strengths and weaknesses of your application.

Some funders (but not all) will then send the reviewers' comments to the applicant to consider and respond to – if this happens, you will normally have a very limited time to produce your response. You should read the reviewers' comments very carefully and craft a suitable response to these. In some cases, there will have been a misunderstanding on the reviewers' part, and you will have the opportunity to clear this up. In other cases (e.g. the necessity for some of the resources you request), you may want to comment on the effects of cutbacks on the scope of the project. It is also quite likely that there will be a spread in scores from the various reviewers – it is the next stage that will try to resolve these.

Next, the awards panel will consider all applications for this funding round (along with the reviewers' reports and your rejoinder, if allowed) and rank all the proposals. Here, it is important to realise that the evaluation panel are not likely to be experts in your particular research field, so your application needs to have been written in a way that is understandable to a non-specialist. The panel will attempt to reconcile the various reviewers' reports, standardise the scoring between the various evaluators and also decide on the amount of funding which should be given to each proposal, if it is successful. The result of this evaluation panel meeting will be a ranked list of proposals for funding along with the amount that should be given to each one.

The final stage is purely administrative. The top-ranked proposal will be allocated funding, then the second-ranked proposal and so on until the pot of money available is exhausted. Those proposals that are left when the money runs out will not be funded.

If You Are Unsuccessful

Normally, unsuccessful proposals will be given feedback from the evaluation panel. Read this carefully, and consider whether it is worth revising and resubmitting the proposal (if the funder accepts such resubmissions – not all do). If the feedback indicates that the proposal was, in principle, fundable but that there was not enough money available in the current round, then resubmission may be in order, especially if there are some improvements which could be made in the presentation or technical aspects of the application.

If the feedback is that your proposal was outside the funder's strategic objectives or priorities, then you will need to consider finding another potential funder, rewriting the proposal and submitting it to this new sponsor.

If You Are Successful

In many cases, a successful application will be awarded less funding than was applied for and often with the expectation that the whole research project will be delivered within this reduced budget. If this happens, you will have to consider carefully whether the project can be delivered with this reduction in resources, or whether the scope of the project will need to be reduced. If there will be significant difficulties in delivering the programme within the reduced resources, you should immediately discuss this with the funder as part of the contract negotiations discussed further in this chapter.

After the formal announcement of your success, there will be a series of negotiations involving the funder, your institution and yourself to draw up a formal contract between the funder and your institution. This will include:

- the conditions of the grant;
- the start and end dates of the project;
- the timing of payments to the institution;
- reporting requirements;
- arrangements for early termination of the project by either party; and
- ownership of intellectual property rights.

Finally, the contract will be signed and you will then be in a position to start the real work!

The Future

In many ways, being an academic researcher involves frequent applications for research funding. Now that you have secured your current grant, you should start thinking ahead and be planning your next applications. To facilitate this, you should be evolving your own long-term research plans and strategy, perhaps looking forward for the next 10 years or so. Of course, your plans need to take account of your potential funders' strategies, and it may need some thought to pull these two requirements into alignment. It will not be long before you will be starting the application process for your next research grant.

9

Who Can Help?

Shirley Coleman

Introduction

Your supervisors will be able to help you in your research, they are the first port of call for ideas, support and encouragement. But if they are not providing the help you need, then you can find help from other academic staff, librarians and technicians. This chapter also looks at a wide range of other sources of help including public services, government departments, trade unions and industrial companies. Even if you feel you have enough help, contact with these sources can be interesting and useful later on in your career. These experiences may add that little extra sparkle which will make your research excellent rather than just good.

Supervisors

The potentially most useful resource is your personal supervisor. His or her task is to help you go through your studies, get the best from them and finish successfully and ready to take up suitable, gainful employment. Some supervisors, however, are not keen to dedicate sufficient time to you and this can be very frustrating.

If you look on the bright side, and assume your supervisor is interested in your studies and conscientious, what can you expect?

You can have regular meetings at which you summarise your work to date or since the last meeting. This means that you keep a good record of your activity throughout your period of study. Your supervisor can add his or her own thoughts and ideas to yours and offer suggestions that spark off further ideas in either of you. As well as you keeping an eye on published work and the advances in your field, your supervisor can be doing the same. This is useful as you can keep tabs on much more work and be more likely to stumble upon relevant information.

Besides providing technical support and inspiration, a good supervisor will deal with all the administration to do with your study. This could include the important question of choosing appropriate (and hopefully sympathetic) external examiners and applying for time extensions.

Research Methods for Postgraduates, Third Edition. Edited by Tony Greenfield with Sue Greener.
© 2016 John Wiley & Sons, Ltd. Published 2016 by John Wiley & Sons, Ltd.

The exact title of your thesis, if you are writing one, or of any publication, is very important; remember, it will follow you around on your CV and job applications for ever more, so a good supervisor will help you choose the best wording which will allow you to adapt your past to suit your future.

A good supervisor can help you to meet academics and others who may be interested in you and your work. He or she may become a life mentor for you, helping you to find employment after you finish studying. If you have to defend your study in a viva voce, a beneficent, supportive supervisor is a great asset as well as a comfort, especially if you are borderline.

Poor supervisors may be unwilling to see you or may not concentrate fully when you try to discuss your work with them.

If your supervisor is poor, what can you do about it? You can try to improve your availability and flexibility about when meetings are held; you can try to improve the quality of the presentation of your work. Your enthusiasm may ignite the interest of your supervisor, if you're lucky. If not, at least you will feel that you've done your best to improve the situation. In the final analysis, if none of these efforts have any effect you just have to try to carry on regardless and look wider afield for people to be involved with. It is worthwhile considering why your supervisor is under-performing. Sometimes supervisors are too busy; it is not always a good plan to go to study with someone whose fame is meteorically rising. They will not have much time for you. Perhaps you or your project bores your supervisor. Another possibility, however, is that your supervisor thinks you are highly competent, are doing a good job and don't need their help. Although this may be a flattering concept, it is also a nuisance as everyone can benefit from the input and interest of an eminent mentor, but again unless you go to great lengths to prove your incompetence there is not much you can do except get on as best as you can on your own. Try to avoid becoming too self-centred. In a few years' time, you yourself will have forgotten most of the vital details you are struggling with now.

In the rare (hopefully) cases where a supervisor is actually incompetent or maleficent, what can you do? You can try to change to another supervisor without becoming too involved in accusations and unpleasantness. Be careful how you choose another supervisor; someone who is keen, conscientious and compatible may be the best even if he or she is not the most renowned. Ask previous tutees how helpful they found their tutor. Try to check whether they are likely to move jobs or retire – racing them up the stairs or challenging them to tennis is not a reliable test! If it is not possible to change your supervisor or it's too late, the best you can do is keep a good record of your work. Don't rely on the supervisor at all, but use the other resources (suggested in this chapter) in case you need to defend yourself later. Be positive and think of your traumatic period of study as good training for coping with later life. After all, if you're careful no one can stop you from doing your work, writing it up and passing successfully. This is more useful than making a big fuss and trying to upset a structure that has so much inertia that it appealed to you as a stable place to study in the first place. If you do need to take matters further, you can always discuss your problems with the head of department, or dean of the faculty. If these people are unhelpful or you wish to go further, seek advice from the students' union and student affairs officers before you do anything rash. Academics have long memories.

Even if you have a good supervisor, other sources of help can provide you with practical examples of where your study is relevant. This makes your work more interesting and allows more unusual questions to be asked when you are interviewed for further study or jobs later on.

Attracting Ideas

Try advertising yourself! If possible, it's a good plan to give a talk or seminar about your work early on in your study. That way you get help in terms of ideas and suggestions from your audience, and you can check what other work is being done in your field and by whom (this reinforces your own literature search).

Academic Sources of Help

Academic members of staff are usually happy to spend a few minutes discussing ideas which interest you, arising from their study. It's often better just to drop in rather than be too formal. You can always ask their secretaries for a good time to catch them.

The authors of relevant papers in journals that you have come across during your study are also likely to be enthusiastic about discussing their work with you. Use the 'address for correspondence' on the papers to contact them. This sort of pro-activity not only provides you with state-of-the-art ideas but also gets you known and may be useful when you are looking for a job. You can combine a visit to more distant authors with the summer holidays, but remember to take some presentable clothes away with you and a notepad!

Conferences often have attractively reduced rates for students and are usually well worth the money and effort of attending. You can cross-examine authors of relevant presentations and make some useful contacts for sharing ideas, help and inspiration. Occasionally, there are conferences especially for post-graduate students. The subject matter will be wide ranging, but they provide a good opportunity for you to make a presentation and for you to share your experiences with other post-graduates.

Universities and colleges usually run a series of seminars and lectures during term time; again, the subject areas covered will be wide but these are often well worth attending.

Other Sources of Help

Libraries

Help may be available from a surprising number of diverse sources as well as the more obvious places. University, college and public libraries are a standard starting point. Besides the books and journals available, try talking to the specialist librarians who have a vast pool of information at their fingertips. Government publications are surprisingly rich in all sorts of facts and figures; the monthly publication *Social Trends* for example, will tell you how many died from falling off ladders in the last month.

Have a look at the reports from government and other research institutes; they may give you some good ideas and contacts.

Government Departments

Government departments may have local offices that you can visit. You can find out which companies are involved in the latest initiatives and whom you should contact to ask about them.

One government initiative is Knowledge Transfer Partnerships in which graduates are employed by a university as a Research Associate for one to three years but work full-time in a company on a specific project. These projects may be relevant to your study, and the list of current programmes could be very useful.

Official statistics bodies may offer help and ideas, for example the Office for National Statistics. There is also the European Commission and the various government departments, for example the Department for Environment, Food and Rural Affairs.

Local Services

Town halls and civic centres have excellent information facilities and are able to supply data on many aspects of the local economic and social scene. Try looking up your city, country and local councils on the internet.

The local Chamber of Commerce is a good source of literature and contact names and addresses. They may be able to put you in touch with appropriate trades societies or trades unions. They are involved in research as well and may be willing to help you. The Chamber of Commerce will also have literature from the Institute of Directors, Confederation of British Industry and other institutions.

Miscellaneous

The financial pages in newspapers give useful information about companies. Keep a collection of any articles that are relevant to your area of study.

Keep a check on the activities of local colleges and schools; they sometimes have project competitions in very interesting subject areas.

Groups of students may work with a local company, and you could make use of the same contact.

Hospitals and healthcare providers have personnel dedicated to disseminating information. They also have staff libraries that you may be allowed to browse through if your study is in an area associated with health.

Technical Support

Laboratory staff know a lot more about the way things work than you can learn from a book, so make sure you are on good terms with them and listen carefully. Materials and equipment supply companies are very happy to furnish literature and demonstrations to post-graduate students. You may expect them not to be interested in you as you have no intention of buying anything, but on the contrary, they see you as the customers and people with influence on the future. They will be very keen to promote their products.

Companies

University or other careers services are probably the best places to learn about companies of interest to you. Usually the information is readily available for you to look at without having to make an appointment or talk to anyone. There is nothing more satisfying when you go to visit someone than knowing more about their company's business than they do! You can try to organise your own visit with a specific person in the company or you can contact their public relations department and try to join a factory tour. In either case, remember to dress the part and take a notebook.

Making a Visit

Arrange the visit when you are sure you are ready – don't risk having to cancel. Read up as much as possible about the place and the people who work there. Be clear in your mind what it is that you want to achieve by the meeting, want to know or want to look at. Write out a checklist to be sure you don't forget anything. Double check your travel arrangements and take a map. Allow plenty of time to get there and plenty of time afterwards in case you're re-invited to stay for lunch, tea, evening meetings or the like. Wear appropriate clothes. Afterwards, write a quick thank you note saying briefly what benefits you gained from the visit. That way, they will be happy to have you back and will remember you if you apply for a job there!

Internet

All of the above can be researched from the comfort or otherwise of your computer via the internet. One of the advantages of surfing the net is the global perspective it gives. One of the disadvantages is that you may feel that the field is expanding out of control. You should find, however, that eventually you start recognising the same things turning up again and again, and that is the beginning of getting to grips with the subject.

Be careful not to spend too many hours looking at the computer screen and playing with the mouse; time flies when you're having fun. Make sure you bookmark the addresses of any pages which interest you, just in case you can't find them again, and print off interesting, relevant information so that you can read it at your leisure, for example in the bath, on the beach or in a bus.

There is still some information which is not public enough to be put on the web but which is not secret enough to be kept from you, so the internet is complementary rather than alternative to the other sources of help.

Family and Friends

Don't forget this valuable collection of committed supporters. Explain your ideas to them and try to cope with their common-sense comments – they often say things that make you think very deeply about what you're trying to do. In most write-ups, you need a layman's introduction, abstract or summary; try reading your attempt to a friend or explaining your project to them, and see whether they can understand it. An outside view will help you to bring out the important points more clearly.

Summary

In summary, there are many sources of help if you have the energy to go out and seek them. The benefits will be widespread in terms of making your study more interesting and relevant and in the advantages it provides in your future career.

10

Information and Library Services

Claire Abson and Alastair Allan

Introduction

You have probably used a university library as an undergraduate. However, as a post-graduate, whether on a taught course or pursuing your own research, you will need to be a more independent and knowledgeable library user. After you have found and interrogated appropriate sources for your subject area, you will need to evaluate their usefulness and accuracy.

Print sources continue to be important, although the degree of importance will vary depending on your subject discipline. However, the vast majority of relevant material will be electronic. It would be impossible to do justice to all the sources available for different subjects in an introductory chapter. Such an introduction would be out of date by the time of publication. Instead, in this chapter, you will be pointed towards ways in which to locate, evaluate and make effective use of the information that is available to you, and how to get the most out of your library, whatever your subject and institution of study.

Know Your Librarian

Most university libraries have dedicated staff responsible for particular subjects, and it is a good idea to find out who your librarian is. The level of support you can expect will vary but, at the very least, there should be someone with the job of ordering library materials for your subject area. That person should also be knowledgeable about information sources, and will probably have some responsibility for training students to develop effective information-seeking behaviours. They may have special arrangements for training staff and researchers, or they may be willing to see you on a one-to-one basis and deal with your particular needs. If they are unable to do this, they should certainly be able to provide you with the information you need to get started.

Research Methods for Postgraduates, Third Edition. Edited by Tony Greenfield with Sue Greener.
© 2016 John Wiley & Sons, Ltd. Published 2016 by John Wiley & Sons, Ltd.

Develop the Skills You Need

Before you begin to search for information, you will need to think about precisely what you are looking for. It is much better to be as specific as you can when you first start, and think carefully about words and phrases you would use to describe your topic as it will save time later. It's also helpful to be familiar with what are known as *search operators*. These are *and*, *or* and *not*, the connecting words you use to string together your search terms. Information databases will be discussed in more detail further on in this chapter, but it suffices to say user interfaces vary, and some provide you with more support than others in how to structure your search, so equip yourself as well as you can before you start. A well-planned search strategy is invaluable.

A search for information about women in the workplace, for example, might look something like this: 'WOMEN and WORKPLACE' would be obvious, but wouldn't necessarily identify all relevant material. Most databases will allow you to use a 'wildcard' where you can replace a letter with a symbol, commonly a question mark, so you could replace WOMEN with WOM?N and find 'woman' as well. In addition, the term 'workplace' is only one way of describing a place of work or a place in the workforce. *Work*, *office*, *factory* and *labour market* are all alternatives (or synonyms) you might use. You may not want to include work in the home. A better constructed search, therefore, might look something like:

WOM?N <u>and</u> (WORK <u>or</u> WORKPLACE <u>or</u> OFFICE <u>or</u> FACTORY) <u>not</u> HOME

This is not a full example, but it illustrates the importance of thinking about the search and trying to structure it before you start. As you begin interrogating different sources, you will need to amend and refine your search further. In addition, as your search generates relevant results, you will find that the sources you are using can also help you. Reference lists are usually included with online references, particularly for journal articles, and these can direct you to other related sources and provide ideas for other relevant search terms.

In addition to thinking about the keywords you use, you also need to ask other questions, for example:

> **Currency**: How up to date does the information need to be? If you are a scientist, the sources you access will need to be current. If your research is in the humanities or social sciences, this may not be as important.

> **Nationality**: Is the country of origin of the material you find important? If you are engaged in educational research, you may be interested in material relating only to the educational system in the United Kingdom, for example.

> **Language**: If your language skills are limited, you will need to limit your search accordingly.

> **Peer review**: You will come across a variety of different types of material as you search, and you will need to be aware of the differences in quality. Evaluating sources will be covered in more depth later in this chapter. In other words, you will need to think about what you want to exclude from your search as well as what you want to include. Most information sources will allow you to refine your search by these criteria, possibly before you even start searching. If you are unsure about how to begin constructing your search strategy, or you encounter problems, library staff will be able to help you.

Tracking down Books

Although the software and interface will vary, essentially most academic library catalogues, or OPACs (Online Public Access Catalogues), are the same. They will generally allow you to search the library collections using a range of options like the author name or the keyword, and there may be other more complex or refined searches you can perform. You will also find that your institution includes other materials in the catalogue, such as student dissertations and DVDs. Electronic books (e-books) are increasingly prevalent, but you will need to use your library to get printed books. The majority of research books are published on paper. Most university libraries buy a range of e-books, but these are generally core undergraduate texts and although you may find them valuable, many of the books you will need to use may not be available electronically.

If you have used an online catalogue of any kind as an undergraduate, you will pick up a new system easily. Innovations called *discovery systems* are becoming increasingly common alongside library catalogues. A discovery system will show you a random selection of content from online sources, such as electronic journals. This will provide you with a good starting point.

However, you will need to become familiar with databases and other sources, and develop your searching skills, in order to obtain a full picture of the available materials on your subject. The value of information databases will be explored later in this chapter.

You will progress beyond the limits of books quickly and, as you make inroads into your research, you will realise that your core reading will be journal articles, conference papers, theses and R&D reports. However, you will probably need to identify and consult books relevant to your topic that your library does not hold. Most university libraries now make their online catalogues available on the web, and your library's own web pages may have links to gateways to enable you to access them easily.

Some university and research libraries working in similar areas (geographically or in terms of their status or the nature of their collections) have put their holdings together to form union catalogues. These will allow you to search several library collections at once by inputting a single search. The best example of this is COPAC (http://www.copac.ac.uk), which brings together the holdings of more than 70 of the largest university research libraries in the United Kingdom and Ireland. These include all sections of the British Library as well as Cambridge and Oxford University libraries. More than 25 million different books or reports in over 450 languages are represented. Briefly, COPAC allows you to search author or title names or keywords and displays a considerable list of results.

There may be other institutions with a strong research profile in your subject area, and you may find it useful to search their catalogues. You can find these easily on the web. There are two ways of getting access to books in other university libraries, and they are discussed in the 'Accessing Other Libraries' section.

Journal Articles and Electronic Sources

The structure of a research strategy has been shown here. It is valuable to keep a note of the terms you use to search and notes about the successes. For instance, in some cases it is valuable to search for biological subjects using taxonomic names but this is not so with other sources. In some places, you must use American rather than British terminology. Notes of these help you to ensure that every set of searches has been thorough. Using reference management software may help you to keep a record, and this will be discussed in more detail later in this chapter.

Once you have formulated your search strategy and considered your priorities, you will need to identify the sources that can direct you towards the most relevant information. If you require current information, for example, you will need to identify any peer-reviewed journals published online and to which your institution subscribes. It is now commonplace for journals to be published in this way as, in addition to cutting costs, the information is available much more quickly.

It is unusual for academic journals to be published only in print form today, and many titles now no longer have a paper version. At an early stage, it is also valuable to identify current awareness databases like Current Contents which provide abstracts of recent articles and index their contents.

If you are returning to study after a lengthy break, you may have used printed abstracts and indexes in the past. The information databases you will now encounter began as electronic versions of these printed abstracts and indexes, but there are very few remaining in print.

Information databases are commercial sources that your library purchases through the web. Services you use on the web like Google are correctly called *search engines*. Large commercial sources like Web of Science should be called *databases*. There is a difference, even though when you use them they have a very similar look and feel. A search engine like Google is something provided on the web free of charge. They work by looking at your search and then finding the same string of letters you have typed. They are never able to interpret meaning. Equally they are never able to tie up searches, and so search engines will find 'carbon dioxide' but not be able to retrieve anything on 'CO_2'. Tips on how to get the best out of search engines are covered in Chapter 11.

Databases, in contrast to search engines, are constructed by experts who will create lists of subject headings and link ones that are the same or similar. So, for instance, a medical database will link AIDS and HIV and will also 'know' that the full term is 'acquired immunodeficiency syndrome'. Databases record all the articles from selected journal titles and allocate key terms to them that describe the content of the article. The result is that different searches can find the same article. Databases also give fuller, and often more accurate, results than search engines.

You can use your carefully composed search strategy to search online for references to journals articles and, often, the full text of journal articles. Databases that contain only references to journal articles will usually give you an abstract or summary of the article content, to guide you as to whether or not the full article is going to be relevant. The subject focus, the national bias and the language coverage will vary from one database to another, even where the databases cover the same broad subject area. The time you have spent thinking about exactly what sort of material you require, and what you do not want, will help you decide which databases to concentrate on.

If you are accessing a database that gives you only references to journal articles, the references it contains may only overlap slightly with the journal holdings of your university library. There are estimates that claim there are over 200,000 academic journals published worldwide, but such an estimate is impossible to verify. The British Library has subscriptions to more than 40,000 major research e-journals in English and knowledge of thousands in other languages. This number excludes free journals, annual publications, government documents, newspapers and journals that are 'difficult to find'. So, therefore, it is possible from this figure to estimate that there are between 75,000 and 90,000 major journals available, and this number excludes ones written in minority languages or provided for a small online audience. In 2013, it was common for most British academic libraries to have access to over 20,000 e-journals. For this reason you will probably be able to find some of the journal articles that appear in your search results, but you will have to obtain some from other libraries. Of the bigger libraries, the Bodleian at Oxford has 38,000 e-journals but this number excludes free open access (OA) journals, a sector of journal publishing which is rapidly developing. In the United Kingdom, it is now a condition of many research grants that resulting articles should be published as OA. This can be either in an OA journal or through a free or OA repository.

Databases are a reliable source of information. You are being referred to a piece of work published in a journal of recognised academic quality. Usually the better databases only include peer-reviewed journals, so they would be preferable to Google Scholar. However, that is not to say that the resources freely available on the internet are not subject to the same rigorous academic processes because, in many cases, web-based articles are just as valid as those published in mainstream journals. You just need to evaluate the sources you use more carefully.

In this chapter, you have already been introduced to ways in which you might evaluate the usefulness of a particular database for your research needs. However, if you are making use of internet resources, you need to assess the quality as well as the appropriateness of the information. Your first port of call should be a subject gateway or other site where evaluative work has been done on your behalf, generally by practitioners, other experts or librarians in that field. Your library may have its own set of web pages linking you to relevant resources in your subject area. Even if it doesn't, you should be able to find relevant subject gateways where that work has already been done.

When you do need to use Google or another search engine, there are additional criteria you should apply to evaluate what you find:

What, if any, research or evidence is the material based on? Does it make reference to any published sources of information?

Where has it come from? Can you identify a body or organisation responsible for it?

Who wrote it? Is there an individual associated with the material? If so, what do you know, or what can you find out, about that person?

Why was it written? Could the author, or the body involved, be biased?

The authority of your own work depends on you asking these questions. If you're unable to answer them, or any of the answers are resoundingly negative, think twice about whether to use the material to inform your work.

Accessing Other Libraries

Your own university library will almost certainly *not* hold all the relevant books and journals you require, so it will be necessary to obtain them from elsewhere.

Inter Library Loan, now often referred to as Document Supply, is one way to obtain material. However, the benefits of searching the catalogues of other libraries have already been described and, where there are other institutions with a strong research profile in your subject area, it may be easier for you to visit their libraries and consult material yourself.

The majority of UK university libraries are members of the SCONUL Access scheme. You can apply online to access your chosen institutions by going to http://www.access.sconul.ac.uk. You may also be able to borrow materials from their libraries, but loan policies will vary from one institution to another. Licensing restrictions will also limit the online materials you can access.

If you do need to make use of Document Supply services, do bear in mind that different institutions' policies vary considerably in this area. You may find that your university library will allow you to request as many items as you require, or you may be limited to a small number. It may be that your department has an allocation to distribute as it sees fit. Familiarise yourself with the local situation, and plan very carefully what you need to request and what materials you can obtain yourself. You should also plan in advance, as far as possible. Journal articles can be obtained quickly. However, books, conference proceedings, theses and R&D reports can take some time. It is often better to feed requests into the system gradually. If you ask for several books at the same time, you may not have enough time to study them all before you need to return them. The British Library Document Supply Centre (in Wetherby, West Yorkshire) is also open to the public, and visitors can use its extensive stock.

Studying at a Distance

It is becoming increasingly common for students to work at a distance from their institution of study. Universities are increasingly flexible with students around their place of study, but don't assume that just because you choose to work remotely your university will realise this. Make sure you are aware of the implications of spending the majority of your time away from your institution. If there is a taught element to your course, this will restrict you. The 'Accessing Other Libraries' section which discusses the SCONUL access scheme is relevant. Access is almost always granted to distance learners.

As databases are generally available through the web, there is more scope for studying at a distance. You will need to discover what passwords you will need to access the relevant resources. Remote access to electronic resources is becoming increasingly common. There may be additional support available to you as a distance learner; for example, your university library may offer to supply postal loans of print materials.

Keep a Record of What You Have Done

One of the most important things to remember when conducting research is to keep an accurate record of anything you use. You will probably, during your undergraduate studies, have been required to produce citations and reference lists in a standardised format. If not, you should familiarise yourself with the style commonly used in your institution or the key journals in your discipline. The Harvard referencing system is the one most commonly used in the United Kingdom in the Social Sciences (see Chapter 38), but the Vancouver system is universal for medicine, for example. Library staff will often be able to give advice on appropriate standards in your area. The library will probably produce online and print guides to referencing and citations. Alternatively, there are Internet sources that will assist you in referencing a range of materials.

Managing and recording the material you use is much easier if you use a reference management package, or piece of software. Your university will probably have one that they recommend and support, such as Endnote or Refworks. A package like this will help you to organise the material you've used in a logical way, and easily convert it into the required format for your reference lists. In addition, there are many videos on YouTube that demonstrate referencing systems and also show you how to use freely available reference management systems like Mendeley or Zotero.

Do not forget to keep a record of searches you have done. Some databases will allow you to save a search strategy you have used so that you can return to it later, or set up an alert so that the database can email you if additional material matching your search criteria is published. This will save you a lot of time and ensure you are completely up to date with developments in your field of study.

Research Data Management

Research data management is the technique whereby researchers ensure that their research data remain available to future researchers over a longer time period than their own project. There is a considerable taxpayer investment in research, and it is reasonable that the fruits of this investment should be permanently available. Output from research in the 21st century is mostly electronic, and only a small proportion of research findings are made visible. The data collected to support one project may well be ideally suited to support others conducted in the future (Managing Research Data, 2012). The responsibilities for research students are that firstly they must conform to the research data management requirements of their funding body and be fully aware of the policy of their own university. The best British organisation to provide advice is JISC (Joint Information Systems Committee). JISC supports the Digital Curation Centre, and its website (http://www.dcc.ac.uk/) is the best available source of information about research data management.

Conclusion

You now have enough guidance for you to begin searching for material to support your research. The means of accessing quality information are changing constantly, and what constitutes the latest technology today will soon be out of date. Electronic journals have now overtaken paper journals in most subject areas. However, as electronic textbooks are now a reality, it may only be a matter of time before access to academic texts is almost wholly electronic. If your computing skills are under-developed, then you must update them. Your supervisor, or your librarian, will point you in the right direction for help, even if they are unable to help you themselves.

Nevertheless, although the means of accessing information may be changing constantly, certain truisms remain. Your research still needs to be:

Well planned – do your thinking before you start, and plan for managing the information you find, and the management and preservation of your data.

Timely – give yourself plenty of time to obtain the materials you need.

Thorough – check out all the information sources at your disposal.

Accurately recorded – you need to save everything you do. Never work from a memory stick. Always use your university's storage to create documents because the security offered is extensive.

If you do all of the above, you will be able to make best use of the wide (and widening) variety of information sources at your disposal.

Reference

Pryor, Graham (ed.). (2012). *Managing Research Data*. London: Facet.

11

Research Methods for Advanced Web Searching

Alastair Allan

Some Background

It is strange to reflect that in 2015, the web is only 21 years old, having first appeared at the end of 1993. In such a short time, its impact is simple because it has changed the world. The first academic web pages appeared early in 1994, and the first e-journals were launched in 1997. The earliest digital theses also appeared in 1997, but the system did not start to build until 2002. Google Scholar appeared in November 2004, which was shortly after the first academic digital repositories had been opened. So in the first decade of its life, it was quickly appreciated by the academic community that the web (and the internet) was going to be very important and have a huge impact on scholarly communication.

Researchers starting their career now have always known and used the web. They saw it when they were in early education, and so their problem lies in the fact that they are still using the web research methods of an eight-year-old and need to develop a mature methodology for searching the web.

> Too much internet usage fragments the brain and dissipates concentration so that after a while, one's ability to spend long focused hours immersed in a single subject becomes blunted. Information comes pre-digested in small pieces, one grazes on endless ready-meals and snacks of the mind and the result is mental malnutrition.
>
> (Hill, 2009: 2)

The full truth about the impact on the mind of the internet or the results of so-called *hyperlink reading*, where readers scan text and only read or follow web links, has not yet been researched.

There are two important but related issues. The first is the size and extent of the web, and the second and central point of this chapter is that many researchers do not understand how

Research Methods for Postgraduates, Third Edition. Edited by Tony Greenfield with Sue Greener.
© 2016 John Wiley & Sons, Ltd. Published 2016 by John Wiley & Sons, Ltd.

to use the web most efficiently as a research tool and, therefore, how to harvest information from the whole of the web.

The size of the web can be described in numbers, but the enormity of its provision never truly sinks in. For instance, the web-based digital journal that is provided free online, *PLOS One* (*Public Library of Science for (Every)One*; see http://www.plosone.org/home.action), published 31,507 articles in 2013. The index to journal repositories, the British OpenDOAR (http://www.opendoar.org/countrylist.php), provides access to 2729 academic journal repositories (as of 2014) that give access to 20.3 million articles. In 2014, the largest open website, Wikipedia (http://en.wikipedia.org/wiki/Main_Page), claimed to have nearly 23 million registered users and 470 million unique visitors in 2012; it now contains more than 35 million articles in 287 languages (Wikipedia, 2014).

A web page that keeps a constant record of the size of the web is WorldWideWebSize.com (www.worldwidewebsize.com). Their calculation for mid-2014, which they acknowledge as an underestimation, varies wildly between 15 and 52 billion pages. They have tracked an increase of around 80% over two years, but their daily peaks and troughs can show differences of 40 billion, and at times their estimation of Google's coverage is greater than their estimation of the size of the entire web. Searching on Google for the English definite article followed by any other word ("the *") finds 25.25 billion pages. If this is repeated for the indefinite article on French, German and Spanish pages, the number of pages retrieved is identical at 25.25. If these figures are true, then that adds to 118 billion. Yet we need to understand that the word will be found more than once on an individual page. But it is possible that many pages in European languages have an English abstract, so this figure is doubtfully high. If one suggests that these languages are only 40% of the web and that another 20% are only images, we reach a tentative maximum figure of 199 billion. The true size of the web is, therefore, somewhere between the two figures: an educated estimate is that it is around 80 billion. This, though, is known as the *searchable* or *open web* because it can be found on search engines and used by anyone. There is also the *closed* or *invisible web*. The invisible web has three sectors. Firstly, the private web or the intranet is owned and used by organisations, and a given intranet is only available to its respective organisation; secondly, there are the subscription sites like journal sites that can only be seen if a fee is paid; and, thirdly, there are the truly enormous websites that are too big to be searched thoroughly by search engines, and so parts of them remain invisible. The size of the invisible web is unknown, so much so that estimates of its size range from 300 billion to 750 billion pages.

Web behaviour has been summarised (Hearst, 2009) as:

- Fact finding
- Information gathering
- Browsing
- Transactions
- Other.

To give these categories some meaning, they can be translated into five common uses. The first use is for personal connection which could be either e-mail or social networking through Facebook or Twitter. The second is using the web like a reference book, so it can find the height of Ben Nevis or provide the address and phone number of the Royal Statistical Society. The third is allowing you to connect to an organisation for either work or leisure. That organisation can provide details of events, images, publications and policy advice and can range from a government department through to a university or charity to a football club. This is the updating and extending knowledge section, and in the early days of the web it was envisaged that users would know which organisations they would wish to monitor, and would bookmark their webpages and then either search or browse their sites. The fourth has grown with the web and is to enable online transactions: to say *shopping* is too narrow. It does include shopping but also includes using government services like paying tax or registering a child for school. It also allows membership in organisations and permits intellectual transactions like engagement with government consultations or using blogs. A last use is study and, of course, the web has a value for study at all levels and for all ages.

The method of website use outlined above, that of bookmarking, is of little use for study. Rarely is it possible to bookmark and return to pages to satisfy academic needs; and so the method of web use that has emerged, and the one that is now prevalent and the usual method of most users, is the use of a search engine as the front end for all enquiry, making the search engine the starting point for all web usage. There is a wide variety of general and specialist search engines. The biggest and most commonly used are Google (www.google.co.uk), Yahoo (uk.yahoo.com) and Bing (www.bing.com). In the United Kingdom, around 91% of all web searches are through Google (Experian, 2012a): in the United States, Google has a 66% share with Bing having 28% of the usage (Experian, 2012b). There are small specialist search engines like SearchEdu.com (www.searchedu.com) that only looks at US university websites. However, bearing in mind its UK dominance, it is Google that will be the focus of this chapter.

Using a search engine is easy. Once any word or series of characters is entered, there are always some results. Below is a good technical explanation of what happens:

> [S]earch results are displayed as a vertical list of information summarizing the retrieved documents. (These search results listings are often known as "search engine results pages," or SERPs, in industry.) Typically, an item in the results list consists of the document's title and a set of important metadata, such as date, author, source (URL), and length of the article, along with a brief summary of a relevant portion of the document. The representation for a document within a results listing is often called a search hit.
>
> (Hearst, 2009: chap. 5)

At this point, it is worth recording that most search engines show 10 results on each page. Sometimes, the first one or two results have an indented list of important sections that point to different subjects on the site.

Once the results have been displayed, then the searcher needs to choose their route through them. An issue is that it is necessary to evaluate the results and maybe to refine the search strategy to produce a more relevant return. One factor that has an influence on both transactions is the number of results that are returned. The way in which the search is entered has a great impact on this: the issue about evaluation will be returned to in this chapter.

When the text is being entered, if one is using the Google Instant page (https://www.google.co.uk/webhp?hl=en&tab=ww) and the terms 'child' and 'abuse' are used, then the logic of the search engine is that it will not just find links to pages on 'child abuse' because the logic of the engine is that it will search for 'child' *or* 'abuse', and so it will find anything about a child or any sort of abuse. It does not link the two terms unless you search for them in a phrase within inverted commas, 'child abuse'. So for most searches that involve a

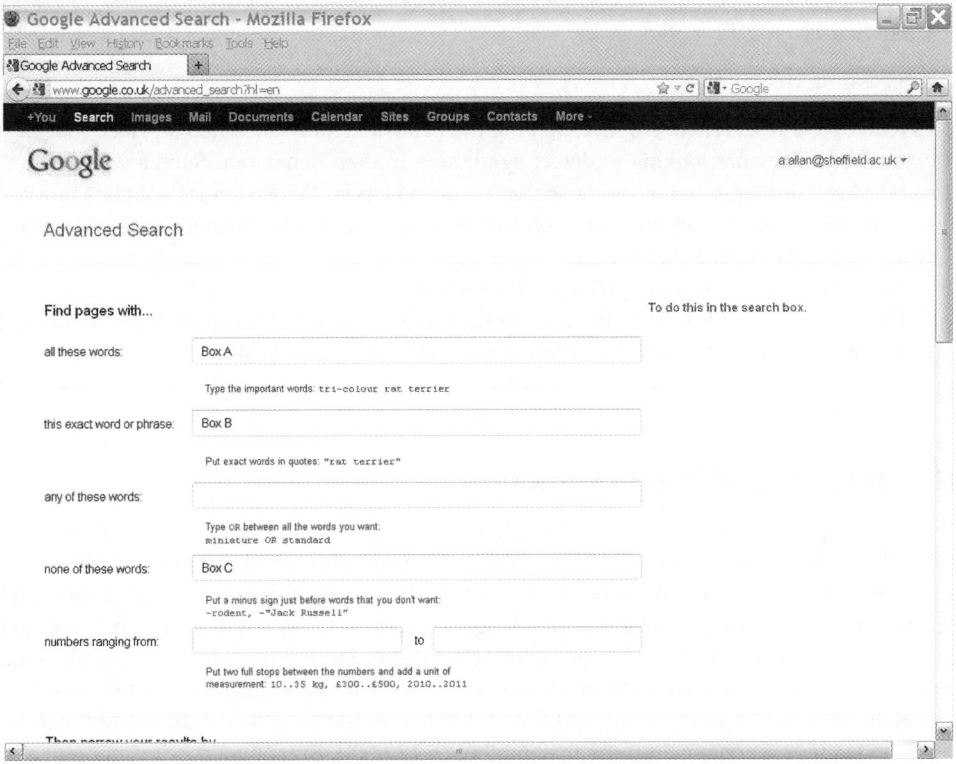

Figure 11.1   Google Advanced Search. http://www.google.co.uk/advanced_search?hl=en.

phrase or a personal name, it is better to use the Advanced Search option (http://www.google.
co.uk/advanced_search?hl=en). This is no longer directly linked from the Instant page but
once you have SERPs, you can go to the cog wheel at the top right and Advanced Search is in
that drop-down menu. So, for names or phrases it is best to use the "this exact word or phrase"
box in "Advanced" (Figure 11.1b).

Many Google searches do use sentences or phrases. In September 2011, Experian (2012b)
reported that even though 26% of Google searches were just a single word, nearly 10% were a
search of six words or more. This appears to indicate that some searches are made by cutting
and pasting a sentence. Passively observing inexperienced Google users, two people used the
following searches:

What are the five most common crimes **in** California	23,200,000 hits
What are the causes **of** global warming**?**	3,590,000 hits

The irrelevant terms are underlined in bold. By splitting the terms into a phrase search plus
another word, the first retrieved 245,000 hits and the second found 32,700. Substituting
'reasons' for 'causes' brought 52,200,000. It is impossible to say which set of SERPs would
have brought the users the best links, but it does show that the results are not just dictated by
the words used but by the way in which the search engine is used.

Case Study

Aggression

Psychological researchers are investigating the behaviour of those who use video games extensively. They are looking to detect aggression in their behaviour. Such a search can use the keyword 'aggression' because there is no synonym. 'Video games' can be known as 'computer games', but the first is preferred because it relates to games on the TV, on computers or on hand-held devices.

   An open search on Google yields 2,420,000 hits.

   The first search finds links to the BBC as the top hit, and on the first page there are links to four news articles, two to US universities and another to a publisher.

How It Works and How to Work It

When a user types a search into the Google search box, Google does not then go out onto the web and look for relevant sites. Google uses its own index which has been built up over years and which indexes and summarises the web. The search (above) for "California" and "common crimes" took 0.32 of a second. Google had checked its own index for that time and found results which it would try to present in a logical order. It would not have searched the entire index but would know how much it had checked and the number of positive hits. If it had gone through 2% of the index and found 3580, it would have then extrapolated that the whole index would contain 50 times that number or 179,000. But, of course, it is not going to let you see all the 179,000 and will, in fact, only ever show you the first thousand.

   This, though, is just too many for most people. Research has demonstrated that more than 90% of people only look at the first SERP (or top 10 results) and that fewer than 0.5% actually get to the fourth page.

> Studies and query logs show that searchers rarely look beyond the first page of search results. If the searcher does not find what they want in the first page, they usually either give up or reformulate their query. Furthermore, several studies suggest that web searchers expect the best answer to be among the top one or two hits in the results listing.
>
> (Hearst, 2009: chap. 5)

> In 93 percent of searches, the users in our study only visited the first SERP.... In only 7 percent of cases did users page onto a second SERP and the number who visited three SERPs for a single query was … likely less than one per cent. Not only did users make do with a single SERP; most of them didn't even bother reviewing the whole page. Only 47 percent of users scrolled the first SERP, which means that 53 percent saw only those search hits that were "above the fold".... On … Google, users can only see four or five results above the fold.
>
> (Nielsen and Loranger, 2006: 39)

Nielsen and Loranger (2006) also quote research that finds that 51% of all users look at the first one but only 2% look at the 10th. Such behaviour is an inadequate response from an academic user. If an academic search is being conducted, then the first one or two hundred results should be examined. It is probable that those from number 300 will be less relevant and that relevance will decline from page to page, but Google will retrieve the more popular pages first and those that have been used much less will be lower in the results. Yet they may have good relevance. It is also relevant to remember that if popularity is being measured, then the most recent results, even though massively relevant, will be less popular because they have not been available for long.

When Google is responding to a search, it is important to realise that it does not have any knowledge. It is finding results from checking for strings of characters. An example is that if one searches for 'AIDS' and 'Britain', then 72,400,000 results are found, but a search for 'AIDS' and 'London' brings 161,000,000. Google does not know that London is part of Britain and, therefore, cannot retrieve results that do not mention Britain but discuss London or Edinburgh or Wales. Similarly a search for 'global warming' and 'carbon dioxide' brings 35,200,000 hits, and one for 'global warming' and 'CO_2' finds a different 32,100,000 results. As time progresses, the designers will teach Google more information to enable it to know and use links. So, therefore, for serious researchers who are using the web to find information, it is important to repeat searches using a different vocabulary. It is worth noting here that it could be valuable to keep a log book of web searches so that important vocabulary can be remembered and reused. Here are examples:

1. aged + alcohol abuse   11,000,000 hits
2. aged + alcohol misuse   415,000 hits
3. aged + alcoholism   8,570,000 hits
4. old people + alcohol abuse   66,000 hits
5. old people + alcohol misuse   54,600 hits
6. old people + alcoholism   2,890,000 hits
7. older people + alcohol abuse   209,000 hits
8. older people + alcohol misuse   828,000 hits
9. older people + alcoholism   316,000 hits
10. elderly + alcohol abuse   749,000 hits
11. elderly + alcohol misuse   310,000 hits
12. elderly + alcoholism   835,000 hits

It is important to understand that in the social work profession, the preferred terms would be 'older people' + 'alcohol misuse', which produced some of the fewest hits. Also, terms such as 'men' and 'women' can be inserted as well as other similar terms like 'pensioners', 'retired people', 'the old' and 'senior citizens'.

There is a Google search method that provides some help in dealing with the issue. It is the use of the tilde ($\sim$) character. If that character is used before the key term (and touching it), then all the known synonyms will be retrieved. Having experimented, it appears to work for the 'alcohol abuse' term (11,800,000, and includes 'alcohol misuse'), and the other three searches still find a mass of results:

1. ~aged + alcohol abuse   11,000,000
2. ~aged + alcohol misuse   415,000
3. ~aged + ~alcoholism   8,570,000

The problem here, though, is that you can still only see a maximum of 1000 results.

So, there is no doubting that the choice of search terms is crucial, but it is also essential to look at alternatives. In biology, the common as well as the taxonomic names must be used. In social sciences, Americanisms like 'teens' for 'youth' and 'mental retardation' for the more correct 'mental disability' should be searched. In the humanities, you need to search 'Richard the Third' as well as 'Richard III'. In medicine (and veterinary sciences), you need to use the Latin names as well as the common names. So, 'TB + cattle' can be 'tuberculosis' + 'cattle' or 'bovine tuberculosis' but should also include '*Mycobacterium bovis*'. In medicine, the search needs to use medical terminology rather than common words, and so the search would be 'myocardial infarction' rather than 'heart attack'. Here, if you search for 'influenza' you get hits intended for the medical professions but the search for 'flu' gets the hits for laypeople.

The placement of the terms in the search engine can also have a bearing. If you search three terms all in different boxes – 'statistics' and 'migration' and 'Britain' – you get 13,300,000 hits: with just the phrase 'migration statistics' (Figure 11.1b) plus 'Britain' (Figure 11.1a), you get just 50,500.

It is, therefore, important to know that the choice and placement of terms in the Google searches are important but, of course, there is no point in increasing your hits from 10,000 to 1,000,000 because you can still only see the first 1000. Yet with a larger and wider set, it is likely that the first hits you get will be more relevant. It is once you are able to find large sets of relevant web hits that one vital skill becomes important. This is web filtering, which is the method of reducing large sets to small sets of less than 1000 relevant hits. It is not yet possible to ask the search engine to narrow down your results or get them sorted into categories or filtered by title, publisher or meaning. Some smaller search engines have attempted this, and certainly it is an innovation that will eventually arrive.

Web Filtering

Aggression and Video Games

The related searches offered by Google to this topic are:

     Aggression and video games psychology

     Video games and aggression statistics

     Video games and aggression in children.

Web filtering is not magic, and it is not really a skill. It is just using the tools that Google provides to enable you to do this. There are some tools provided. These change because new tools can appear and some can disappear. An important one is Related Searches. If you search for Migration Statistics, at the bottom of the first SERP you are offered a list of related searches and it offers you searches for terms like 'migration statistics by country'. These related searches are built by Google using the searches done by other people, and so it is probable that those offered will be popular searches that retrieved good results.

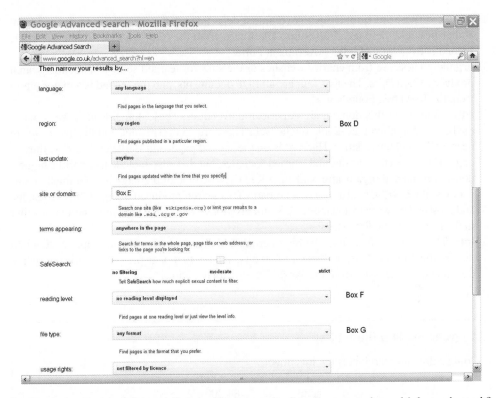

Figure 11.2 Google Advanced Search: filtering results. http://www.google.co.uk/advanced_search? hl=en.

Domain Filtering

One of the best filters can be part of a regular search. This is *domain filtering*. On the Advanced Search page, this is seen in Figure 11.2e. A point not yet made is the value of the box that is called 'none of these words' (Figure 11.1c) because if the search has been cast too wide, a single irrelevant term can reduce the hits significantly. The lower half of the Google Advanced Search page is 'Then narrow your results by …', and these are the filtering boxes. The one we are concerned with is 'Site or domain' (Figure 11.2e). Every website address contains a domain and a country code, and so the University of Sheffield website is www.shef.ac.uk. 'Shef' is obviously the name and 'uk' is the country code (the USA does not have one, Germany is 'de' and Spain is 'es'). The domain for Sheffield University is 'ac', which shows it is a university. By putting such domains in the 'site or domain' box, you are asking Google to restrict the results it gives you to the websites that belong to that group.

Let us take the example above of 'old people + alcoholism'. If we go back to the search and implement domain filters, we will be given a far fewer number of hits. There were 2,890,000 hits:

1. **ac.uk** – These are results just from British Universities, and there were 2720 results. The first was from the Royal College of Psychiatrists, and the second from the journal article repository UK PubMed Central.

2. **gov.uk** – These are from British government departments, government agencies and local authorities, and there are 275 results. The first is from the Norwich City Council, and the second is from 'gov.uk', the government portal.
3. **org.uk** – These are from charity websites or the sites for learned societies, and this search produces 12,600 hits. The first four are all PDF documents, and the second is a major report from the Rowntree Foundation.
4. **edu** – Sites from this domain are American universities. These domains produce some of the best research work available on the web. There are 35,600 results, and the first page are nearly all small universities. Harvard is on the second page, and Berkeley is on the third.
5. **org** – These are US charities and learned societies, but they also include the world's biggest non-governmental organisations like the Red Cross. Some of the large international organisations like UNESCO also use this domain. This search produced a large 620,000 results, and mostly they were from smaller US organisations except an Alcoholic Anonymous link in the middle of the first SERP.
6. **int** – These are the sites from the big international organisations, although many others do not use the domain. This produces 13,600 hits, and the first page presents a series of policy documents on the issue from the World Health Organization.

Aggression (44 million hits)

Doing a domain search cuts the hits down:

ac.uk	84,300
gov.uk	36,200
org.uk	125,000
edu	565,000
org	21,900,000
int	24,800

Evaluation

A connected point here allows us to refer back to the issues about evaluating the web pages that are found. What has not been firmly stated before is that no organisation has any responsibility for the web. Site owners are allowed to post any content they deem appropriate. It has never been possible to accurately index or organise the web, and this is part of its beauty. The serendipity aspect of the web is an integral part of its value. So pockets of the web are invaluable for research and the whole of it is a reflection of society, but much is of lower value and is not research material. As website owners have total freedom, it follows that no one reviews the content in the same way that journal articles are peer reviewed. For this reason, it is not possible and not advisable to have a high level of confidence in the material found on the web. The process of evaluation is the one that needs to be developed when using web-based information. It should be easy to know that information coming from Cambridge University, the Royal Society and the US government should be of the highest value and fully trustworthy, but not all judgements are this easy.

It has already been commented that it is not possible to sort the results on the SERPs to enable easy consideration of groups of results. One of the ways in which results might be sorted would be to put them into domain groups, but this is not yet an available filtering technique. An important comment about the quality of searches comes from Hearst (2009):

> There is a general concern that Web search obliterates the distinction among sources, and provides searchers with few cues about the quality or reliability of retrieved information sources.
>
> (Rieh, 2002, quoted in Hearst, 2009, chap. 7)

One of the features on the SERP is the presentation of the URL, and from this the user can detect the domain and the country of origin. This information is not so valuable for inclusion, but it helps to exclude sites because they are commercial ones or they emanate from overseas sources that might not be able to provide appropriate insights. Obviously, as well, it will be possible to learn the URLs of important organisations, and that will help with evaluation. There are, though, some guides on the web that are designed to help with evaluation. These do outline the procedures that are recommended and also indicate how to identify a page without credibility. There are two that have existed for several years. The first is by Elizabeth Kirk from Johns Hopkins University and is called 'Evaluating Information Found on the Internet'; see http://guides.library.jhu.edu/evaluatinginformation. The second by Robert Harris is from the VirtualSalt website and is 'Evaluating Internet Research Sources', at http://www.virtualsalt.com/evalu8it.htm.

Site Search

An extension of the domain search is the search through a nominated site and uses the same 'site or domain' box. If you enter the homepage URL of any website, then Google will just retrieve results on your subject from that site. There are three advantages. The obvious first one is that you are taking information from a website that you know to be of high value. The second is that your search can be very general and can retrieve a number of valuable sources on the topic generally which you are able to sort by reading through them. The third is that, of course, the results might not have featured in a full search of that topic but also Google thoroughly retrieves every relevant reference which it might not do when responding to a general search. In this example, although still interested in the issue of 'elderly' and 'alcohol abuse', the search will just be for the term 'alcohol abuse' and the nominated site will be the Joseph Rowntree Foundation which is the premier charity that researches social policy issues in the United Kingdom. Their URL is www.jrf.org.uk. This search brings 75 hits, and it immediately becomes apparent that there is extensive research on alcohol and youth but also on issues that might relate to older people like homelessness and poverty. This is an example of a search where the quality of the results almost compels the user to inspect all the results. However, the downside of this approach is that if you conduct a search on the Rowntree site with their internal search facility, you get 218 hits.

At the start of this section, the 'related searches' were referred to. Only a single example was mentioned, but in this context the 'reading level' can be considered (also Figure 11.2f). This filter can be applied to any set of results that has been found. Once it is used, it separates the results into three reading levels which are Basic, Intermediate and Advanced. Basic is a reading age of 9 years, and advanced is one of 15 years. Once the domain filter (Figure 11.2e) has been used, the Reading Level filter can be a most valuable way of further pinpointing the best results because the 'advanced' reading level highlights those written in academic language. So, with the search of British charities (org.uk) in list item 3 above that found 12,600 hits, by using the Reading Level filter it reduced the hits to 1638 advanced-level ones and 6048 intermediate-level ones, with several thousand results being lost which does help to concentrate the search results. Once you have a page of results, you can go to the top, under the search box, and you can apply the Reading Level filter by going to Search Tools and All Results, and by selecting Reading Level under that.

Format

One of the most popular filters is the Format filter (Figure 11.2g). The Format box cannot be applied to a completed search. This is a drop-down box, and some of the options are very general like asking for a Word document or for specific ones like Google Earth documents. Over the years, it has been discovered that there are three valuable options under Format:

1. **PDF** – Looking only for PDF documents results in only research reports and journal articles being retrieved. This, together with the Reading Level filter, can often reduce a large set of results to a manageable set of relevant articles. Again, for the British charities search (org.uk) on the elderly and alcohol abuse, there are 12,600 hits, which with the pdf filter slims down to 389 hits. On the first page, there are six out of 10 relevant articles when the results are limited to pdf.
2. **XLS** – Obviously, this retrieves statistics and there are often few hits. For this search, there were only three results, but the URLs of the providers indicate that they could be valuable.
3. **PPT** – These are PowerPoints and are often documents that accompany presentations. For the elderly search there were only 18, but nine seemed to be exactly on the required topic.

If we change the search to the broader one on 'global warming' + CO_2 and use the US university domain of 'edu', we get 119,000. Of these, 59,800 are PDFs, 686 are statistical XLS files and 2960 are PowerPoint slides. This is a worthwhile way for any researcher to filter their results because it puts the emphasis on the academic style of presentation.

Aggression and video games (957,000 hits)

Using the format filters, the results are also reduced:

PDF	57,000
XLS	1370

Country Filter

There are two ways of filtering by the country to which the web pages refer. The first is to enter the country code in the domain box. (A list of country codes, this one produced by NASA, is at http://goes.gsfc.nasa.gov/text/web_country_codes.html.) The country code means that the website has been produced in that country rather than being a page about it. The second way is to use the second filter box which is Region (Figure 11.2d). This provides a drop-down box with a list of the world's countries. This filter does not necessarily just use the pages from that country, but the data about the webpage might indicate the page is about that country.

Doing a search on 'AIDS' and 'co-infection', there are 183,000 results. These are the results of investigating three countries with a high AIDS infection using the country codes:

Botswana	10
Kenya	4710
South Africa	17,200

Then using the Region drop-down box:

Botswana	48
Kenya	157
South Africa	291

The use of country codes and the Region box is a valuable way of finding information that emanates from a particular country.

Aggression

Using the two different country filters for the 'video games' and 'aggression' search does again reduce the sets significantly.

If we select 'United Kingdom' in the Region drop-down box, then the number of hits is:

United Kingdom   810

This is a much reduced set, and there are nine news articles with one hit from a charity.
Using the .uk suffix in the domain box, one retrieves:

.uk   27,200

The first 10 hits are almost identical.

Finding the Academic Websites That Search Engines Cannot Find

It is not illogical to be aware that there are pages on the web that search engines cannot find.

In the 'Some Background' section, there is an explanation of the invisible web, but the invisibility of certain pages can be explained. Google is a vast search engine and has been programmed to see everything. Other search engines are smaller and simply do not trawl the web to the extent that Google does. Search engines run systems called *bots* that look at pages and their links and remember them and create an index from them. In this context, Google is being studied and there are pages that it cannot see. In many cases this may be because they are too new to be linked by another site or possibly they may never have been noticed by any service and so have always remained in the shadows. In the 'Some Background' section on the invisible web, another reason was suggested, and that is that some websites are simply too big to be trawled thoroughly or accurately.

Let us take the World Bank as an example. It is a first-rate site, and a major sub-section is its publications at http://www.worldbank.org/reference/. The site is so well organised that it is possible to calculate that there are 17,722 records in a sub-site that is their Open Knowledge Repository. The Google response, though, is that if you use a split search of 'World Bank' + 'publications', it finds more than 48 million documents: a phrase search for 'World Bank publications' gets 356,000. These hits will have found sites that link to World Bank publications, and so a domain-specific search for 'publications' seems to be the best route but that finds 241,000 results, which this author judges is at least over 200,000 records too many. This demonstrates that for a whole range of reasons, Google is unable to be as specific as a user might require. For that reason, it is valuable to go directly to the individual website and conduct a search from their homepage.

Finding Journal Articles

The point that academic journal articles are lodged in repositories has been mentioned in this chapter. A repository is a website (or part of one) that collects and freely distributes the articles published by members of an organisation. The important feature here is that the articles are freely distributed whereas they would usually be only available from a journal for which a subscription is necessary. This is known as 'open access' publishing, and it is part of a wider movement that drives towards universal, freely available academic research information.

General Repositories

A general big repository is WorldWideScience.org (http://worldwidescience.org/index.html) which is a 'global science gateway comprised of national and international scientific databases and portals' and is run by the US Government Office of Scientific and Technical Information which is a sub-body of the US Department of Energy. This gives the whole site a great deal of status. Much of the data come from government organisations, and this is a good method for finding obscure reports. This database also has a fluid definition of 'science' and includes articles on archaeology and social welfare. A search for 'tuberculosis' finds 922,004 results which it filters down to 4374 'top results', all of which are research papers, but then only lists 3981 references; similarly, a search for 'elderly' + 'alcohol abuse' brings 23,913 hits but presents only the 1256 top results.

Another excellent US government site is science.gov (www.science.gov). This claims to search over 60 databases and 200 million pages of scientific information and also acts as a gateway to more than 2200 federal websites. This site is actually sponsored by 17 US government science organisations within 13 federal agencies and, again, often finds government information not easily available elsewhere. This site has an advanced search feature and, again, searches find very large swathes of information. A search for 'elderly' + 'alcohol abuse' finds 29,537 articles and presents 1235 of them: the search for 'tuberculosis' finds 527,106 and presents 2561. Searchers are enabled to limit their search by general field (like 'Earth and Ocean sciences'), and there is an award-winning mobile search part of the site. It also features an alphabetical subject index, but this is very much limited to applied, medical and pure science topics. Experience proves that both of these US sites are rarely used by academic researchers; this is an oversight because they do give access to research work done by government agencies outside universities.

There are other large repositories or search features that also give access to research articles or research reports that cannot normally be found by search engines. BASE (www.base-search.net) contains more than 68 million records and links into repositories. It is a multi-disciplinary search engine for scientifically relevant web resources and harvests data though the Open Archives Initiative. The service is run by the Bielefeld University Library in Germany, which provides a full English language set of pages in addition to its German ones. BASE provides good access to European research published in the English language, but its strength is the links to works in European languages.

OCLC WorldCat (www.worldcat.org) is the biggest library catalogue in the world. It shows the library holdings of more than 10,000 libraries worldwide. A sub-section is OAIster (oaister.worldcat.org), which indexes more than 15 million records that WorldCat has found that are articles from repositories, digital libraries and online journals. It provides a collection of freely available, previously difficult-to-access, academically oriented digital resources that are easily searchable.

Academic Repositories

Open access publishing is strongly supported by universities worldwide. National repositories exist in several countries, but the important element is the individual repositories built by individual universities. In the United Kingdom, there is a series of university repositories. They normally serve two functions, and the first is to store and distribute completed PhD theses as they are accepted for a degree. The second is to store copies of articles, book chapters and some reports published by members of the university's academic staff and their research staff and students. The ideology is that the articles have been written in university time, and so the copyright belongs to the author and the university and not the journal publisher. By making the articles available 'open access' on the university repository, it means that the university does not need to subscribe to a particular journal in order to use the intellectual content that it already owns. A second result is that the university is able to demonstrate its research excellence and also is able to mark the research territory in which it works. A third result is that making the article widely and freely available increases the readership and so increases the impact of the work and increases the number of citations. The last result is that the university can share its research harvest with the rest of the world's research community and, in particular, allow access to researchers in poorer developing countries who could not otherwise afford the journal fees necessary.

The name for the largest group of British repositories is Sherpa (http://www.sherpa.ac.uk/repositories/). Here there is a list of the 22 partners, and there is also a full list of all the 176 British ones at the Directory of Open Access Repositories (DOAR) (www.opendoar.org). There is also a search facility for DOAR at http://www.opendoar.org/find.php, but this covers all the repositories that it has found worldwide, whereas the Sherpa search facility at http://www.sherpa.ac.uk/repositories/sherpasearch.html) only recovers articles from Britain.

It is now usual for a major US university to have a digital repository that showcases the institution's research work. Quite normally, this is indexed on OAIster. There is, however, at least one list of links to these repositories. This is Digital Commons, which is a list of institutions using the Digital Commons platform; it can be found at http://digitalcommons.bepress.com/institutions.html. There is a list of Australian repositories at http://arrow.edu.au/repositories.php. Other countries also have such lists.

Organisational Repositories

BioMed Central was founded in 2000 and is a profit-making company that specialises in online open access journal publishing (www.biomedcentral.com). Rather than levy a subscription, it charges fees to the article authors and now is host to 270 peer-reviewed open access journals. It has the support of the US National Institutes of Health (NIH) and the Wellcome Trust. In early 2015, an estimate was that there were about 190,000 articles on the site. The main corporate sponsors welcome the fact that the research that they fund is published and publicised on an open website. Although it is possible for other repositories to harvest BioMed Central, there are more than 1.6 million hits registered on Google, and Google Scholar can find over 82,000. Both these figures demonstrate that this is another site that cannot be satisfactorily accessed from a search engine.

Many of the larger websites maintain publications pages, and many international sites maintain large collections. Here again, the search engines will be able to see such pages, but in the many hundreds of thousands of hits to relevant sites that they retrieve, it is likely that most of the valuable sites will be hidden lower in their finding lists. In this context, only a few demonstrative examples are given. One of the oldest journal repositories is arXIV, which existed before the web was born. arXIV is a website that contains over half a million articles on physics, mathematics, computer science, quantitative biology, quantitative finance and statistics; it can be found at http://arxiv.org. Large international organisations have large collections. The UN's Food and Agricultural Organization (FAO) at http://www.fao.org/documents is one such collection, and another one with 360,000 physics articles is CERN at cdsweb.cern.ch.

Aggression and Video Games

There is often a variation between the numbers of hits that individual sites can find. Investigating the results, the following often present unique results:

OAIster	60 hits
Worldwide Science	2107 hits
BASE	335 hits
Science.gov	10,357 hits
OpenDOAR	386,000 hits
Sherpa	347,000 hits
Biomed Central	29 hits

It is fairly certain that the Sherpa search engine has not found 'video games' AND 'aggression' but 'video games' OR 'aggression'.

Open Access Journals

Although their articles will be found by Google Scholar and by Google search, there is value in knowing where to find open access journals. The reason for open access journals has been explained, but this philosophy was given official approval in 2012 by the Finch report

(Research Information Network, 2012). Briefly, the report addresses the issue of how to enable fast access to research articles for all those who need to read them. The online environment now dictates that users want or need instant access to articles that are free at the point of use. The barrier of articles being in print or only available to subscribers needs to be overcome. The report looks at both repositories and open access journals to investigate how either can answer the problem. They recommend that governments and research funding agencies should support open access journals, and they look at ways of enabling a transition from the present system.

The central website that records and archives open access journals is the Directory of Open Access Journals at www.doaj.org. This is an extensive listing that is arranged by subject. Some of them are new titles that have been published because the 'free-on-the-web' model is attractive and some are new open access titles, but some are academic journals that have now decided to publish their journals free of charge and sometimes they have been subscription journals in the past. Often they can be peer-reviewed and of good standing. The largest journal in the world is identified as *PLoS One*. PLoS (www.plos.org) has seven journals together with current awareness services and blogs. It is better to use the journal website than rely on search engines for their articles because the websites can provide tailored services like newsletters, bulletins and RSS feeds.

Google Scholar

Now that the issues of repositories and open access journals have been covered, it is the time to discuss Google Scholar (http://scholar.google.co.uk/schhp?hl=en). Google Scholar is Google's academic site. This search engine delivers only information about scholarly articles and does so by trawling academic repositories rather than visiting publisher's journal websites. A biased, unscientific and subjective judgement is that Google Scholar is the best development in scholarly communication over the past 25 years. Scholar's coverage is magnificent, and the bottom line is that it is completely free. Search results are retrieved, and on the right-hand side of the page links are given to free copies of the article which can be read or downloaded. It is vital to note, though, that Scholar only delivers articles (or abstracts of articles); it does not find web-published research reports, and it only finds conference papers if they are published on the web as though they were a journal article.

The Scholar SERP is different from that of the Google search.

When the results are delivered, they are not ordered by popularity, but they are put in approximate order of the number of citations that the article has had. The result is that the first result should be the one that has had the highest impact, but another result is that the articles that are delivered first are the ones from peer-reviewed journals. The logic is that only good articles from good journals will get a large number of citations. This means that Google Scholar is nearly always delivering the best.

In Figure 11.3, the Advanced Search box is shown. (The page has recently been abridged.) To find Advanced Search, you click on the down-pointing triangle in the search box. It is a far better approach than the simple instant search on the Scholar page. This is because it allows the searcher to look for a named author if that is required, and it also allows filters to be applied to the search. There are date, publisher and author filters. The effect is that it is possible to keep the number of results delivered down to a low number and, therefore, to enable all the

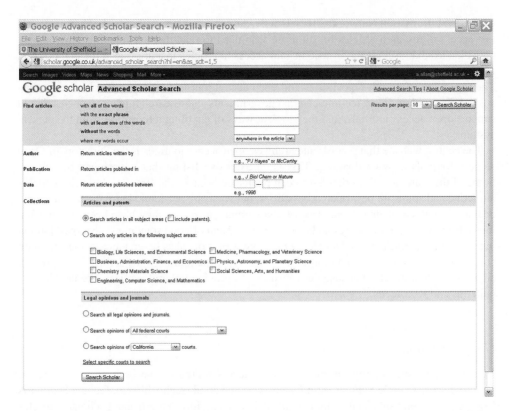

Figure 11.3 Google Scholar Advanced Search page. http://scholar.google.co.uk/advanced_scholar_search?hl=en&as_sdt=1,5.

results to be investigated because fewer than 1000 pages are being delivered. If we research the topics covered above:

Tuberculosis	1,450,000, 1,810,000
(with a post-2009 date filter)	311,000
Elderly and alcohol abuse	106,000
(with a post-2009 date filter)	17,300
Tuberculosis + medicine + 2014	95,000
Elderly + social science + 2014	31,100

Another valuable feature is given on the SERPs pages. Every record has, on the left, a list of citations. This link can be opened, and the citing articles can be seen and chosen. Similarly, there is a link to 'Related Articles' which is an extension of the Scholar search. Under this link, there are 101 records that Google Scholar has found that use the same terms as the one selected. Using both these links it is, therefore, possible to find one article and then discover several other articles that are important for the same topic.

Electronic Theses

When the academic repositories were discussed, it was mentioned that many also delivered completed PhD theses. It is normally difficult to find these on a Google search, and they are not delivered by Scholar. A full title in the phrase box (Figure 11.1b) plus the author name will usually find them, but they will not normally emerge after a subject search. Obviously a source can be the repositories themselves, but now there are national websites that try to capture the whole of a country's PhD output.

There are listings for countries all over the world, but in this chapter only three such services will be mentioned. The oldest and now the biggest is the National Digital Library for theses and dissertations which is at the appropriate URL of www.theses.org. This was originally set up in 1997 and is now very large. It is a site that is difficult to browse, but there are two different search services that both work well. Additionally, the home page gives links to digital thesis services from Australia, Brazil, South America, Europe, Germany, Scandinavia, Britain, South Africa, Portugal and Canada.

The British service that collects directly from British universities rather than from websites is called EThOS, which is a service from the British Library at http://ethos.bl.uk/Home.do;jsessionid. Not all the theses have been digitised there, but it is possible to register and ask for one to be digitised. The last one is the Australian service on Trove (trove.nla.gov.au) which collects all aspects of Australian information. There are many parts, but there is a thesis search and the majority are digitised.

Conclusions

This chapter not only describes the best ways of using a search engine but also tries to explain the best methods for using the web as a research resource. This involves explaining how information on the web can be best found and also explaining extensive sites that can be missed by search engines. Obviously, for many people there are just far too many options in this chapter to use on a regular basis. That in itself indicates that the web contains an almost limitless amount of information and highlights the importance of developing good research methods.

The only downside of developing first-class web-searching skills is that a series of first-class searches will reveal far more information than one person could ever read.

List of Websites Referred to in the Text

arcHIV	arxiv.org
Arrow repositories	http://arrow.edu.au/repositories.php
BASE	www.base-search.net
Bing	www.bing.com
BioMed Central	www.biomedcentral.com
CERN	cdsweb.cern.ch
Digital Commons	http://digitalcommons.bepress.com/institutions.html
Directory of Open Access Journals	www.doaj.org
Directory of Open Access Repositories	www.opendoar.org

EThOS	http://ethos.bl.uk/Home.do;jsessionid=
'Evaluating Information Found on the Internet' by Elizabeth Kirk	http://guides.library.jhu.edu/evaluatinginformation
'Evaluating Internet Research Sources' by Robert Harris	http://www.virtualsalt.com/evalu8it.htm
FAO documents	http://www.fao.org/documents
Google	www.google.co.uk
Google Scholar	http://scholar.google.co.uk/schhp?hl=en
Joseph Rowntree Foundation	www.jrf.org.uk
National Digital Library for Theses and Dissertations	www.theses.org
OCLC WorldCat	www.worldcat.org
OCLC OIAster	oaister.worldcat.org
Open DOAR	www.opendoar.org
Public Library of Science (PLoS)	www.plos.org
PLoS One	http://www.plosone.org/home.action
Science.gov	www.science.gov
SearchEdu	www.searchedu.com
Sherpa repositories	http://www.sherpa.ac.uk/repositories/
Trove	trove.nla.gov.au
University of Sheffield	www.shef.ac.uk
Web country codes	http://goes.gsfc.nasa.gov/text/web_country_codes.html
Wikipedia	http://en.wikipedia.org/wiki/Wikipedia
World Bank publications	http://www.worldbank.org/reference/
WorldWideScience.org	http://worldwidescience.org/index.html
WorldWideWeb size	www.worldwidewebsize.com
Yahoo	uk.yahoo.com

References

Experian. (2012a). Data center – top sites & engines. Costa Mesa, CA: Experian Hitwise. http://www.hitwise.com/uk/datacentre/main/dashboard-7323.html

Experian (2012b). Google share of searches at 66 percent in September 2011. Costa Mesa, CA: Experian Hitwise. http://www.experian.com/hitwise/press-release-google-share-of-searches-sept-2011.html

Harris, R. (2010). Evaluating internet research sources. VirtualSalt. http://www.virtualsalt.com/evalu8it.htm

Hearst, M. (2009). *Search User Interfaces*. Cambridge: Cambridge University Press. (Available free online). http://searchuserinterfaces.com/book/

Hill, S. (2009). Howard's End Is on the Landing: A Year of Reading from Home. London: Profile Books.

Kirk, E. (2012). Evaluating information found on the internet. Johns Hopkins University. http://guides.library.jhu.edu/evaluatinginformation

Nielsen, J., and Loranger, H. (2006). *Prioritizing Web Usability*. Berkeley, CA: New Riders.

Research Information Network, Working Group on Expanding Access to Published Research Findings. (2012). *Accessibility, Sustainability, Excellence: How to Expand Access to Research Publications* [the 'Finch report']. London: Research Information Network. http://www.researchinfonet.org/wp-content/uploads/2012/06/Finch-Group-report-FINAL-VERSION.pdf

Wikipedia. (2012). Wikipedia. http://en.wikipedia.org/wiki/Wikipedia

12

Searching in Unfamiliar Fields

Tom Bourner and Sue Greener

A Strategy for Search

While many readers of this book will be pursuing research following their undergraduate academic studies, others may have found their ways into research through their professional occupations, such as history, engineering, law, medicine, social services, chemistry and human resources. We call the latter *practitioner researchers*. The two routes have contrasting approaches to literature search. One starts from what is known and seeks to describe a detailed background to the research area, possibly reviewing the history of what has gone before in order to establish a wide context, and to support arguments for a new theory. This search is *intensive*. The second route usually seeks to discover current knowledge that will help to answer a well-defined question. It is *extensive*. This latter is reflected in present-day methods and teaching of *evidence-based medicine* (Sackett *et al.*, 1996) and *evidence-based management* (Rousseau, 2006). Evidence-based practice requires the practitioner to obtain evidence of the up-to-date knowledge in the field, and that field may be unfamiliar territory.

The two approaches are contrasted in Table 12.1.

Table 12.1 suggests that the sort of intensive literature review that professionals learn how to do during their time in higher education does not fully prepare them for the kind of extensive search that would help them most with problems they are likely to encounter in professional practice.

Professional practitioners often face problems for which their academic background and professional studies provide them with little or no familiarity. For example, a doctor may have to solve a problem based on employment law, a designer may have to explore materials science to produce a new product design and many professional practitioners need to solve problems connected with information and communication technologies. In these cases, researchers will be unfamiliar with the journals in the field, and generally lack the tacit knowledge that would help them to find and evaluate information in the new field. This makes it difficult to get started. The search to establish 'what is already known about the problem' is not likely to

Research Methods for Postgraduates, Third Edition. Edited by Tony Greenfield with Sue Greener.
© 2016 John Wiley & Sons, Ltd. Published 2016 by John Wiley & Sons, Ltd.

Table 12.1   Approaches to literature searching contrasted.

	Literature search in higher education	Literature search in professional practice
Intensive or extensive search	Intensive search (drilling down)	Extensive search
Relationship to academic disciplines	Intra-disciplinary	Inter-disciplinary and search outside of current disciplines
Starting point of search	The state of current knowledge in the field	A problem or an opportunity
Key question for search to answer	What are the gaps in the current literature on this topic?	Who has ever encountered this problem, or some variant of it, and what possible solutions have been suggested or tried?

involve drilling deeper within a known subject discipline. The search must be far and wide for where the problem, or some variation of it, has been encountered in other fields. A structured approach is needed. One which is straightforward to follow, can be remembered for future use and offers rigorous results on which newcomers to a field can rely.

We describe here a recursive process for literature search by people researching in fields where they have little or no familiarity with the literature. The elements of the strategy are:

1. Focus on key writers and core journals as well as seminal articles.
2. Identify potentially significant works, journals and people by applying the principle of duplication of references.
3. Ensure convergence on the set of articles that are most relevant to the issue under investigation by appreciating the principle of diminishing returns to search.

Searching in Unfamiliar Territory

We focus here on just one form of literature to present the strategy: *journal articles*. In our experience, this is the most useful source of up-to-date thinking on a subject. Reasons for this include:

1. Journal articles usually contain brief reviews of the relevant literature in addition to an account of the method, research findings and how the findings add to what was previously known.
2. Academic journals are refereed and that provides some quality control.
3. The journals are usually published several times each year, and many provide open access online, so they are likely to be relatively up-to-date.
4. Journals are the vehicles through which researchers have traditionally shared their findings with other researchers, so they are published in forms designed to enable researchers to find them through a literature search. Thus, for example, they often contain keywords and abstracts designed to enable other researchers to find them in the process of a literature search.

Although we focus on journal articles, this search strategy can also be applied to other forms of literature.

We present the search strategy as a set of stages as we feel that will be most helpful to the practitioner researcher.

Stage 1: Search Profile

Summarise the starting point for the literature search in the form of a *search profile*. A search profile defines the scope of the search. It provides a preliminary answer to the following questions:

- What is the topic of this research?
- What is the research trying to find out? What is the research question?
- What is the aim of the research investigation?

We illustrate the method in the appendix at the end of this chapter with a specimen search profile. Producing a search profile will concentrate your mind and force you to clarify what is within the scope of your investigation and what lies outside it. As the literature search proceeds, the scope of the research is likely to change, and it is important to acknowledge that and not regard the search profile as a straitjacket. Its purpose is to provide sufficient clarification of the research topic and definition of the scope of the research to enable the process of the literature search to start.

Stage 2: Keywords

What words best describe the subject matter of the research? Keywords can be words, combinations of words or short phrases. The keywords will be the way into the literature of the research topic. How many keywords should you start with? The more you use, the more likely you are to find your way into relevant literatures. However, a scattergun approach will leave the researcher reeling with variable or patchy results. Again, a systematic approach will help. The basic procedure here involves three steps:

1. Generate as many plausible keywords as possible.
2. Rank them in terms of their relevance.
3. Record those you use and where you used them.

Later on in this chapter, there is a section on keywords with ideas for ways to generate keywords.

Stage 3: Starter References

Use the highest ranked keywords to search for relevant journal articles.

You are likely to use databases, search engines or repositories to find journal articles, although of course you can still use library reference catalogues if you have access to these.

The latter are helpful because they often come with personal support to answer those annoying questions to which you probably should know the answer, but do not. Collections of periodical/ journal indexes come in a variety of forms: printed, CD-ROM or more usually via the web.

Using an electronic database enables the researcher to search through years of journal indexes easily, rather than having to scan volume after volume manually. The database of journal indexes that you select will depend on the area of your research, provided you have already identified the relevant subject discipline; examples would be LexisNexis for law or PubMed for medicine. Keywords, on the other hand, enable you to search across subject disciplines, and you can use search engines and repositories as well as interdisciplinary databases (examples of the latter would be the Academic Journals Database and Web of Knowledge). It is easy to start with an Internet book supplier such as Amazon or a search engine such as Google Scholar, which searches only academic texts.

Use the information in your search profile in order to focus your search. Most databases and repositories, and all search engines, provide a *keyword search* facility. You may also have the opportunity to specify parameters, such as year(s), language(s) and whether you want to do a national or international search. Focus your search at this stage to obtain a list of journal articles that will be of direct, rather than vague, relevance to the issue you wish to research.

Once you have produced a list of journal articles, read the abstracts and identify and obtain the 10 articles that look most promising, judging by their titles and abstracts. Be ruthless; unfamiliar fields can be enticing, and much time can be wasted by trying to read whole articles that look vaguely interesting at this early stage of the search.

Stage 4: Seminal Works, Core Journals and Key People

Identify works that have been particularly influential in the field, the *seminal works*; the journals that are most likely to publish articles on the topic, the *core journals*; and the people who are also working on the issue, the *key people*. Core journals may not all lie within the same field of study as categorized by traditional academic disciplines. For example, a search on 'managing professional change through action research' is likely to yield core journals covering a range of academic disciplines.

Use the references cited in the 10 articles to identify:

- articles and books that are mentioned more than once – these are the potential *seminal articles*;
- journals that are mentioned more than once – these are potential *core journals* for the topic of interest; and
- authors who are mentioned more than once – these are likely to include the *key people* researching and writing about the topic of interest.

In other words, you are using the principle of duplication to identify seminal articles, core journals and key people who can help with your search.

This stage can be completed with various degrees of meticulousness. At one end of the scale, you can simply look through the references for duplication of published work, journals or authors. At the other end of the scale, you can key all the references listed in all the articles into a spreadsheet and then sort to produce an alphabetical list of titles of the works cited, names of the authors and titles of the journals. Duplications become conspicuous in this process. This may seem a time-consuming approach, but a significant proportion of the references will need to be keyed in at some stage, if only to provide a list of references for the final report of the research. One way to save time will be to key this information directly into a reference management package such as EndNote or Mendeley. In some cases, the software package will import key information directly without manual labour, still allow you to sort on any field including keywords, and import references directly into your research document.

A short cut would be a preliminary scan across all the references, highlighting only those that look at all plausible and thereafter working only with the highlighted articles.

Stage 5: Seminal Articles

Assemble the articles and other works, such as books and reports, that you identified while finding the duplications. If some are accessible only online, you will need at this stage to record the date of access and URL for later referencing. Check the frequency of citation using Google Scholar; the larger the citation number, the more likely that you have found a seminal article or at least an influential one.

Stage 6: Core Journals

Each of the core journals should publish an index of contents at regular intervals. Look at the indexes of contents of the core journals for the last few years to identify the most recent articles addressing the topic of the research, or closely related issues.

Stage 7: Key People

Use a *citation index* to look for other published work by the authors you have identified as potential key people in the field. A 'citation index' is a regularly updated index that lists works cited in later works (examples are the Science Citation Index and Social Science Citation Index). It includes a list of the sources from which the citations were gathered. Researchers use it to locate sources related by subject to a previously published work. For example, a search in a citation index for Dr P.W. Smith's work will reveal all the articles that have cited or referred to Dr P.W. Smith. This will enable you to identify what else these authors have written that is relevant to the topic in question. You may also wish to send some authors a note, perhaps by email, to explain your interest in the topic. You could, for example, enclose a list of your top 10 references to date and ask them to add any notable omissions from your list, including anything else that they have written on the topic. You may get back copies of work that is so new that it has not yet been published.

Another way of finding key people is to use social media; search for communities of learning within a practitioner area, and consider using Twitter or Linked-in to find them. Many professional periodicals will carry references to such communities or groups which are sometimes easily found under a specific hashtag (examples include #copyright and #elearning). Virtually 'follow' the key people for a while to find out whether they have something to offer your research, and consider connecting directly with them.

Table 12.2 The literature search process.

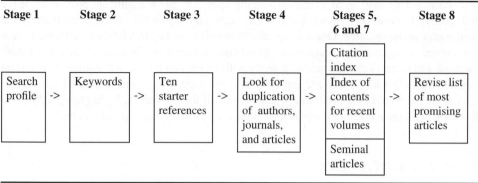

Stage 1	Stage 2	Stage 3	Stage 4	Stages 5, 6 and 7	Stage 8
Search profile ->	Keywords ->	Ten starter references ->	Look for duplication of authors, journals, and articles ->	Citation index; Index of contents for recent volumes; Seminal articles ->	Revise list of most promising articles

Stage 8: Further Iterations

Stages 5, 6, and 7 will have generated additional references. Select, from them, a handful that look most promising, and repeat the process from stage 4 onwards to identify additional potential seminal works, core journals and key people. This process can then be repeated until no new works, journals or people are found. The principle of diminishing returns to search should ensure that the process will not continue indefinitely.

The eight-stage process is summarised in Table 12.2.

Keywords

Generate as many plausible keywords as possible, and then rank the most plausible ones in terms of their relevance to the research topic. The following list of ways to generate plausible keywords is a selection of ideas resulting from a brainstorm of MBA students on a creative problem-solving course on this theme.

1. Use a dictionary.
2. Use a dictionary of synonyms and antonyms.
3. Use known academic journals/trade or professional journals.
4. Use the contents/index pages of a relevant book.
5. Do an internet/intranet search.
6. Use the contents pages and bibliography of other dissertations.
7. Look at keywords in other people's published articles in journals to get an idea of what makes a good set of keywords.
8. Talk to practitioners.
9. Check for a regulatory body.
10. Use one keyword and Amazon.com.
11. Annual reports of organisations in the field.
12. Mindmaps/brainstorming.
13. Rich pictures.
14. Journal articles/references in articles.

15. Newspapers and other media.
16. Focus groups/action learning set.
17. Library catalogues.
18. Newsgroups.
19. Trade conferences.
20. Conference proceedings.
21. Random words.
22. Use metaphors.
23. Free association.
24. Word association.
25. Ask someone else to do it.

Conclusions

When you need to search the literature in an unfamiliar field, or across unfamiliar fields, you need a structured approach to avoid the frustrations of a haphazard hunt. With a simple focus on how to find seminal articles, core journals and key people, and recognizing the principles of duplication and diminishing returns, we have offered a quick start to such a search. We believe you will succeed in that search if you follow our eight-stage process. You will find further guidance in Egger *et al.* (2001).

Appendix: Specimen Search Profile

Topic: Development of professional doctorates in England during the 1990s.

Research question: What is the rationale for the development of professional doctorates?

Aim of the investigation: Discover the thinking behind the development of professional doctorates in England in the 1990s.

Parameters:

Time:	1990 to present
Place:	Unrestricted; anywhere in the world
Languages:	English only in the first instance.

Keywords:

1. Doctorate
2. Doctor of
3. Professional doctorate
4. Practitioner research
5. Taught doctorate
6. Continuing professional development
7. Work-based learning.

References

Egger, M., Davey Smith, G., and Altman, D. (eds.). (2001). *Systematic Reviews in Health Care: Meta-analysis in Context*. London: BMJ Books.

Rousseau, D. (2006). 2005 Presidential address: is there such a thing as "evidence-based management"? *Academy of Management Review* 31(2): 256–269.

Sackett, D.L., Rosenberg, W.M., Muir Gray, J.A., Haynes, R.B., and Richardson, W.S. (1996). Evidence-based medicine: what it is and what it isn't. *British Medical Journal* 312(7023): 71–72.

13

Sources of Population Statistics

Keith Dugmore

Why Are Statistics about Populations So Important?

Many people and organisations seek counts and classifications of populations, and, with easier access to information over the web, interest continues to grow. Academic researchers across many disciplines use population statistics for a multitude of different projects. Here are just a few examples: ethnic minorities in the labour market; social polarisation 1971–2001; carers, employment and services; health in Scotland; ageing and migration trends; labour market outcomes in Northern Ireland; defining Travel-to-Work Areas; health inequalities in England; a geo-demographic classification of mortality; and migration and social change in rural England.

Such research is usually interested not only in total populations, but also in classifying them in various ways. Some of the most common classifications are age, sex, marital status, ethnicity, education, and occupation, but these are only the tip of the iceberg, as will be illustrated in this chapter.

One particular use of population statistics is not always immediately appreciated: they are often vital in providing denominators for other counts, so as to create percentages or rates per 1000. Unemployment rates and crime indices are ready examples.

Populations and Classifications

The Total Population can be defined in various ways. The most common definition is the Usually Resident Population, used for the 2011 Census and also the annual Mid Year Estimates (MYEs). Other definitions include the Population Present (which includes visitors and short-term migrants) and the Workplace Population (people who are in employment and whose usual place of work is in the area).

People can be classified in a multitude of ways. The 2011 Census is a good starting point for examples, giving classifications (which are sometimes referred to as *breakdowns*) of people

Research Methods for Postgraduates, Third Edition. Edited by Tony Greenfield with Sue Greener.
© 2016 John Wiley & Sons, Ltd. Published 2016 by John Wiley & Sons, Ltd.

by age, sex, marital status (including civil partnership status), relationship to others within the household, student status, usual address one year ago, country of birth, citizenship, month and year of entry into the United Kingdom and intended length of stay, national identity, ethnic group, religion, language, general health, long-standing illness or disability, provision of unpaid personal care, educational and vocational qualifications, second residence, economic activity, employment status, hours worked, occupation, workplace and means of travel to work. Some of these topics are also used to derive further classifications such as the National Statistics Socio-Economic Classification (NS-SEC) and the Social Grade.

Researchers also often wish to analyse at the 'Household' level. A household is: one person living alone, or a group of people (not necessarily related) who live at the same address, share cooking facilities and also share a living room or sitting room or dining area.

Again, the Census provides examples, gathering information on the households' tenure, type of accommodation (detached, terraced, purpose-built flat etc.), whether they are self-contained, central heating, and the numbers of rooms, bedrooms and cars. Households are also classified according to the people in them, for example household size, and type (e.g. lone parent or two-pensioner households).

Which Population Counts Would You Ideally Like for Your Project?

When tackling a particular research project, it's best to start by specifying what population and classifications would be ideal. Here's a checklist of the key features:

Population definition? The most obvious populations are individual people; the focus is usually on where they are resident, but there is increasing interest in the numbers of people working in or visiting an area. Some projects, however, may be better tackled by analysing households.

Classifications of that population? Research is usually focussed on particular subgroups, and these can be quite detailed, for example a four-way tabulation such as 'sex, by age, by ethnic origin, and employment'.

Geographical area? Are you interested in the United Kingdom as a whole (England, Wales, Scotland and Northern Ireland), or Great Britain (which excludes Northern Ireland), or just one of these countries? Or, perhaps you are focussing just on Sheffield, or a local ward?

Geographical level? This is another vital decision. Some research is best carried out by studying a single area, but in many cases there is value in dividing it into finer geographical areas, such as Local Authorities or, at the lowest level for most population statistics, Census Output Areas. However, you may find that you cannot get such detailed cross-tabulations for smaller areas with their smaller populations.

Timeliness? Is this an important issue for your project? Datasets vary greatly in their frequency of collection and publication, from monthly economic indicators to the decennial Census. It is also necessary to be alert to the delays between collection and publication. The first results of the March 2011 Census, for example, were published more than a year later.

Confidentiality and disclosure control? Government has always rightly been concerned not to release confidential information, but in recent years there has been a much more active and restrictive policy of disclosure control. This can result in some data not being released even as small-area statistics, or in inflicting damage to the counts by such procedures as rounding or suppressing numbers.

Format and accessibility? Lastly, there are considerable variations in the ease with which datasets can be obtained, ranging from downloads of simple files in popular formats from open websites, to much more detailed datasets, complex file structures, and restrictions on access to individual records, as typified by major government social surveys.

You should also keep an eye on the following considerations, although, with luck, these are rarely show-stoppers in the academic environment:

Coverage and quality? Government bodies such as the Office for National Statistics (ONS) make considerable effort to achieve good coverage or response rates, and responses of high quality.

Cost and licencing? The good news is that the great majority of population statistics produced by government are free, and do not have onerous licencing restrictions.

Having specified your ideal dataset, you are almost certain to find that you will need to compromise, for example weighing the merits of very detailed cross-tabulation of a minority population (but only at the national level) against very detailed local geography (but with only separate counts of each characteristic). Again, you may find that a survey has rich information on 'income', but that you have to accept a less than ideal classification of 'ethnicity'. Do consider your alternative options carefully.

Which Organisations Produce and Supply Data?

The most important population statistics are produced and supplied by central government, but there is also an important academic infrastructure in the United Kingdom that helps distribute data to researchers. There are also commercial distributors whose data are sometimes used in academic research.

Government as Producer and Supplier

It is important to remember that the United Kingdom comprises four countries, and that practices can vary between them.

Government datasets have some obvious attractions: they usually cover a whole country, rather than just odd regions; collection methods are uniformly applied; collection is often compulsory, giving coverage of >95% of the particular population; and quality is usually high. This is, however, sometimes at the expense of timeliness. A further attraction is that most government datasets are free to end users.

Table 13.1 The key national producers of population statistics.

Country	Agency
England and Wales	Office for National Statistics (ONS): http://www.ons.gov.uk/ons/index.html
Scotland	The Scottish Government: http://www.scotland.gov.uk/Topics/Statistics/ Browse/Population-Migration
	General Register Office for Scotland (now part of the National Records of Scotland): www.gro-scotland.gov.uk
Northern Ireland	Northern Ireland Statistics and Research Agency (NISRA): www.nisra.gov.uk

It's worth noting that it is often quicker to locate information about specific data sources – such as surveys – within the ONS's website by using Google, rather than using the site's Search facility, or by hunting through the menus.

Other parts of government also produce statistics about particular populations. Some obvious examples are the Department of Work & Pensions (e.g. the numbers of people claiming benefits such as retirement pensions), the Department for Education (pupils) and the Home Office (migrants).

In recent years, governments have put an increasing emphasis on transparency and making data freely available, and the website https://data.gov.uk is a goldmine of datasets from many parts of government. These include not only statistics but also digital maps, which can be valuable to researchers undertaking demographic projects.

Academic Suppliers

Academic researchers are particularly fortunate in having a well-established data infrastructure which makes available major datasets such as Censuses and sample surveys.

The UK Data Service (UKDS) (http://ukdataservice.ac.uk) is a comprehensive resource funded by the Economic and Social Research Council to support researchers, teachers and policy makers who depend on high-quality social and economic data. It provides a single point of access to a wide range of secondary data including large-scale government surveys, international macrodata, business microdata, qualitative studies and Census data from 1971 to 2011. All are backed with extensive support, training and guidance to meet the needs of data users, owners and creators. All documentation and resources are freely available.

Commercial Suppliers and Producers

There are also commercial *value-added resellers* which supply existing population statistics simply and swiftly, primarily to commercial users. They often produce their own proprietary new datasets too, most notably geodemographic classifications of neighbourhoods, such as Mosaic (produced by Experian) and Acorn (produced by CACI). Both have been used in a variety of academic research projects.

What Are the Major Sources of Population Statistics?

These can be grouped into five broad categories. This summary focusses on the sources of statistics for England and Wales that are available from the ONS. In many cases similar information is available for Scotland and Northern Ireland too, but inevitably there are some differences.

Censuses of Population – 2011, 2001 and Earlier

The most recent Census of Population, which was carried out in March 2011, continued the tradition of providing the most fundamental source of population statistics. Its major strengths are: the range of questions asked, and hence classifications available (see above) and the ability to analyse these together; consistent information on most topics across the United Kingdom; detailed Output Area geography down to circa 120 households; compulsory coverage resulting in a response rate of circa 94% (with missing records being imputed); and it is freely available in popular formats. The main drawbacks are that it is carried out only every 10 years, and that there are inevitably limitations on the number and type of questions that can be asked.

The United Kingdom's 2011 Censuses offer several different categories of statistical products:

> **Resident population: tables** – These published tables classify the population by many different topics. These have traditionally represented much of the greatest part of Census output, with key statistics appealing to the popular market, and more detailed cross-tabulations catering for specialist users' needs.

> **Resident population: area classifications** – These have proven popular, with classifications of Local Authorities being produced for several recent Censuses, and an Output Area Classification (OAC) at the neighbourhood level across the United Kingdom being produced for the first time in 2001, and again, using the 2011 Census.

> **Workplace population** – These tables classify the population at their place of work by selected topics.

> **Origin and destination tables** – Commuting (journey from home address to work address) and migrants (address 1 year ago and current address). Such datasets are more complex to analyse, but academic specialists in particular have used them for important projects for government, such as defining Travel-to-Work Areas and analysing migration patterns.

> **Commissioned special tables** – These can be ordered (at a modest cost) by users who require more detail than is available in the published tables.

> **Microdata files** – These are of two types: individual Samples of Anonymised Records (SARs) and the Longitudinal Study, which links records of anonymised individuals across two or more censuses. Both of these products appeal to specialist users, particularly academic researchers.

The Census operation also enables the creation of two further products which are of great value to users: digital boundaries for Output Areas for mapping, and a Postcode / Output Area directory, which enables the linking of various postcoded address datasets to Census area statistics.

Extensive documentation about the Census is available on the ONS's website at http://www.ons.gov.uk/ons/guide-method/census/2011/index.html.

Administrative Statistics

The enormous growth of computerised administrative records in recent decades is providing new opportunities for the creation of statistics for various populations.

These are usually available for a single country, and sometimes Great Britain or the whole of the United Kingdom. The widespread adoption of address postcodes, when linked to grid references, enables the creation of statistics for very small areas. Such statistics also have the major advantages of being timely and frequently updated.

> **Births and deaths** – Statistics on births and deaths have been produced for many years, and they are a fundamental element of population statistics, not only for analysis in their own right (e.g. comparing life expectancy in different areas) but also for producing updated estimates and projections of future populations.

> **Neighbourhood Statistics** – The ONS's Neighbourhood Statistics website (NeSS) provides a one-stop shop for numerous statistics, including many derived from administrative files. The particular value of NeSS as a supplement to the core Census data is as a source of information about topics not included in the Census, and more recent information.

NeSS datasets can be downloaded from the website http://neighbourhood.statistics.gov.uk which also provides interactive mapping. Scotland and Northern Ireland have similar Neighbourhood Statistics initiatives.

Navigating the NeSS site using the 'Topics' option, the datasets are grouped using domains. Table 13.2 gives pointers to the topics and populations that are available.

NeSS also acts as a shop window, pointing to the original sources of the datasets, and gives a lead for pursuing more that might be available from the originating departments. Table 13.3 illustrates some of the major departments and the data that they collect.

Many thousands of government administrative datasets are now available at https://data.gov.uk.

Updated Estimates – Mid Year Estimates

The ONS's MYEs provide the most recent and best available estimates of the resident population, and they become increasingly important, particularly as denominators, as the Census statistics become more out of date. The MYEs classify the population by age and sex, and they are produced at national, sub-national, and small-area levels. There are also estimates of the very elderly, marital status and ethnicity.

Table 13.2 Neighbourhood Statistics – domains and examples.

Domain	Example topics
Census	Tables on most Census topics
Crime & Safety	Local crime data
	Fire and rescue statistics
Economic Deprivation	Datasets based on Department of Work and Pensions' Claimant records
	Income estimates and business demography
Education, Skills & Training	Key Stage and General Certificate of Secondary Education (GCSE) statistics
	Pupil absences
Health & Care	Estimates of healthy lifestyle behaviours
	Hospital admissions, life expectancy, mental illness and standardised mortality ratios
Housing	Changes of ownership (sales by price)
	Dwelling stock by Council Tax band
Indices & Classifications	Indices of Multiple Deprivation
	National Statistics area classifications
People & Society	Births, deaths and population estimates
Physical Environment	Commercial & industrial floorspace and rateable value, energy consumption, and land use
Work Deprivation	Businesses: enterprises and local units

The MYEs are of the estimated resident population of an area including all people who usually live there, whatever their nationality. Members of UK and non-UK armed forces stationed in the United Kingdom are included, and UK forces stationed outside the United Kingdom are excluded. Students are taken to be resident at their term-time address. The MYEs do not include Short Term Migrants (staying for less than 12 months) or Visitors (staying for up to 3 months).

ONS uses a components-of-change method to update the MYEs since the Census. The general procedure for updating for both national and local estimates is to add one year to ages of the resident population, add births and subtract deaths. The remaining components, in- and out-migrants, are by far the most difficult to measure, and this difficulty increases when estimating for local (subnational) areas.

Sample Surveys

Surveys are usually based on a relatively small percentage sample of people or addresses drawn from lists or registers. They have the advantage of including many detailed questions but the drawback that, because the sample is comparatively small, estimates are only robust nationally or regionally, and not for small areas. Surveys can, however, provide valuable input to modelled data for small areas and individuals.

The results of government surveys are usually presented as standard tables in published reports. It is often possible to also request ad hoc special tables (although the service may not

Table 13.3   Government departments – examples of population information collected.

Government department	Database	Possible topics
National Health Service	National Health Service Central Register: GP Patient Registers	• Date of birth • Sex • Address and changes • Births • Marriages • Deaths • Health conditions
Department of Work and Pensions, HM Revenue & Customs	Customer Information System (this includes children at birth, as well as the adult population)	• Date of birth • Sex • Marital status • National Insurance – working population • National Insurance – workers from overseas • Income • Benefits (various, including child allowance, retirement pensions and disability) • Household structure
Department for Education	Annual School Census	• Date of birth • Sex • Language • Ethnicity • Free school meals • Travel to school • Educational attainment
Home Office	e-Borders	• Passport details • Citizenship • International migration

be speedy). A further option for academic researchers is to carry out their own analyses of anonymised individual records using the facilities available through the UKDS.

Government bodies carry out many sample surveys of people and households. Some of the most important ones are listed in Table 13.4.

Projections of Future Populations

As well as looking at recent population counts, or estimates of current populations, researchers sometimes wish to look to the future, and anticipate numbers and classifications 10, 20 or more years ahead.

It is important to appreciate that government produces 'projections': these are mechanistic, based on stated assumptions, and are usually trend-based. They are not 'forecasts' or 'predictions', which would imply more judgement and belief. Projections provide an indication of the

Table 13.4 Major sample
surveys carried out by
government.

Annual Population Survey
British Crime Survey
British Social Attitudes Survey
English House Condition Survey
Family Resources Survey
General Household Survey
Health Survey for England
Labour Force Survey
Living Costs & Food Survey
Scottish Household Survey
Survey of English Housing
Survey of Personal Incomes
Wealth & Assets Survey

---

impact that changes in demographic patterns might have on the size and age structure of the population in the future, and they become increasingly uncertain the further they are carried forward. To help understand the uncertainty, a number of variant projections are produced based on alternative demographic scenarios.

One important set of population projections published by ONS are the national projections for the United Kingdom and its constituent countries, They are based on the MYEs and a set of underlying assumptions of future levels of fertility, mortality and net migration, and they classify the projected populations by age.

ONS also produces subnational population projections. Past trends are used to project forward the population to give an indication of the future population for 25 years from the base year. Subnational population projections are available for all England local authority areas by age and sex. The projections are consistent with the national population projections for England, and are used as a common framework for local area resource allocation, planning and policy making in a number of fields such as health and education.

In addition to these, projections of the numbers of households in England are published by the Department for Communities and Local Government. These are consistent with the ONS's sub-national population projections and indicate the number of additional households that would form if recent demographic trends continue.

Conclusions: Hunting Population Statistics

In conclusion, researchers nowadays have the great advantage of being able to search for datasets by using the web. This might lead to exactly what you want. If so, that's fortunate, but even then it is always worth seeking contact with people who have expert knowledge of these datasets. A swift email or phone call might save you much effort in pursuing something that is likely to prove worthless for your project, or open up hitherto unknown rich data sources.

14

An Introduction to Bibliometrics

Silvia Salini

Why a Chapter on Bibliometrics?

In recent years, bibliometrics has become a primary tool for the evaluation of scientific research. Such an evaluation is commonly used to support promotion decisions, or to allocate grants. Researchers may move from being the subjects of evaluation to those doing it, and then back and forth between the two roles. However, while there is a large literature on research performance indicators, there is little consensus on the best methods of bibliometric measurements.

An early study presenting the difficulties in constructing research performance indicators was made by Martin and Irvine (1983). They defined three concepts related to the evaluation of a publication: quality, importance and impact. It is, however, very difficult to obtain an absolute or direct measure of the quality, importance or impact of a publication. The evaluation criteria and the assessment depend on the goals, the evaluator (reviewer, national agency, director of a research institution), the stakeholders who commissioned the assessment, what is being evaluated (publication, research, set of publications, set of researchers) and at what level (macro, meso, micro). What is certain is that, whether we like it or not, bibliometrics is pervasive, and at some point we have to deal with it.

Introduction

Bibliometrics is a relatively new science. It started out as a statistical tool for analysing bibliographic data made necessary by the large increase in the number of journals and scientific papers. Only lately, due to a change of perspective, bibliometrics has become a common tool for the quantitative evaluation of scientific research.

This new perspective is not without risks and controversy. While it is considered an indispensable tool for the evaluation of the academic productivity of groups and institutions, in most cases it is criticised as a tool for analysis of individual publications and ranking of individual scholars. What is actually measured through bibliometrics is also somewhat unclear, as we

Research Methods for Postgraduates, Third Edition. Edited by Tony Greenfield with Sue Greener.
© 2016 John Wiley & Sons, Ltd. Published 2016 by John Wiley & Sons, Ltd.

shall see in this chapter. The basic unit of bibliometric indices is the number of citations, so it might appear unreasonable to equate bibliometric indices with 'research quality'. In fact, these indices measure the diffusion and, to a lesser extent, the actual impact of research, which is not the same as quality.

In this chapter, we give a definition of bibliometrics. Then, following Glänzel,[1] we retrace the historic routes of bibliometrics from its origins to the present day, focussing on databases and the new challenges and opportunities they offer. After that, we look at the pros and cons of two alternatives for the evaluation of research, peer review and bibliometric indices, including the most recent indices based on web usage. Finally, we present an overview of the main bibliometric indices and offer some conclusions.

The Definition of Bibliometrics

The Oxford English Dictionary defines *bibliometrics* as 'The branch of library science concerned with the application of mathematical and statistical analysis to bibliography; the statistical analysis of books, articles, or other publications'. The term *bibliometrics* is often confused with similar terms:

- **Scientometrics** – The application of mathematical and statistical methods to scientific and technological communication products designed to determine the relative contribution of scientists, institutions and nations in the advancement of knowledge.
- **Infometrics** – The study of the quantitative aspects of information in any form (not just information recorded in books, articles etc.) and in any social group (not only academics).
- **Webmetrics, cybermetrics and netmetrics** – Extensions of the methods of biblio-, scientoand infometrics to flows of information that materialise on the internet (not just the web, but also e-mail, FTP, P2P and networks).

In 1969, Pritchard defined the term *bibliometrics* as 'the application of mathematical and statistical methods to books and other media of communication'. In the same year, Nalimov and Mulchenko defined *scientometrics* as 'the application of those quantitative methods which are dealing with the analysis of science viewed as an information process.' Hood and Wilson (2001) supply more details about the precise meanings of these terms.

The History of Bibliometrics

Origins

The systematic counting of publications originated in the field of psychology in the early 1900s (Godin, 2006), but the scholars regarded as the founders of bibliometrics – D. de Solla Price, E. Garfield and R.K. Merton – come from different disciplines: history of science, information science and sociology of science, respectively. Conceived as a statistical approach to the management of an increasing flow of academic information in the field of psychology,

---

[1] An overview of the history of bibliometrics presented to the European Summer School on Scientometrics (Vienna, September 2011), at www.scientometrics-school.eu.

bibliometrics was thus born into the rapidly expanding domain of science studies. Through the measurement of the quantitative aspects of these processes, bibliometrics has enabled analysis and understanding of the cognitive processes of communication in science.

Originally, bibliometrics aimed primarily at monitoring, describing and modelling the production, use and dissemination of (academic, especially scientific) knowledge. The first applications were developed to optimise library access and circulation, improve bibliographic databases and extend information services. It was in this context that, throughout the 20th century, the *laws of bibliometrics* originated. These are:

- Lotka's Law (Lotka, 1926) characterises the frequency of publication by authors in a given field. It states that 'the number of authors making n contributions is about $1/n^2$ of those making one; and the proportion of all contributors that make a single contribution is about 60 per cent'.
- Bradford's Law (Bradford, 1934) states that 'If scientific journals are arranged in order of decreasing productivity on a given subject, they may be divided into a nucleus of journals more particularly devoted to the subject and several groups or zones containing the same number of articles as the nucleus when the numbers of periodicals in the nucleus and the succeeding zones will be as $1:n:n^2\ldots$'. Formulated by Bradford after studying a bibliography of geophysics, this law serves as a general guideline to librarians in determining the number of core journals in any given field.
- Zipf's Law (Zipf, 1949), originally proposed in quantitative linguistics for the analysis of James Joyce's *Ulysses*, is often used to predict the frequency of words within a text. This law states that in a relatively lengthy text, if one lists the words occurring within that text in order of decreasing frequency, the ranking of a word on that list multiplied by its frequency will equal a constant. The equation for this relationship is $rf = C$, where r is the ranking of a word, f is the frequency of occurrence of the word and C the constant.

D. de Solla Price (1963, 1976) is considered the founder of scientometrics. His rather timely contribution was to revise the laws of bibliometrics according to the new system of science communication. We owe to de Solla Price the first systematic approaches to the study of modern science, as we are indebted to him for the development of methodology for the analysis of science, based on the foundations of research in scientometrics. It was also de Solla Price who first addressed the analysis of co-citations between papers and who proposed models for the study of these networks. Moreover, he laid the foundations of modern techniques for the evaluation of research, studying, among other things, its policy implications.

Eugene Garfield was the founder and chairman of the Institute for Scientific Information (ISI; now part of Thomson Reuters). In the early 1960s, Garfield developed the Science Citation Index (SCI), the world's first large multidisciplinary citation database. Although the SCI was specifically developed for advanced information retrieval and for science information services – 'The SCI was not originally created either to conduct quantitative studies, calculate impact factors, nor to facilitate the study of history of science' (Garfield, 2009) – it has since become the common source for scientometric studies.

The journal *Impact Factor* (IF) was introduced as a measure for the comparison of journals independently of 'size' and to help select journals for the SCI (Garfield and Sher, 1963). The co-citation-based *Atlas of Science* developed and issued by the ISI was considered a new kind of 'review literature', also suited to help students in their choice of career in science. Garfield later recognised the power of the IF for journal evaluation and considered it also a journal performance indicator (Garfield, 1975).

Robert K. Merton, meanwhile, came to represent the sociologists' view of scientometrics. Among his most well-known ideas related to science and its measurement are the Matthew effect,[2] according to which eminent scientists will often get more credit than a comparatively unknown researcher, even if their work is similar; and the notion of citation as a reward system (currency of science). According to the sociologists' view, communication in science is not merely linked to cognitive processes (cf. information science), but also characterised by the positions that scientists hold in the community (Merton, 1968, 1988).

Evolution

In the 1970s and 1980s, scientometrics/bibliometrics developed rapidly and found a new orientation. In what can be considered a 'perspective shift', bibliometrics evolved from an invisible college, from a sub-discipline of library and information science, to a major instrument for evaluation and benchmarking. New fields of application and challenges opened to bibliometrics as a result – but many tools were still designed for use in scientific information, information retrieval and libraries, and thus became used in a context for which they were not designed.

The development of a specific *scientometric methodology* took place during this period. Co-citation analysis was proposed for the structural mapping of science, and ISI issued the co-citation-based *Atlas of Science* (Marshacova, 1973; Small, 1973). A decade later, Callon *et al.* developed another cognitive mapping procedure, Leximappe, based on co-word analysis (Callon *et al.*, 1983). Later on, these methods were supplemented by and combined with other text-based (term frequency) and citation-based (bibliographic coupling, direct citation link, author co-citation) techniques. During this period also, a consistent system of scientometric indicators was developed for the evaluation of research performance, at the Hungarian Academy of Sciences Information Science and Scientometrics Research Unit (ISSRU; Budapest, Hungary) and Centre for Science and Technology Studies (CWTS; Leiden, the Netherlands) (Braun *et al.*, 1985; Moed *et al.*, 1985; Schubert and Braun, 1986; Braun and Glänzel, 1990).

After the 1990s, a new, important trend in bibliometric application emerged. The level of aggregation began to decrease from the national level through the institutional level down to the level of the evaluation of research groups or even individual scientists. This was allowed for by the rapid development of both scientometric methods and information technology (IT). The spectacular recent evolution of bibliometrics is due in no small part to the IT revolution we have witnessed. The subsequent developments and their synergies have facilitated both breakthroughs in and the popularisation of bibliometrics.

In 1997, Almind and Ingwersen published the first studies of problems related to the area defined by the term *Webmetrics*. The science of Webmetrics – or Cybermetrics, or web metrics (metrics web) – attempts to measure the World Wide Web in order to obtain knowledge about the number and type of hypertext links (hyperlinks) on the various web structures and the (still expanding) patterns of use of resources that populate the cyberspace.

---

[2] The term takes its name from a line in the biblical Gospel of Matthew (25:29): 'For unto every one that hath shall be given, and he shall have abundance: but from him that hath not shall be taken even that which he hath'.

The Principal Databases

In the past, access to electronic versions of bibliographic databases suitable for bibliometrics usage was the privilege of a very few institutes. This changed, firstly, when multidisciplinary and other databases became available to university and institutional libraries, and then again, as the internet mushroomed. Today, such information is readily accessible to all.[3] The most relevant multidisciplinary databases are, at present:

- **Web of Science (WoS)** – Edited by the ISI and distributed by Thomson Reuters, this has a selective coverage of most relevant journals (and other literature sources).[4]
- **Scopus (SCO)** – The major competitor to WoS, sponsored by Elsevier, Scopus covers a higher number of journals than its rival, especially for social sciences and humanities.[5]
- **Google Scholar (GS)** – The academic research version of the ubiquitous search engine (one possible interface for querying is Publish or Perish, developed by Anne-Wil Harzing),[6] GS is more extensive than the WoS and SCO databases, but its data quality is poorer (GS is, of course, based on web search and is not a database as such).[7]

Further to these generalist sources, databases with a specific disciplinary focus provide alternative citation data sources. These offer better coverage and classifications in their field, as well as a variety of special features. Examples include MathSciNet, SciFinder (Chemical Abstract Service (CAS)), PubMed, INSPEC, BIOSIS and RePEC. There is also potential in citation indexing in the cyberworld, as demonstrated by open access sources such as arXiv.org, Citebase, CiteSeerX and Citations in Economics.

Much of the bibliometric literature discusses the characteristics of different databases, especially the largest, most popular ones, and the correlations between and among bibliometric measures. Primarily interested in medical databases, Falagas *et al.* (2008), among others, have compared the content coverage and practical utility of Scopus, the Web of Science, Google Scholar and PubMed. Bakkalbasi *et al.* (2006) used citation analysis in an observational study examining GS, SCO and WoS to test the hypothesis that the three search tools lead to different citation counts. They observed considerable variability: the question of which tool provides the most complete set of citing literature may depend on the subject and publication year of a given article. Norris and Oppenheim (2007) compared WoS, SCO, GS and CSA Illumina (Cambridge Scientific Abstracts) in order to analyse the social science literature. They found Scopus to offer the best coverage in this area of the four databases researched, and suggested that this might be used as an alternative to WoS as a tool to evaluate research impact in the social sciences. Archambault *et al.* (2009) used macro-level bibliometric indicators to compare results obtained from WoS and SCO from geographical and rankings perspectives. They showed the correlations between the measures obtained with these databases for the numbers of papers and the number of citations received by country, as well as their ranks, to be

---

[3] The third is accessible to all; for the first two, universities have to pay license fees; those that currently do not have access are few.

[4] http://isiwebofknowledge.com.

[5] www.info.scopus.com.

[6] http://www.harzing.com/pop.htm.

[7] Which rather brings us full circle, given that the development of search methodology by Google was based on bibliometrics.

extremely high. Franceschet (2010) made a comparison of bibliometric indicator scores and citation-based rankings as computed on WoS and GS, and provided some advice on their use.

With regard to databases, it is important to distinguish between commercial and open access. The purpose of the first type of database is the profit and not the growth of science. This is also demonstrated by the requirements that the journals must meet to enter, for example, in WoS. These formal requirements have nothing to do with the quality, relevance or impact of scientific research, but they are actually essential to control costs related to updating the database. These arguments are the basis of the current debate on open access (Guédon, 2007).

Bibliometrics as a Research Field

After information science and the sociology of science, *science policy* became the third driving force in the evolution of scientometrics. The application of science policy has brought a new perspective and resulted in a re-interpretation of bibliometric conceptions and conventions. The need for metrics in information services, science policy and research management has become widely recognised, with successful application of scientometric methods largely contributing to their increasing popularity. Importantly, electronic communication, the web and open access have paved the way for the democratisation of bibliometrics, a process that carries many risks alongside its benefits. The development of appropriate methods for different levels of aggregation is necessary, while quick and dirty statistics and evaluations, as well as uninformed application and misuse of bibliometrics, could be very dangerous. Peter Hall in the September 2011 *IMS Bulletin* states the following:

> As statisticians we should become more involved in these matters than we are. We are often the subject of the analyses discussed above, and almost alone we have the skills to respond to them, for example by developing new methodologies or by pointing out that existing approaches are challenged. To illustrate the fact that issues that are obvious to statisticians are often ignored in bibliometric analysis, I mention that many proponents of impact factors, and other aspects of citation analysis, have little concept of the problems caused by averaging very heavy tailed data. (Citation data are typically of this type.) We should definitely take a greater interest in this area.

The new situation implies several new tasks and challenges in bibliometric research and technology that require new methods and instruments. The high quality of data is, of course, vital. Extending bibliometrics to the technical sciences, social sciences and humanities is a continuing problem. One other important challenge is to integrate bibliometrics with other data (micro and macro level) related to the efficacy and effectiveness of research, and to the quality, effectiveness and efficacy of education. This is closely connected to ranking issues, funding issues (Geuna and Martin, 2003) and 'multi-dimensional' assessments of higher education systems.

Bibliometrics and scientometrics have become important and hot topics of research for scientists from many disciplines, including computer science, statistics, sociology, mathematics and so on. (Glanzel, 2003). A non-exhaustive list of the main journals on which one can keep oneself updated on the many problems and ongoing discussions on bibliometrics comprises *Scientometrics, Journal of Informetrics, Journal of the American Society for Information Science and Technology* (JASIST), the electronic journal *Cybermetrics, International Journal of Scientometrics, Bibliometrics and Infometrics* and others. There are also journals that do not focus primarily on bibliometrics but often address the themes of bibliometrics, like *Social*

Studies of Science, Research Policy, Research Evaluation, PLoS One, The Journal of Higher Education and others.

New centres for science, technology and innovation research have also been founded world-wide that have a variety of links to the bibliometrics field. Among these, we might mention research centres with a long tradition like ISI (United States), ISSRU (Hungary), VINITI (Russia), CWTS (Netherlands) and NISTADS (India), along with newer ones like OST (France), REPP (Australia), NIFU STEP (Norway), SOOI (Belgium), KAWAX (Chile), WISELAB (China) and IFQ (Germany), along with the companies Science-Metrix (Canada), Evalumetrics (UK), Evindence (UK) and others. Special courses are also offered by universities and research centres. In particular, the University of Vienna, IFQ, Humboldt-Universität Berlin and KU Leuven jointly organise the annual European Summer School for Scientometrics (esss).

Bibliometrics and Evaluation of Research

As mentioned in this chapter, the evaluation of scientific research requires considerable caution, owing to several methodological and technical issues and to their policy implications. The open access journal *Ethics in Science and Environmental Politics* (ESEP) in 2008 published a special issue titled 'The use and misuse of bibliometric indices in evaluating scholarly performance'.[8] The papers in this special issue give an idea of the criticisms of and caution required in using bibliometrics indicators in order to evaluate research. A detailed discussion, to which I am indebted in this short paragraph, of criteria of evaluation of scientific research is Frosini (2011). Here, reference is to the evaluation of a single product of research, which is still the main basis for the evaluation of researchers and institutions.

It is important to stress that evaluation criteria of any type are relative, not absolute. In particular, this means that even if it is to evaluate a single product of research, any evaluation criterion adopted must be able to place the publication in a ranking on an overall publications scale referring to the same subject area at a given historical moment. The criteria for the evaluation of individual products of research are classifiable into two broad categories:

- **Qualitative** – peer review; and
- **Quantitative** – bibliometric criteria (citations and consultations).

The evaluation criterion of peer review is the more traditional and most used criterion for assessing the quality of a research product, normally provided by the reading of a product of research by one or more experts (referees) in the same subject area to which the product to be evaluated relates. One of the major strengths of this criterion is that it can be applied to all types of products and not only to articles published in certain journals. Books, book chapters, articles, patents and working papers may be evaluated thus. Working papers deserve a special comment. These are the typical research products of young academics, such as post-graduates, cognisant of the fact that the process to acceptance of a work by journals often requires more than a year. The best known and most well-established research evaluation, conducted in the United Kingdom as the Research Assessment Exercise (RAE) (since 2014 replaced by the Research Excellence Framework (REF)), used both peer review and bibliometric criteria

---

[8] http://www.int-res.com/abstracts/esep/v8/n1/.

(informed peer review), but includes working papers among the research products counted and rated.

This criterion of peer review is obviously not without limitations (Martin and Irvine, 1983: 72). To simplify, these can be summarised as the subjectivity of assessment (which is not always objective and transparent) and the prohibitive costs (which makes it impracticable in the case of large-scale evaluations).

The bibliometric criteria proposed and actually realised to date can be divided into two main categories:

- Criteria dependent on citations made in articles published in a given set of journals (print or electronic); and
- Criteria-dependent data items in consultation websites (usage log data).

Recently, whole books have been written about bibliometric criteria for research assessment (Moed, 2005; Andrés, 2009; Vinkler, 2010). As noted, Garfield initially proposed to establish the operation of bibliometric indices dependent on citations made when he founded the SCI in 1955. The critical point of these indices is the role of the citations. Supporters of extensive bibliometric indices argue that the citations contained in a paper constitute admission of 'scientific debt' to cited authors who have previously worked on the same field or related fields, with contributions of some basic form. There is no doubt that such admissions of positive influence (impact) may affect some citations; it is yet to be verified, however, whether this affects all, or even most, citations.

There are many weaknesses in this approach, highlighted for example by Martin and Irvine (1983: 68). Some criticisms focus on the unpredictability of the timing of citations (Spiegelhalter and Goldstein, 2009); some focus on the existence of negative and rhetorical quotes (Cozzens, 1989); others speak of *self-citations*, *in-house citations*, *in-group citations* and the *halo effect* (Adler *et al.*, 2009); while still others discuss the different *citation habits* of different disciplines (Seglen, 1997). Unfortunately, a gradation of *degree of influence* of the various citations cannot be performed in bibliometric studies: all citations are valued on an equal footing. Furthermore, considering that it has been customary to use citations – at least for the last 30 years – and to cite wherever reasonably possible, a judgement of quality on a paper with a large number of citations cannot be derived.

It is true, however, that an important consideration for the use of bibliometric criteria is that the papers published in journals of high editorial position have already undergone a rigorous peer review process that should have guaranteed the fulfilment of quality criteria. If more are cited, then the publication as well as high quality also have a high impact and are therefore important for the growth of knowledge. Highly relevant to the issue of citation is the percentage of rejection of papers by journals and the factors that influence this (topics, citations, fashion, groups, countries etc.).

The indices of the second type given – those based on consultations of papers that can be read and downloaded from the web – have been extensively investigated in recent years by a group of scholars at the Los Alamos National Laboratory (LANL), led by Johan Bollen (Bollen *et al.*, 2005, 2008). Notably, a project called MESUR (Metrics from Scholarly Usage of Resources) was established which has produced the construction of a giant database, comprising about 1 billion events, processing and producing several metrics (indices) of impact; also, there are numerous articles that have released this new bibliometric criterion, as well as the metrics connected. The *usage factor* (UF), defined as a bibliometric indicator quantity of new generation, is complementary to the traditional pattern known as *impact factor* (IF).

Another important initiative is due to the Public Library of Science (PLoS). They publish transparent and comprehensive information about the usage and reach of published articles onto the articles themselves, so that the entire academic community can assess their value. They call these measures for evaluating articles 'Article-Level Metrics',[9] and they are distinct from the journal-level measures of research quality that have traditionally been made available until now.

Bibliometrics Indicators

As measures of different aspects of research output, indicators generally concern scientific journals, single researchers or groups of researchers (departments, scientific disciplinary sectors, universities etc.). Most basic indicators are determined for publication output (as a measure of productivity), co-authorship (as a measure of collaboration) and citation rates (as a measure of impact) or a combination thereof.[10]

Most indicators are derived from simple counts of items extracted from various bibliographies and databases. Advanced measures are *network indicators*, derived from the analysis of co-authorship and citation networks. The more commonly used bibliometrics indicators are based on the ratio of citations per paper (indicator of average impact/diffusion). Among this group, the well known and already mentioned IF is no more than the arithmetic average of citations of papers in a journal in a given period (typically two years), while the immediacy index (II) calculates the average number of citations of those papers published during the same year (and thus represents a measure of how quickly the average article in a journal is cited). It is important to note that these two indices are regularly referred to journals. It is not correct to attribute the IF or II of a journal to papers that appear in it. The citations distribution is strongly asymmetric to the right, following the Pareto distribution (Seglen, 1997; Adler *et al.*, 2009): a few articles have a lot of citations, the mode and the median for many journals are much lower than the mean (in some cases, even 0), and it is unlikely that the impact factor for any paper will be equal to the impact factor of the journal in which it appears. As an example, Table 14.1 reports the frequency distribution of citations for the papers published in the *Journal of the American Statistical Association* in 2010. Figure 14.1 shows the typical distribution shape of the citations. In this example,[11] 31% of the papers have zero citations, 22.3% have only one citation and so on. The mean (total citation/total paper = 335/148) is 2.26, the median is 1 and the mode is 0.

However, IF provides *a priori* information about the 'expected quality' of an article. In general, the use of citations should concern large aggregates, for appropriate time intervals and with appropriate normalisation.

The other well-known and widely used bibliometric indicator is the *Hirsch* (h)-index. This is defined as the number such that, for a general group of papers, h papers received at least h citations while the other papers received no more than h citations (Hirsch, 2005). Typically

---

[9] http://article-level-metrics.plos.org.

[10] Glänzel presents some methodological challenges for the bibliometrics measures to the European Summer School on Scientometrics (Vienna, September 2011); at www.scientometrics-school.eu.

[11] Data was downloaded from Web of Science in July 2012.

Table 14.1   JASA 2010 – frequency distribution of citations

Citations	Frequency	Percent	Cumulative Percent
0	46	31.1	31.1
1	33	22.3	53.4
2	27	18.2	71.6
3	14	9.5	81.1
4	10	6.8	87.8
5	4	2.7	90.5
6	2	1.4	91.9
7	3	2.0	93.9
9	3	2.0	95.9
11	2	1.4	97.3
12	1	.7	98.0
16	1	.7	98.6
18	2	1.4	100.0

Data from Web of Science 2012.

used to evaluate individual researchers, the h-index is a number defined over an ordinal scale with only equivalence and ordering properties. The distance between two consecutive classes in terms of 'average effort' is gradually increasing. This is one of the most criticised aspects of this index. Moreover, it is not additive (if one researcher has an h-index of 5 and another of 7, the h-index calculated on the publications portfolio of the two researchers is not 12). Some

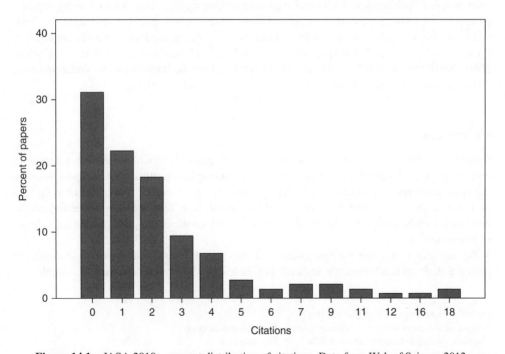

Figure 14.1   JASA 2010 – percent distribution of citations. Data from Web of Science 2012.

modifications of the h-index are present in the literature, such as the *individual h-index*, which takes account of the number of authors (Batista *et al.*, 2008); the *m-quotient*, which considers the age of the individual (Hirsch, 2005); and the *contemporary h-index*, which includes the age of the papers (Katsaros *et al.*, 2006). The packages that calculate the h-index and its variants are Publish or Perish,[12] Plug-in Fire Fox[13] and TritaH-Scholar H-Index Batch Calculator.[14]

More recently, indicators that consider the reputation of citations have been proposed (as opposed to the indicators seen before, in which one citation has exactly the same prestige as another). Other more complex indicators weighing citations depending on the impact of the citing papers have been developed (here, a paper is important if it is 'endorsed' by important papers).

The last point to be made related to bibliometrics indicators is the need for normalisation. Indicators have tended to ignore unsurprising but important variations: the fact that journals have different levels of productivity, scientists have different seniority, research institutions have different staff numbers, papers have different numbers of co-authors and, last but not least, articles from sub-fields have different citation propensity. The problem of field normalisation is particularly relevant in the benchmarking of institutions. Until now, indicators have had to regard a set of papers within the same field or discipline, but different scientific fields may have different propensities regarding the practice of citation (citation rates) and differences of maturation of citation impact (citation accumulation).

A common feature of field-normalised indicators is that they are based on comparisons between the amount of citations received by the group of publications examined and a comparison term given by the expected number of citations received/made by analogous publications in the specific discipline(s) of interest. The comparison term should represent the citation propensity of the publications of interest. It could be determined using a selection of a reference sample of publications with similar citation propensity (e.g. same journal, superimposed classifications and neighbourhood); an indicator based on references or citations received; and/or selection for the central tendency indicator (e.g. mean, median, harmonic mean etc.). Examples of indicators of this type are Fractional Citation Counting (Leydesdorff and Opthof, 2010), SNIP (Moed, 2010), Audience Factor (Zitt and Small, 2008), MNCS (Waltman *et al.*, 2010) and *Success*-index (Franceschini *et al.*, 2011; Kosmulski, 2011).

Conclusions

Considering the current state of play, what kind of approach should a researcher follow in order to guarantee a good assessment for himself and for his institution? A simple answer to this question is not easy to provide. As always, however, a reasonable rule of thumb is that the best route is probably somewhere in the middle – in this case, between the extremes of, on the one hand, merely adapting to bibliometric criteria, and, on the other, just ignoring the science of bibliometrics.

Researchers who adopt the first course may attempt to consider all possible strategies to increase their own bibliometric indices. For example, they might increase the number of

---

[12]http://www.harzing.com/pop_hindex.htm
[13]https://addons.mozilla.org/en-US/firefox/addon/scholar-h-index-calculator/
[14]http://www.tisreports.com/products/14-The_H_IF_index.aspx

co-authorships to increase the number of published papers, they may try to get into strong groups working on hot topics to increase the likelihood of being quoted, they may submit papers only to the top journals for their field and so on. This behaviour can definitely be rewarding in the short term, but in the long run it represents a risk both for individual researchers and for academia as a whole. Firstly, there is not, in fact, an optimal behaviour that maximises all bibliometric criteria; and, secondly, the evaluation criteria, as we have seen, change as a function of the goal, the scientific field, the country and time (development of the field, academic trends etc.). The possible effects on the scientific community of such behaviour being adopted by a critical mass could be very serious. If all researchers were to conform to the impact factor policy, for example, many important papers that have shaped the history of science would have gone unwritten, unfashionable topics would be abandoned, and small research centres and institutions and US universities on the top of international rankings would be even more crowded than they already are. Indeed, this book would not even have been written!

Researchers who adopt the second approach remain woefully unaware that any reviewer, even if free to follow a qualitative approach, can now, thanks to the wide dissemination of bibliometric databases and accessibility to all, derive a first impression of a scholar from his or her bibliometric indices (informed peer review). Indeed, this is increasingly likely. The exponential growth in the number of papers to be evaluated in recent years has made this a common strategy for first screening.

In terms of advice on how to be at least reasonably bibliometrics savvy, it is also important not to be just self-referenced. To have a consensus in one's own small scientific community is not a sufficient guarantee for the true potential of research. Submitting your work to the judgement of the most prestigious journals of your field is, of course, also a way of conveying knowledge, an opportunity for your ideas to be read by people with experience and expertise, and an incentive to challenge one's mental schema. In other words, a thing of value becomes really important when it is communicated in an efficacious and effective way, and this will inevitably be recorded bibliometrically, whatever the methodology adopted. Simon Day in his Editorial of *Journal of Royal Statistical Society (Series A)* (2012) writes:

> Some used to say, 'publish or perish', and perhaps they still do. Certainly it is important for university-based scientists to publish their work. There are many reasons for this including gaining (or maintaining) an appropriate level of self-recognition, and the same for that of one's research institution. Thoughts of research assessment, league tables, and so on, come to mind. Some of this need or desire for recognition crosses over to industry-based scientists also. However, there are gains and losses. We all know that publishing takes time—not just the wait for referees and editors to give comments and to come to an opinion, but the time to write, edit, respond to editors' comments (often requiring more detail of certain aspects while reducing the overall length of a paper), and so on. The option of giving up and spending time (and perhaps money) on other more fruitful or interesting projects often lures us. Yet we all know that it is important to disseminate the results of all studies to a wide audience—if nothing else simply to prevent the waste of other people's time repeating work that has already been done (even if shown to be futile) [...]. Let us be clear for whom research and publication are for: not for us (the scientists) but for society as a whole. Sadly, this is sometimes forgotten. We should keep in mind such questions as 'Is what we are doing useful, or simply interesting (to us)?'. This does not bar 'theoretical' research in favour of 'applied' research. Our research should be interesting to us; perhaps some fun also. But we have a duty to our scientific peers, and to society as a whole, to disseminate our findings.

We must assume that everyone has specific features and capabilities. The best way to become a good researcher in life is to be the best of ourselves. Homologation does not bring innovation, creativity and excellence, but the lack of judgement and comparison. Following the guidelines of this book is definitely good advice to produce good research. Combined with tenacity, commitment and open-mindedness, research will always be rewarded.

References

Adler, R., Ewing, J., and Taylor, P. (2009). Citation statistics (with discussion). *Statistical Science* 24: 1–28.
Andrés, A. (2009). *Measure Academic Research: How to Undertake a Bibliometric Study*. Cambridge: Chandos Publishing.
Archambault, E., Campbell, D., Gingras, Y., and Larivire, V. (2009). Comparing bibliometric statistics obtained from the Web of Science and Scopus. *Journal of the American Society for Information Science and Technology* 60(7).
Bakkalbasi, N., Bauer, K., Glover, J., and Wang, L. (2006). Three options for citation tracking: Google Scholar, Scopus and Web of Science. *Biomedical Digital Libraries* 3–7.
Batista, P.D., Campiteli, M.G., Kinouchi, O., and Martinez, A.S. (2006). Is it possible to compare researchers with different scientific interests? *Scientometrics* 68: 179–189.
Bollen, J., Van de Sompel, H., and Rodriguez, M.A. (2008). Toward usage-based impact metrics: First results from the MESUR project. arXiv :0804.3791v1.
Bollen, J., Van de Sompel, H., Smith, J.A., and Luce, R. (2005). Toward alternative metrics of journal impact: a comparison of download and citation data. *Information Processing and Management* 41: 1419–1440.
Bradford, S.C. (1934). Sources of information on specific subjects. *Engineering: An Illustrated Weekly Journal (London)* 137: 85–86.
Braun, T., and Glänzel, W. (1990). United Germany: the new scientific superpower? *Scientometrics* 19(5–6): 513–521.
Braun, T., Glänzel, W., and Schubert, A. (1985). *Scientometric Indicators. A 32-Country Comparison of Publication Productivity and Citation Impact*. Singapore: World Scientific.
Callon, M., Courtial, J.P., Turner, W.A., and Bauin, S. (1983). From translation to problematic networks: an introduction to co-word analysis. *Social Science Information* 22: 191–235.
Cozzens, S.E. (1989). What do citations count? The rhetoric-first model. *Scientometrics* 15: 437–447.
de Solla Price, D.J. (1963). *Little Science, Big Science*. New York: Columbia University Press.
de Solla Price, D.J. (1976). A general theory of bibliometric and other cumulative advantage processes. *Journal of the American Society for Information Science* 27: 292–306.
Falagas, M.E., Pitsouni, E.I., Malietzis, G.A., and Pappas, G. (2008). Comparison of PubMed, Scopus, Web of Science, and Google Scholar: strengths and weaknesses. *The FASEB Journal* 22: 338–342.
Franceschet, M. (2010). A comparison of bibliometric indicators for computer science scholars and journals on Web of Science and Google Scholar. *Scientometrics* 83(1): 243–258.
Franceschini, F., Galetto, M., and Maisano, D. (2011). The success-index: an alternative approach to the h-index for evaluating an individual's research output. *Scientometrics* 1(21).
Frosini, B.V. (2011). *La valutazione della ricerca scientifica: scopi e criteri di* valutazione *[in Italian]*. Rome: Vita e Pensiero.
Garfield, E. (1975). ISIS Atlas of Science may help students in choice of career in science. *Current Contents* 29: 5–8.
Garfield, E. (2009). From information retrieval to scientometrics – is the dog still wagging its tail? Paper presented at the Fifth International Conference on WIS, Tenth COLLNET Meeting, Dalian, China, 13–16 September. http://garfield.library.upenn.edu/papers/dalianchina2009.html
Garfield, E., and Sher, I. (1963). New factors in the evaluation of scientific literature through citation indexing. *American Documentation* 14(3): 195–201.
Geuna, A., and Martin, B.R. (2003). University research evaluation and funding: an international comparison. *Minerva* 41(4): 277–304.
Godin, B. (2006). On the origins of bibliometrics. *Scientometrics* 68(1): 109–133.
Guédon, J.C. (2007). Open access and the divide between "mainstream" and "peripheral" science. http://eprints.rclis.org/12156/

Hirsch, E. (2005). An index to quantify an individual's scientific research output. *Proceedings of the National Academy of Sciences of the USA* 102: 46.

Hood, W., and Wilson, C.S. (2001). The literature of bibliometrics, scientometrics, and informetrics. *Scientometrics* 52(2): 291–314.

Katsaros, C., Manolopoulos, Y., and Sidiropoulos, A. (2006). Generalized h-index for disclosing latent facts in citation networks. http://arxiv.org/abs/cs.DL/0607066

Kosmulski, M. (2011). Successful papers: a new idea in evaluation of scientific output. *Journal of Informetrics* 5(3): 481–485.

Leydesdorff, L., and Opthof, T. (2010). Normalization at the field level: fractional counting of citations. *Journal of Informetrics* 4(4). http://leydesdorff.net/opthofleydesdorff/rejoinder.pdf

Lotka, A.J. (1926). The frequency distribution of scientific productivity. *Journal of the Washington Academy of Sciences* 16(12): 317–324.

Martin, B.R., and Irvine, J. (1983). Assessing basic research: some partial indicators of scientific progress in radio astronomy. *Research Policy* 12: 61–90.

Merton, R.K. (1968). The Matthew effect in science. *Science* 159(3810): 56–63.

Merton, R.K. (1988). The Matthew effect in science, II: cumulative advantage and the symbolism of intellectual property. *ISIS* 79: 606–623.

Moed, H.F. (2005). *Citation Analysis in Research Evaluation*. Dordrecht: Springer.

Moed, H.F. (2010). Measuring contextual citation impact of scientific journals. *Journal of Informetrics* 4(3): 265–277.

Moed, H.F., Burger W.J.M., Frankfort, J.G., and Van Raan, A.F.J. (1985). The use of bibliometric data for the measurement of university research performance. *Research Policy* 14: 131–149.

Norris, M., and Oppenheim, C. (2007). Comparing alternatives to the Web of Science for coverage of the social sciences literature. *Journal of Infometrics* 1: 161–169.

Pritchard, A. (1969). Statistical bibliography or bibliometrics? *Journal of Documentation* 25: 348–349.

Schubert, A., and Braun, T. (1986). Relative indicators and relational charts for comparative assessment of publication output and citation impact. *Scientometrics* 9(5–6): 281–291.

Seglen, P.O. (1997). Why the impact factor should not be used for evaluating research. *British Journal of Medicine* 314: 498–502.

Small, H.G. (1973). Co-citation in the scientific literature: a new measure of the relationship between two documents. *Journal of the American Society for Information Science* 24: 265–269.

Spiegelhalter, D., and Goldstein, H. (2009). Comment to the article 'Citation Statistics'. *Statistical Science* 24: 21–24.

Vinkler, P. (2010). *The Evaluation of Research by Scientometrics Indicators*. Cambridge: Chandos Publishing.

Waltman, L., van Eck, N.J., van Leeuwen, T.N., Visser, M.S. and van Raan, A.F.J. (2010). Towards a new crown indicator: some theoretical considerations. *Journal of Infometrics* 87: 467–481.

Zipf, G.K. (1949). *Human Behavior and the Principle of Least Effort*. Cambridge, MA: Addison-Wesley.

Zitt, M., and Small, H. (2008). Modifying the journal impact factor by fractional citation weighting: the audience factor. *Journal of the American Society for Information Science and Technology* 59(11): 1856–1860.

15

Choosing and Using Software for Statistics

Felix Grant

I assume that you already recognise the need for statistical analysis at some level, and that you already know, or will soon decide, what techniques are appropriate to your needs and purposes. So, I shall not advise you on what statistical methods you need. That advice is in other chapters. I shall, instead, advise on how to choose the software aids most appropriate to you for applying those techniques once they have been selected. Questions to consider are:

- What can you afford?
- What sort of person are you?

What Can You Afford?

Can you afford to buy (or persuade someone else to buy) new software, or must you rely on what is already available? At the time of writing, the cost of statistics software ranges from under £100 (about €140 or US$100) to over £1000 (about €1400 or US$1400). You will be wise to investigate your position in this regard, at an early stage. There is no point in identifying an ideal solution if it is not accessible to you. If your institution, employer or sponsor has a standard program in place, they may be reluctant to buy something else. In that case, you have to decide whether the price of purchase is justified by the advantages offered. Even if you persuade someone else to make a purchase, or if you are factoring in the cost of software for a funding body bid, you will still need to provide a cost–benefit case to justify the outlay in terms of time savings and/or improved productivity.

If you cannot match your ideal package with realities of supply, all is not lost. You may be able to borrow facilities. If even a restricted range of options is available to you, the advice here will equip you to choose between them. If all else fails, knowing what you want will at least help you to make best use of the software provided.

Research Methods for Postgraduates, Third Edition. Edited by Tony Greenfield with Sue Greener.
© 2016 John Wiley & Sons, Ltd. Published 2016 by John Wiley & Sons, Ltd.

With the questions of cost decided, the ultimate check must obviously be whether or not a software package embodies, or enables, the techniques you have chosen or, perhaps, parallel equivalents. This is a matter of detailed examination of the manufacturer's literature, application of your own subject area knowledge, and follow-up enquiries which we will consider in more detail towards the end of this chapter. Before you can make sound decisions about the capacity of a package, though, you must first make a realistic assessment of your own relation both to statistics and to software in general.

What Sort of Person Are You?

All software is trying to provide a perfect balance between power, ease of use and affordability. But where that perfect balance lies varies from person to person. What sort of person you are is an important factor in choosing the best tools for your purposes. Don't let anyone tell you that only the mathematics or statistics wizard can do valuable research. There are as many approaches to research as there are researchers or study areas, and using software to complement your strengths in other areas is as valid as using it to amplify your reach in one where you are already secure. Also, depending on the work you are doing, very different blends of rigour and intuition may be your best ally. With that in mind, ask yourself some honest questions – and answer them candidly.

Do you delight in statistical methods, or fear them? Are you happy with detailed mathematical work at university level, or does it fill you with dread? Do you willingly delve into the workings of the software tools that you already use, or do you just want them to work smoothly and not bother you with detail? Do you enjoy programming, or is it a closed book to you?

The answers to these questions will take you a long way towards your final choice. If you can cheerfully call yourself confident in all four respects, you may prefer products that emphasise power and flexibility. If, on the other hand, you feel confident in none of them, your priority will be maximum support and ease of use. In most cases, of course, you will be somewhere between the two extremes and looking for some intermediate balance of power with usability.

How much statistical work do you really need to do? And are you a visual thinker or a symbolic puzzle solver?

The most efficiently powerful tools for a strong mathematician with confidence in both underlying statistical methods and programming practice are those which overtly present themselves as a high-level language and offer a Command Line Interface (CLI) at which you type your requirements. If you are a visual thinker from the arts or humanities, however, with little or no background in any of these areas, a far better solution is a program built around a Graphical User Interface (GUI) which mimics or uses the methods and software which you already know – and you even may be better off with one of the more extensive technical graphics packages.

Very few products these days are entirely within either camp. Most offer at least a basic form of both methods – but there is a great variation in the emphasis which is placed on one or the other. I will cover the two ends of the spectrum first, then move on to look at the packages which occupy the middle ground in between. Finally, I will look at some special cases which do not fit into these neat compartments.

I will use specific products as examples; I have selected them because I know them well and can recommend them all, but you should look beyond the specifics to general principles which will help you to best match available tools to your particular tasks and needs. Remember, too, that this is a fast-changing area: by the time this book is published, changes and developments will have occurred in many products. I shall assume that you are using the increasingly dominant Wintel platform. This is a personal computer with an Intel or compatible processor chip, running Microsoft Windows. If that is not true in your case, the general considerations in this chapter will still be valid, but not all of the specific products will be available to you. Many products are available on one or more alternative platforms; some are not.

Emphasising the Graphical

Starting from the GUI end of the range, we will first look at the technical graphics option. You should seriously consider whether such software might do all you need. This sector of the market is well suited to exploratory statistics, and visualisation is a large part of that approach.

The current generation of scientific graphics packages includes various basic statistical functions. Some (e.g. StatGraphics Plus from Manuguistics Corp.) regard this as part of their core function, while for others (e.g. Axum from MathSoft and SigmaPlot from SPSS) analysis is a value-added addition to a primarily visualisation function. For many purposes, graphic visualisations will tell you what you want to know. All the graphics packages are slightly different, so it is worth examining them all for particular techniques that might meet our needs. Diamond from SPSS, to take a particularly striking example, offers a fractal foam tool that allows the most intuitive exploration currently available of relatedness between numerous aspects of your information.

If your analysis needs are more than can be met by such software, consider the other side of the same coin: will a statistical analysis package provide the graphics you need? You may need both types of software, but usually not. It makes little sense, for example, to acquire both Axum and MathSoft's analysis package S-Plus (discussed further in this chapter), when both include the same graphics capability. Think carefully about your actual needs; consult with your supervisor and others with experience in this area.

Moving on from these *visualisation crossover* solutions, we come to a body of statistics tools that emphasise one or other of the *window, icon, mouse, pointer* (WIMP) approaches to analysis rather than a programming or command line.

You may be familiar with a spreadsheet program, most likely Microsoft's Excel spreadsheet, part of the Microsoft Office suite of programs that have swept the board as far as standard products are concerned. You may want to continue using this spreadsheet. You may even *have to* continue using it, if you need to exchange data and results with other people. Spreadsheets are valuable tools, and they can be used to great effect for basic statistical work if you know how to get the best from them; but you should be aware of their limitations. In particular, some statistical spreadsheet functions become unreliable if you push them too far; and you will, in many cases, work harder for the same results than you would in a proper statistical package. Nevertheless, most analysis and visualisation products allow you to leave your data within your spreadsheet for general management purposes, import it for analysis and perhaps even export the analyses back to Excel.

You must organise your spreadsheet in the way that the receiving product will expect to find it, with variables in columns and cases in rows (or, in database terms, each column is a field and each row is a record), and some spreadsheets are better supported than others. Excel generally gets the best support for current file versions. Legacy data, for example from Lotus 123, often has to be saved in older file formats such as WK1. Users of Corel's otherwise excellent Quattro Pro have the hardest ride; although a few packages will read this format directly, most require the mild inconvenience of an intermediate file save in some other supported format. Nevertheless, whatever your spreadsheet, you can share its contents with your analysis software. For users of Excel, probably the easiest package to use (at the time of writing) is Unistat, from the company of the same name.

Unistat is actually a separate, standalone product with its own built-in datasheet (a *datasheet* is what most stats packages call their specialist internal spreadsheet array). It is also, however, specifically designed to allow hiding it within other products and comes out of the box all ready to integrate seamlessly with Excel. Once Unistat is installed, you have three options: run Excel alone, run Unistat alone or run Excel with Unistat as an add-in extension; the third option does not use a separate datasheet at all, and you need do no conscious importing or exporting. Highlight the Excel columns you want to analyse or visualise, select a function from the three new items that Unistat has placed on your Excel menu and Unistat will kick in automatically to operate on the highlighted data. When you have what you want, click one of four output buttons on a new toolbar at the top to export it to your choice of destination: back to Excel, out to a Word report, into an HTML-formatted web page or onto the Windows clipboard for use in some other suitable program. Statistical work from spreadsheet data doesn't get much easier than this.

Another package that makes the most of the GUI approach, but in a completely different way, is Statview. This is one of a range of products tailored to different needs from the giant SAS Institute. Statview originated on the Apple Macintosh platform (where it is still available) and brings a view of things that is somewhat different from the Wintel norm, but the differences are all positive.

To Statview, everything is an **object**. Each of your variables (e.g. age, intensity, salinity and colour) is an object. A graph of a relationship between those variables is an object. An analysis of the way in which a variable is distributed is an object. At first sight, this is strange (especially if you are used to spreadsheets), but it has tremendous advantages once you look more closely. The objects have a physical presence on your screen, and they can be picked up and dropped on each other. If you want to investigate the relation between two variables, for example, just open a scattergraph object with a mouse click, then drag the two variables into it. Statview is particularly popular in the medical and health sciences, but it also has a good record of success for users with no prior experience of statistical work at all. The object-based approach is a good analogy for everyday, hands-on experience of working with things or ideas. Statview also allows you to save an analysis you have done as a new object, so you can use the same method again, just by dragging and dropping new data into it.

If you are not afraid to deal directly with statistical ideas, prefer a free-standing package with a conventional worksheet approach to your data, but want to stay in the WIMP category, a strong contender is Stat-100 from Biosoft, which offers remarkable value for money at a price well below its specification. Probably the lowest priced package around at the moment, it demands very few system resources, will run under older versions of Windows and on older machines or small crowded notebooks, yet will do most of what more expensive competitors can do. It is also a very effective partner to some of Biosoft's other, more specialised tools (see the 'Special Cases' section). There is a more sophisticated sibling, Stat-200, that is still low priced at double the cost, but that belongs in the later *middle ground* section (see 'Going for the Best of Both Worlds'). JMP from SAS also provides a versatile and richly analytical service with detailed statistical options.

Taking Direct Command of Your Software

At the other extreme from these graphical environments, we find the overtly programming approaches.

Perhaps the purest and best known of these is Genstat, a statistical programming language developed from the more general Fortran specifically for statistical analysis. Marketed by NAG (Numerical Algorithms Group; www.nag.com), it concentrates on analysis, giving graphical and text output in very stripped-down form, with presentational aesthetics taking a back seat to clarity. If you are confident in Fortran and want to build your own highly efficient routines focussed on your own needs, this is a good place to start the investigation of software for your chosen route. Genstat does have a Windows interface, for interactive exploration of data, but its users usually see this as a tool for preliminary investigation before getting down to serious batch processing.

At this end of the scale, my warnings about use of statistical functions in spreadsheets apply with even greater force, but the spreadsheet program itself can still be used to great effect as a container and low-level organiser of your data. As before, most programs will import data from widely used spreadsheets, particularly Excel. If your preferences, constraints or working software environment predisposes you to remaining within Excel for the analysis itself, despite the problems, NAG again have a solution in the form of their Statistical Add-ins for Excel, a package of Fortran routines which integrate as an Excel add-in to provide command line function call access. JMP can also work in an integrated manner with Excel.

Many products in this category could, perhaps, be placed in the *middle ground*, since their WIMP interfaces are so highly developed. Mathsoft's S-Plus, for example, is built around Lucent's powerful S programming language, and it is through direct access to that aspect that a user will use the facilities available to maximum effect. Despite this, however, S-Plus installs on your computer to present you with a graphical exploration interface as powerful as any; the language is hidden behind the WIMP front end until wanted. Despite appearances, S-Plus would be an expensive and overpowered purchase for most users who want a graphical program, but it is an excellent choice if you want exploratory and visualisation capacity on top of a programming environment. The freeware R, which has evolved from S-Plus, also has a variety of front-end supports such as R Studio to help the user visualise their data and organise their analyses and output.

If you are mathematically as well as statistically confident, you may want to investigate the use of generic mathematical tools with programming or command line facilities as an alternative to dedicated statistical ones. I leave these until the 'Special Cases' section.

Going for the Best of Both Worlds

Stat-200, Biosoft's big brother to Stat-100 in the graphical group, is a good introduction to the middle ground where both GUI and programmed approaches coexist on more or less equal terms in the same package. It is, primarily, an interactive exploratory package that makes full use of the sophisticated WIMP facilities in 32-bit versions of Windows; it has, nevertheless, a procedure language that enables a considerable degree of batch automation.

The range covered by this, the most popular category, is vast. It includes, for example, Statsoft's Statistica: an immensely powerful modular package that integrates with other Statsoft products to provide anything from basic descriptives to full-blown control of a factory.

Statistica is a WIMP-driven package, conceived from its very earliest days as a graphic tool. It has, however, two programming languages of its own which enable any degree of batch automation, allow control of other programs or subservience within them and can be used to build specialised *turnkey* systems.

Another interesting and widely used member of this group is Minitab, which has in many ways taken the opposite development route to Statistica. Minitab originated as a command line program, has evolved towards the WIMP model and now straddles the two approaches. In the version current at the time of this writing, the screen is divided horizontally with the WIMP menu and tool bars at the top, a session window below and a datasheet at the bottom. Operations selected from the menu are echoed in the session window as commands; these can be copied, pasted, edited, and executed or stored as batch processes from a command line. Command sequences can also be written from scratch in the session window and executed directly without recourse to the menus at all. SPSS by IBM is popular, in particular, amongst social science and medical students and similarly has a divided screen to help the user visualise their data and organise their analyses and output.

Special Cases

I have concentrated primarily on general-purpose statistics software, but there are specialised tools out there for many more focussed purposes, either in their own right or alongside the more general products.

Design of Experiments (DoE)

One of the big growth areas in recent years has been the computerised design of experiments. Industry is the economic engine here, and agriculture was the crucible where the techniques evolved, but DoE is applicable to every study where any control at all can be exercised over the collection of data: from literature to physics and everything in between. DoE is essentially a particular application of standard methods, but purpose-designed software support simplifies its application considerably.

Increasingly, DoE tools at some level or other are built into general products (particularly the high-end ones), and these may well suffice for your needs. If you will be making intensive use of these methods, however, it is worth considering the dedicated packages that are available. DesignEase and DesignExpert from Stat-Ease Corp., and the DoE-Fusion products by S-Matrix, are both excellent. Both companies offer two levels of power to suit your needs, and both achieve their ends in different ways to suit different user psychologies.

Closely related to DoE is sample design software, such as the excellent nQuery Advisor from Statistical Solutions.

Mathematics in the Round

Useful as statistical analysis software is, it does not answer every need. In certain circumstances, and if you are sufficiently relaxed with mathematics, both analysis and modelling can be more efficiently and rapidly productive using general mathematical methods. Products such as Mathematica, Mathcad, Maple and Matlab (all in different ways) can provide invaluable support or even replacement for their dedicated statistical cousins. On the one hand, you may want to work exploratively with formulae in natural notation, using the speed of a computer to rapidly build up an intuitive model for data features that resist conventional methods. Or, you may find that building your own numerical approaches from fundamental mathematical building blocks enables you to shake off preconceptions and attack (and solve) intractable problems in new ways.

Other Special Purpose Tools

If your study is particularly concerned with a particular aspect of your data, there may well be an established way to deal with it that offers greater efficiency than a general analytical product. Search the web, and/or the available literature, for references to your particular concerns in relation to software.

For example, if you are concerned with highly peaked data (which occurs in fields as diverse as aeronautics and art history), a suitably framed search may turn up two very different but equally valuable avenues to explore. One relies on focussed use of a generic package, S-Plus, using modified kernel routines. The other uses a specialist program designed for the purpose, Peakalyse, from scientific software supplier Biosoft.

Peakalyse is one example of the many specialist software tools that are available, not only from Biosoft but from plenty of other companies. There may be one company that already produces exactly what you want. In Biosoft's range, you will find (apart from Peakalyse) products for purposes such as assay analysis, curve fitting, data extraction from published plots, and mixture characteristics, amongst others.

Suitability for Your Purpose

General Considerations

What I have been able to give here is no more than a broad sketch of the software market within which you must seek the best tools for your needs. With that sketch in mind to orientate you, decide the *type* of software product you would ideally like to use and the types that you would prefer to avoid if possible. Then, from detailed knowledge of your own study before you, start by making a list of the essential statistical techniques to be employed, and a subsidiary list of the desirable ones.

At this point, as you survey the market and make a detailed examination of feature lists on websites and in manufacturers' literature, you will find either that several products serve your purpose or that none matches your needs perfectly and you must seek to make a second choice fill the bill. The first, happy state of affairs is most likely if your needs match your skills closely; the second is more probable if there is a mismatch in either direction. Either way, you should delve more deeply in order to choose between the candidates.

Ask to try out the software, to get a feel for it in your own hands and context. Increasingly, companies are making time-limited demonstration copies available for web download. Ask for the names of satisfied customers, and then approach those customers with a request for a brief hands-on experience. Talk to colleagues and others in your area, or who are like you as a person. Join mailing lists (see the 'Help and Support' section).

If you are unable to match needs with techniques on offer, are there ways to achieve your ends through other techniques that *are* available? If you have the confidence, can your required techniques be synthesised from others, achieved by modifying others or built from scratch by programming them yourself with the facilities provided? Can you pair up a main package with a specialised one (or more) to get the combination you need?

It is worth approaching the publishers of those products that come closest to what you are seeking. They have a tremendous store of accumulated in-house experience, and will often come up with a solution that had not occurred to you. In some cases, they will even make modifications to provide what you want. After all, if you want it, then so perhaps do other potential customers.

Help and Support

Minitab and Statistica, though about as different as they can be, both originated in the academic system, and this is still very visible in their approaches to users. When you buy these products, you are also buying into a system of user support which knows how people like you learn their statistical tools. They are not the only ones. Systat, for example, is another. It has one of the best manual/tutorial sets around for the non-statistician.

This is one of the many issues you should consider and which have nothing to do with statistics: how much help and support are you going to need, and who will provide it? If you are secure in your own abilities, within the context of your subject, you are free to choose from the whole gamut of software that the market can offer you. If you will be relying on sources of help amongst those around you, you will need to consider the need to stay within their range of experience and/or adaptability. If you are inexperienced, unsupported and nervous, you should look carefully at the support resources available to you.

Support resources do not have to come from the manufacturer or publisher of the product (though they often do so). There are internet news and discussion groups, and mailing lists for some widely used products (e.g. Genstat has a thriving user list). Some products (Minitab, again, is notable in this area) have also spawned a large number of textbooks. Look for books that mention your particular area of study in connection with particular products. There may be valuable support from those who have gone before. Search the web, that most prolific source of the latest information and contacts.

And Finally …

Never forget that statistics software is only a tool, no more useful than what you do with it. Allow statistics to strip away the undergrowth and reveal the true form of your data, and allow software to take the drudgery out of the process, but never, ever allow either of them to usurp your own understanding, experience, intuition and independent powers of thought. Like all power tools, they are dangerous if they are not fully in your control but liberating as long as you remain in charge.

16

Computer Support for Data Analysis

Clifford E. Lunneborg

Introduction

In this 21st century, no post-graduate student will complete the analysis of his or her research data without turning to a computer. The computer has moved from being a widely available and convenient tool to being both ubiquitous and essential.

Data should never be analysed by computer. The computer provides invaluable support to the research student in the analysis of data. But it is the researcher who must design, monitor, refine, report and interpret the analysis, guided at all stages by the questions that motivated the research.

The researcher must remain in control of the analysis. Beware of statistical computing packages; they invite the over-analysis of research data. Recent growth of statistical computing resources makes this warning imperative. Statistical packages are fast and easy to use, at least in the sense that they are user-friendly. They span many statistical techniques and offer many options in the implementation of any one technique. All of this makes it tempting to try out all-too-many analyses.

The increase in amount of computing support for data analysis prompts a second warning: statistical computing resources should be identified while designing the research, *not* after the data have been collected.

Five Stages of Statistical Method

A paper by R.J. MacKay and R.W. Oldford (2001) provides an accessible account of how data analysis needs to be integrated into an over-reaching statistical method. The method is represented as a series of five stages:

Research Methods for Postgraduates, Third Edition. Edited by Tony Greenfield with Sue Greener.
© 2016 John Wiley & Sons, Ltd. Published 2016 by John Wiley & Sons, Ltd.

- Problem
- Plan
- Data
- Analysis
- Conclusion.

In the problem stage, the researcher identifies, at least by name, the variates and attributes that must be assessed in the research as well as the aspect of the problem to be studied. Is the goal of research to describe a set of relations, to establish causal links or to predict future events? In the plan stage, measurement instruments or techniques are selected for the named variates and attributes, and decisions are taken as to where and in what amount data are to be collected. The data stage requires continuing data monitoring as well as data storage.

The evolution of the statistical method through the first three stages affects the analysis stage, not least the computing support that will be required. You do not need any unpleasant surprises when you reach the analysis stage. Ensure that computational support will be available for the descriptive, causal or predictive analysis of the particular amount and kind of data your measurement techniques and research design call for. Make certain that, as data are collected, they remain appropriate to the planned analyses and are stored in a manner compatible with those analyses.

Planning the analysis, including its computational aspects, as you plan your study protects against nasty surprises and reduces the chances that you will expend your energies on fruitless or needless analyses.

Confirmatory and Exploratory Analysis

Most research programs will profit from exploratory as well as confirmatory data analyses. Briefly, confirmatory analyses are those that answer the questions that drove your research. Exploratory analyses provide clues about how better to design your next study. Exploratory analyses have been described as unplanned or dependent upon the data, once collected and examined. They may be of that kind, but even so, their value will be the greater if attention is given, in the problem and plan stages, to what might go wrong. Are there additional attributes that could be worth measuring? In what detail should the data be saved for subsequent analyses?

Throughout, and particularly at the conclusion stage, you should distinguish confirmatory from exploratory analyses.

Computing Resources

There are *statistical procedure packages* and *statistical languages*. The former typically offer a limited number of statistical procedures, either standard or specialised, in an environment easily traversed by the user. The latter provide an extensive set of tools that a knowledgeable user could deploy to implement an extremely rich range of procedures.

To an impressive extent, the distinction between statistical procedure packages and statistical languages has dimmed, though it certainly has not vanished. The language programs have added pull-down menus, permitting researchers to carry out standard analyses with a series of mouse clicks. And the procedural packages have added scripting or macro capabilities, giving researchers greater flexibility in carrying out analyses.

Every major statistical computing resource provides far greater capabilities than it did five years ago. Furthermore, most statistical packages have become far more cooperative; data can be saved in a variety of formats from one package, say SPSS, to be readily input to a second package, such as S-Plus, or be analysed using the freeware R. As a result, the researcher can often move from package to package, analysing data on one package, for example, and then turning to other packages to prepare presentation graphics or papers for publication.

New versions of statistical packages are released quite often. The capabilities of any one package will have increased between the time this chapter was written and the time the book reaches you. Fortunately, it is quite easy to learn the capabilities of the latest version of a package. The website provided by the package's publisher can be consulted. Web addresses for several statistical packages are provided later in this chapter. The websites of the Royal Statistical Society (www.rss.org.uk) and the American Statistical Association (http://amstat.org) provide links to the websites of package publishers. At the time of this writing, an impressive list of statistical packages with links to their websites is maintained on the website of the publishers of the statistical package Stata (see http://www.stata.com/support/links/stat_software.html).

Standard Statistical Packages

Many universities have adopted one of the major statistical packages for use across faculties, and this package, naturally, will be the one to which you will turn first. Four of the more popular packages are referenced in this section. There is a wider discussion in Chapter 15 of how you might go about choosing a statistical package, in terms of the nature of your work, your own nature and the level of the analysis that you might do.

Genstat (an acronym for *general statistics*) is the product of researchers at the Rothamsted Experimental Station and is distributed by the Numerical Algorithms Group (NAG; www.nag.com). Owing to its origins, Genstat is strong on the analysis of designed experiments and on related regression models. Genstat also provides facilities for a range of multivariate analyses such as cluster analysis, principal components, correspondence analysis, discriminant analysis and multivariate scaling, as well as the study of time series, survival analysis and the analysis of geospatial data.

SPSS (Statistical Package for the Social Sciences; www.spss.com) boasts a host of standard parametric and nonparametric statistical tests and is widely used for standard analyses of variance, for regression modelling and for factor, cluster and discriminant analyses. The base-level package can be supplemented with add-on modules for, amongst other applications, categorical predictive models, conjoint analyses, exact small-sample statistical tests and data mining.

Minitab (www.minitab.com) has a loyal following amongst teachers of statistics and may be familiar to you as a student. Minitab shares much of the statistical coverage of SPSS and Genstat. It has a worksheet orientation to data and, thus, may appeal to those students who are comfortable with spreadsheets but find spreadsheet programs limited in their statistical capabilities. See the 'Specialised Statistical Packages' section on Excel.

S-plus, the augmentation of the statistical language S distributed by Insightful (formerly MathSoft; www.insightful.com), is the tool of choice of many academic statisticians. The Windows version, S-Plus 2000, now supplements a command line interface with pull-down windows for common analyses and for the preparation of statistical graphics. S-Plus, though strong on graphics, is best known for its extensibility, as is the freeware R. S-Plus and R are the platforms used by statisticians for developing new procedures. If yours is a novel statistical application, they may be the best environment for your analyses. Venables and Ripley (1999) give an excellent survey of S-Plus with a good indication of its adaptability. If you become serious about developing new applications, Venables and Ripley (2000) provide valuable assistance. There are many useful books about R and also a strong internet following which supplements the rather sparse and concise help facilities within R.

Specialised Statistical Packages

The Microsoft spreadsheet program, Excel, is increasingly popular as a statistical analysis platform, particularly in business schools. The statistical capabilities of Excel (including those provided by the Analysis ToolPak supplied with the program) cover the range of procedures associated with an introductory course (e.g. Berk and Carey, 2000). For more advanced analyses, third-party add-ons may be available. For example, XLSTAT (www.xlstat.com) provides a set of 25 tools to facilitate data management and to implement a variety of nonparametric tests, analyses of variance and the more popular multivariate analyses. The XLSTAT website provides links to other Excel statistical products and to Visual Basic programming resources for those who need to create new statistical applications for Excel.

BUGS (Bayesian Inference Using Gibbs Sampling; http://www.mrc-bsu.cam.ac.uk/bugs/) provides the first platform for the study of Bayesian conditional independence models that is relatively easy to use. If you think that models of this kind might be of interest to you, look at the examples at the website; this package may be invaluable.

Resampling Stats (www.resample.com) provides a platform for quickly developing applications of permutation (randomisation) tests and bootstrap inference. These nonparametric resampling techniques, which effectively replace difficult if not impossible mathematical analysis with repetitive computations, have grown in importance as the speed of your desktop or notebook computer reached a level where the applications became feasible. Lunneborg (2000) provides an introduction to these techniques.

StatXact (www.cytel.com) provides implementations of the more common permutation tests. Some of these features have been imported from StatXact to SPSS.

BLOSSOM (http://www.mesc.usgs.gov/blossom) is another package of specialised permutation tests. The package features permutation applications to multiple-response data and linear models. BLOSSOM is available at no cost from the US Geological Survey.

Conclusion

Only a very few of many statistical packages have been mentioned. The Stata website link mentioned here runs to 12 pages and provides access to the websites of 170 software publishers.

If the computing resources of your university or laboratory are not appropriate to your research:

1. Search the list for more appropriate candidates.
2. If the search is not successful, you may need to develop a statistical application for your own needs. This will seem somewhat more difficult than the prospect of running a mainstream

statistical application, but do not be daunted. A statistical computing platform like S-Plus or R, with thousands of data-handling functions to build upon, should permit you to meet your goals.

3. If your computational requirements are not only novel but also demanding, you may have to develop your application in a lower level (but faster executing) language. Fortran and C++ have been the standards for developing statistical algorithms. S-Plus, among other statistical packages, provides a well-documented pathway for incorporating Fortran or C routines. In this way, you need code only those parts of your application that are computationally most sensitive while using standard functions for the balance of the analysis.

As a research student, you are gaining skills and knowledge of lifetime value. Each data analysis is an opportunity to learn more about an ever-expanding range of computational resources.

Good learning and good computing!

References

Berk, K.N., and Carey, P. (2001). *Data Analysis with Microsoft Excel*. Pacific Grove, CA: Duxbury.

Lunneborg, C.E. (2000). *Data Analysis by Resampling: Concepts and Applications*. Pacific Grove, CA: Duxbury.

MacKay, R.J., and Oldford, R.W. (2001). Scientific method, statistical method and the speed of light. *Statistical Science* 15: 254–278.

Venables, W.N., and Ripley, B.D. (1999). *Modern Applied Statistics with S-Plus*, 3rd ed. New York: Springer-Verlag.

Venables, W.N., and Ripley, B.D. (2000). *S Programming*. New York: Springer-Verlag.

Part III
Measurement and Experimentation

17

Instrumentation in Experimentation

Anand D. Pandyan, Frederike van Wijck and Garth R. Johnson

Introduction

Measurement instruments can be used in all sorts of experiments, ranging from studies during space exploration to mother–infant interactions or the recovery of dexterity after a stroke (Table 17.1 has a few examples).

In this chapter, we shall introduce you to the use of instruments in experimentation. We shall discuss the concept of measurement, consider *what* you might choose to measure and, then, *how* you can measure using instrumentation. We shall describe the basic components of a measurement system and explain their successive functions. We shall illustrate different types of measurement systems with examples of how instrumentation has been used in the measurement of human performance. We shall present criteria on which to evaluate measurement instruments, including aspects of scientific robustness (e.g. sources of error) as well as practicalities. In the discussion that follows, we encourage you to think about *why* you might want to use measurement technology by considering the balance of potential benefits and drawbacks of using instrumentation in your research. Finally, we shall suggest where you might find help, guidance and resources.

Concept of Measurement

What Is Measurement?

Among many definitions of measurement, we find:

> Measurement refers to the use of a standard (such as a metric ruler) to quantify an observation.
>
> Kondraske (1989)

He adds,

> Assessment is the process of determining the meaning of a measurement or collective set of measurements in a specific context.

Research Methods for Postgraduates, Third Edition. Edited by Tony Greenfield with Sue Greener.
© 2016 John Wiley & Sons, Ltd. Published 2016 by John Wiley & Sons, Ltd.

Table 17.1   Some examples of measurement instruments, as they may be applied in different areas of research

Topic of interest	Examples of instrumentation used
Cardiac rehabilitation	Heart rate monitor electrocardiogram (ECG)
Alcohol content in the blood stream	Breathalysers
Relation between brain and behaviour	Positron emission tomography (PET)
Density profiles across a lipid membrane	Reflectometers
Strength (materials, animals including humans, and plants)	Strain gauges and signal amplifiers
Forensic applications (lie detectors)	Heart rate monitors
	Blood pressure monitors
	Impedance monitors
Environmental hazard monitoring	Geiger and scintillation counters
Obstetrics & Gynaecology	Ultrasonic transducers and measurement
Non-destructive testing in engineering	systems
Thickness of polar ice caps	
Breast screening	Infrared thermal sensors and cameras
Insulation protection (buildings and transport)	
Archaeology	Carbon dating with mass spectrometers

Thus interpreted, measurement is instrumental in arriving at objective and accurate inferences from observations. Part VI of this book offers guidance about analysis and interpretation of data.

What Can We Measure?

Measurement is the process by which you allocate numbers to a quantity or an event. Measurement leads to the generation of data. Your data may be categorised as: nominal, ordinal, interval or ratio levels of measurement. Table 17.2 provides the characteristics of each level.

- **Nominal-level data** pertain to information that can be ordered only into categories which are different but which lack a hierarchy (e.g. male–female). Also called *categorical*.
- **Ordinal-level data** can be classified hierarchically, although the magnitude of the difference between each step in the hierarchy is either unequal or unknown (e.g. the level of

Table 17.2   Key characteristics of nominal, ordinal, interval and ratio levels of measurement. $\sqrt{}$, Condition has to be satisfied; ×, condition need not be satisfied

Characteristics level	Hierarchical order	Quantitative measurement	Equal interval lengths	True zero point
Nominal[#]	×	×	×	×
Ordinal	$\sqrt{}$	×	×	×
Interval	$\sqrt{}$	$\sqrt{}$	$\sqrt{}$	×
Ratio	$\sqrt{}$	$\sqrt{}$	$\sqrt{}$	$\sqrt{}$

Adapted from Pandyan *et al.* (1999).

consciousness according to the Medical Research Council (MRC) Grading for Strength) (Riddoch *et al.*, 1943).

- **Interval-level data** can be organised into a hierarchy with the magnitude of the difference between each successive step being known and equal, but which lacks an absolute absence of the quantity (e.g. temperature in °C).
- **Ratio-level data** have all the characteristics of interval-level data, in addition to an absolute zero (e.g. velocity measured in m/s and temperature measured in °K).

It is essential that you are able to categorise your measurement into one of the above levels of measurement as this classification will dictate which statistical methods you can use for data analysis. The general rule is to use parametric tests when your measurements provide ratio or interval levels of data that are continuously distributed, non-parametric tests when your data are either ordinal or nominal. *In certain circumstances, when an ordinal level of measurement has many levels (e.g. a score of greater than 40 levels) and is based on a hierarchical method of assessment, one could treat the ordinal-level measure as equivalent to an interval or ratio scale.*

Measurement, as defined by Kondraske, usually involves interval or ratio-level data entailing standard units. Opinions differ: some authors (see Wade, 1992) prefer to use the concepts of 'measurement' and 'assessment' interchangeably. This implies that measurement may also include nominal and ordinal-level data, using counts and frequencies. But caution is urged when these terms are used interchangeably as assessment is the interpretation of measurement within ecological and environmental constraints. As an illustration, a body builder will often have a body mass index (BMI) in the obese range but will not be classified as obese.

Before discussing how to measure, you should think carefully about what you wish to measure and why that should be relevant. What is the construct you wish to measure, and how does your dependent variable relate to this?

Measurement Systems: Basic Principles

A measurement system is a configuration of elements that interact with each other to quantify a specific phenomenon for assessment, analysis or control purposes. Figure 17.1 is a simple diagram of a generic measurement system. This consists of the following components: a signal source, a transducer (or electrodes) and amplifiers, data display and storage devices and in some cases a control device. We will describe the components related to measurement in the following sub-sections.

Signals normally result from an energy conversion process and can be broadly classified into two types: analogue signals and discrete signals. Analogue signals are continuous in time (e.g. monitoring hear rate), whereas discrete signals occur as discrete events in time (e.g. maximum and minimum temperature in a 24-hour period). The primary input to the measurement system is the signal: the phenomenon to be measured. Measuring any phenomenon without the assistance of instrumentation generally leads to data of nominal or ordinal levels.

Measurement instrumentation is needed to obtain interval or ratio-level data. Elements of this are given throughout this section.

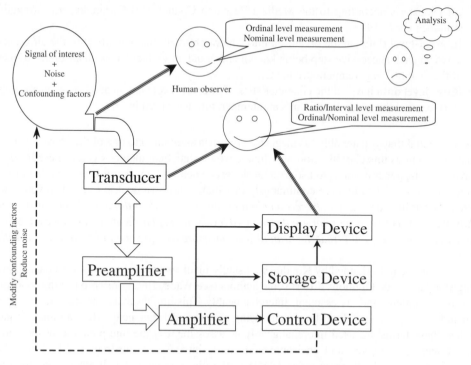

Figure 17.1   Simple diagram of a generic measurement system.

Transducers

A transducer converts energy from one form to another. The transducers in measurement systems normally convert changes in any form of energy to changes in electrical energy. This process of energy conversion can be passive (e.g. a thermometer) or active (measuring displacement using a potentiometer). Passive transducers are often used when the magnitude of the signal is large when compared to the levels of noise (e.g. the measurement of the boiling point of water). However, much of the time the magnitude of the desired signal is significantly lower than the surrounding noise signals (e.g. human biological potentials) and an active transducer, one that is energised by an external power source, is needed for recording purposes.

Preamplifiers and Amplifiers

In many applications that use active transducers, the output signals often have a low amplitude and are contaminated by noise and other confounding factors. Also, it may not always be possible to take measurements at the site of recording, such as when monitoring harmful radiation or measuring rapid eye movement (REM) during sleep. In these circumstances, the signals need to be amplified prior to transmission for display and storage. In general, amplification is carried out in two (or more) stages. In the first stage(s), preamplifiers are used to amplify the input signals; this amplification in addition to amplifying the signal of interest will also amplify the associated noise. Whilst the signal is being preamplified, the system

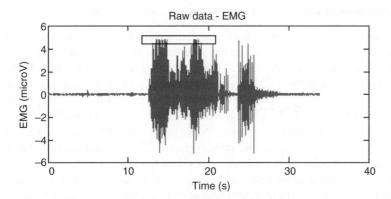

Figure 17.2 An example of a recording of human muscle activity. The signal was recorded using bipolar electrodes (a method of improving the signal-to-noise ratio) with a sampling frequency of 1980. There was evidence of data saturation (amplification of the signal was too high) as identified in the boxed section. If the EMG was quantified (using a root mean square procedure) prior to filtering of the means frequency noise, the mean was calculated as 0.299 μV; after the removal of the means frequency using an eighth-order Butterworth notch filter, the mean reduces to 0.295 μV.

also tries to reduce the levels of unwanted noise by filtering (this is described in the 'Filters' sub-section). Once the ratio of the signal to noise has been improved (i.e. signal of interest is amplified and the noise is filtered out), further amplification can be carried out on the signal of interest. For example, in space exploration and radio transmission, the signals collected by the transducers and microphones need to be amplified many times before they can be transmitted without contamination. This transmission of the signal between stages of amplification may be by wires or wireless (e.g. via Bluetooth technology). At the amplifier, the signals are further amplified and processed before being displayed or stored. It is important to note that if the signal amplification is too high, useful data can be lost due to saturation (Figure 17.2).

Filters

Transmitted signals often become contaminated with unwanted variation. This random signal contamination, known as *noise*, often arises from interference from the mains power supply but may also originate at the point of observation (Figure 17.1). For example, any movement of a sleeping person whose REM is being monitored may appear as noise in the recoded signal. The unwanted noise signals need to be removed for a meaningful interpretation of the recorded signal. This is usually done by filtering. Commonly used filters are:

- **Notch filters**: to block selected frequencies in the signal (a 50 Hz notch filter is used to eliminate the mains interference from signals – in the United States, a 60 Hz notch filter is used).
- **Band pass filters**: to allow only signals in a selected band to pass through (when measuring muscle activity, signals with frequencies below 20 Hz and above 200 Hz are not normally required, so a band pass filter with a band width of 20–200 Hz is used). Please be aware that any useful information at the filtered frequencies will also be lost.

Data Storage on a PC

Measurement systems are often used in association with personal computers. If you use a personal computer in your research, you can take data in the form of an analogue signal from an amplifier and convert it into a digital signal that is a series of bits. You do this with an *analogue-to-digital converter* (ADC) card. Most commercial ADC cards provide a 12-bit or 16-bit conversion. The 12-bit ADC is less sensitive than the 16-bit ADC. Suppose you have a signal that varies from 0 to 10 V. A 12-bit ADC card has a resolution of 4096 (2^{12}). Therefore, the 0 V signal will be equivalent to 0 on the ADC, and 10 V will be equivalent to 4095. Thus, a change of one volt from the amplifier will cause an ADC signal to change by approximately 409 units, and one ADC bit will measure approximately 2 mV.

The interface between an amplifier and your PC can be established either with a serial or with a parallel port. Many commercial programs are available for data collection on a PC. However, commercial programs may not have the flexibility your need for your research. Instead, you can develop your own data collection program using a high-level programming languages such as HpVee, Labview, or Matlab or a general purpose programming language such as C++. Whatever your approach, you should test that the ADC card matches the data collection program to be purchased or developed.

Performance of Measurement Systems

Although the diagram (Figure 17.1) identifies the individual components of a measurement system, you must remember that there is a high degree of interaction between them.

An ideal transducer (or amplifier) will not draw any energy from the signal source (or transducer). In reality, all transducers extract some energy during the measurement process. For example, a thermometer will absorb heat when it measures temperature. If a transducer or amplifier extracts a disproportionate amount of energy from the signal source, it will affect the quantity being measured and thus bias the outcome. When you select a transducer, you should select one that will minimise this bias. This is usually achieved by selecting a transducer with a *high-input impedance* (*impedance* is the technical term for a complex resistance). The higher the input impedance of the transducer, the less energy it will draw from the signal source. At the output from the transducer, the process has to be reversed: the impedance has to be low so that the preamplifier or amplifier will be able to extract the signal from the transducer without any contamination. The general rule for all the components in the diagram is that the input impedances should be much higher than the output impedances of the preceding stage.

Confounding effects corrupt the signal but cannot be eliminated by filtering. For example, the ambient temperature may influence the output from a strain gauge. Therefore, ideally you should choose a transducer whose output will not be influenced by any factor other than the signal of interest. However, if a confounding effect cannot be avoided, you may be able to compensate for its influence or you may be able to measure it separately and so allow for it in data analysis. A common example is that of correcting for the influence of ambient temperature on measurements of interest.

Evaluation of Measurement Instrumentation

In this section, we discuss several criteria that may help you to choose instrumentation for your research.

Health and Safety

A risk assessment of both the measurement system and the process of measurement is important before implementation. If your research involves human beings, pay special attention to their safety and comfort. Verify with the manufacturer if there are any health hazards, called *contra-indications*, for a particular type of equipment. For example, the use of an electromagnetic measurement device may be dangerous for anybody who has a pacemaker. In Europe, any piece of equipment should bear the CE mark. This tells you that it meets European safety standards.

Accuracy and Precision, Sample Frequency, Threshold and Resolution

Threshold and **resolution** indicate the sensitivity of a measurement system. When the input to the measurement system is increased from zero, at one particular value of the input the measurement system will record an output – this is the threshold of the measuring system (Doebelin, 1983). Any signal with a magnitude below the threshold cannot be measured. Resolution is the smallest detectable change in input that can be measured by the measurement system (Doebelin, 1983).

Accuracy and **precision** express the quality of data that a system will measure. See also Chapter 27. Accuracy is the difference between the measured and true value of a variable (Doebelin, 1983). The manufacturer will usually provide information about the accuracy of a system. However, it is important to understand that the accuracy of an instrument refers to the system as a whole, as it functions in a specific environment using specific methodology. Therefore, accuracy may change over time, through use or by damage, and may also be subject to factors such as room temperature (Durward *et al.*, 1999). Because of this, it is important to *calibrate* your measurement system before data collection. Calibration is the procedure in which the input to the system is systematically varied over the range of values expected in the experiment and compared with the output.

Precision indicates the ability of a system to yield the same value for repeated measurements (Durward *et al.*, 1999) and thus indicates how much the value of a discrete sample varies from the mean of a group of measurements. The smaller this variation, the more reliable the system. Since the precision of a system may vary within a measurement range, you should estimate precision throughout the full range required for the experiment.

The **sampling frequency** of the measurement system is the number of times per second at which data are collected. If the sampling frequency is equal to or less than the frequency of the input, the reconstructed signal becomes distorted. This phenomenon is known as *aliasing*, defined as an error between the sampled and the original signal (Cohen, 2000). The minimum sampling frequency that can be allowed, the so-called *Nyquist frequency*, is based on the *sampling theorem*. This theorem postulates that the sampling frequency should be greater than or equal to twice the frequency of the original signal (Cohen, 2000). However, exact application of this rule has serious limitations, and we advise you to sample at about four times the signal frequency (if this is possible). However, care is urged in over-sampling too. Increase in the sampling frequency will lead to an overestimation of the measured quantity in certain rare circumstances (Figure 17.3).

Linear measurement systems – in which any change in input over a defined range produces a proportional change in the output – are generally recommended. However, although some

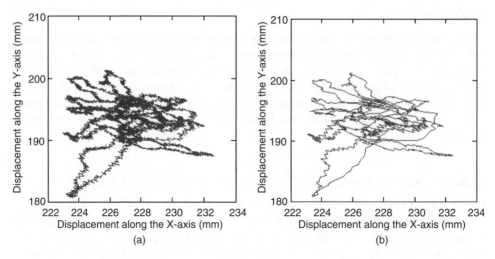

Figure 17.3   (a) The measurement of human sway in 2D when measured at 1000 Hz and the estimated path length is 1483 mm. (b) The same measurement of human sway in 2D when down-sampled to 100 Hz and the estimated path length is 154.46 mm.

transducers may not be linear, this should cause no problem provided that the relationship between the input and the output can be quantified and is reproducible.

Sources of Error

Before you use the measurement system to collect experimental data, you should establish any sources of error and their magnitudes. For example, when you use an opto-electronic movement measurement system, stray infrared light, caused by reflective material, may cause errors. Fluctuating air temperature may disturb measurement systems that involve sound.

Crosstalk, which is said to occur when the output of one signal is contaminated with that from another, may be another source of errors. It can occur when different measurement components are at work at the same time (Durward *et al.*, 1999). Hence, before you collect data, check if there are any signs of crosstalk. By systematically changing input to one channel only and monitoring the output of all other channels, you can examine whether only the channel to which you changed the input has registered a change. During this test, you should ensure that channels with no active inputs are grounded.

Hysteresis is the phenomenon where the output of a system depends on its immediate past history of perturbation. For example, in measuring the stiffness in a joint such as the elbow, a therapist will move the joint about its full range of movement. If the therapist were to move the joint twice with an interval of 15 seconds between measurements and record the stiffness, then the stiffness encountered during the first test would be higher than the stiffness in the second. Read Durward *et al.* (1999) for a detailed discussion of hysteresis. Hence, if you suspect that your measurement instrument is susceptible to hysteresis, you should test it repeatedly through its entire measurement range with values ascending and descending.

User-friendliness

Is your instrument practicable? Questions that you should ask about any proposed equipment are:

- **Set-up**: Is it easy to set up? This is important, especially if you intend to use it routinely or if you want to use it in several different places.
- **Calibration**: Is calibration automatic and quick?
- **Automatic data collection**: Is data collection automatic? If it is, it will be fast and help to avoid errors caused by the operator. It will also ensure that all trials are automatically indexed so that there is no confusion between them later.
- **Real-time display**: Will it display data in real time? This will enable you to check data integrity during capture. If you detect errors, you can correct them in the following trial.
- **Automatic analysis**: How much of the data analysis process can be automated? But be aware that analysis by system software might not approach the analysis in the way that you would like; it may obscure interesting features that more careful examination could reveal.
- **Data storage**: Is data storage easy and efficient? Does it automatically identify each trial?
- **Synchronisation**: When you have several instruments to collect data, can they be synchronised automatically? An example of a situation where you may need several instruments is in lie detection, where heart rate, blood pressure and galvanic skin response are measured simultaneously.
- **Wider database**: Will your system easily enable you to input the data into a wider database?

Other Practicalities

- **Cost**: Estimate the costs of purchase, maintenance and running of equipment. Also remember you will need to account for training costs as well as costs involved in data collection and analysis. For example, in some clinical centres, expensive devices are purchased for routine measurement; however, due to resources constraints associated with trained personnel and data analysis, many of these device are rarely used.
- **Environment**: Where will you have your equipment? Is there enough space? Is there other equipment nearby, or are there other activities that could interfere with your process and introduce errors into data collection?
- **Sturdiness**: Is your equipment robust enough? Where this is required: is it portable enough?

Discussion

Why Measure?

Before you start to implement measurement systems in your own research, you should weigh the potential benefits against the drawbacks. You may wish to consider both the outcome of measurement (the quality of the information obtained) as well as the process of measurement. Table 17.3 summarises the main benefits and drawbacks of implementing measurement instrumentation in research.

Table 17.3   A summary of the main benefits and drawbacks of implementing measurement instrumentation in research

	Potential benefit	Potential drawback
Outcome	• Additional information • Greater accuracy • Greater precision • Greater resolution • Data of higher level (interval/ratio)	• Information overload; additional information is not necessarily relevant. • Differences are not necessarily relevant.
Process	• Data acquisition and analysis: standardised to a high level • Using a computer enables data to be stored, accessed and analysed post hoc.	• Calibration if often required, and repeated calibration will need to be done. • Costs • Time • Training • Dedicated space

Outcome of Measurement

A benefit of using instrumentation may be that it provides additional information. For example, in agriculture, measuring soil acidity provides farmers with additional information on which to base decisions regarding the types and amounts of fertilisers.

The downside of this may be that – unless you carefully think about what you need to measure – this additional information may not be relevant and may only lead to information overload.

Instrumentation allows you to improve accuracy and precision. Without instrumentation, your measurement system would be just the assessor (Figure 17.1); it would depend only on his or her observational skills and interpretation of the observations.

Most sensing, processing and presentation of experimental data are now done with standard routines that yield quantitative outputs, as depicted in Figure 17.1. Measurement technology does not eradicate measurement errors. Rather, the sources of error are of a different nature, as discussed in this chapter. The main issue is: when only quantitative measurement methods are employed, can the magnitude of such errors be determined, corrected for or taken into account (Tourtellotte and Syndulko, 1989)?

Measurement tools with quantitative scales provide much greater resolution, a much higher level of detail, than qualitative scales (which are widely used in the humanities, medicine and allied professions, and environmental studies). Thus, smaller changes may be registered which would otherwise remain undetected. For example, measuring the concentration of toxins in river water allows local authorities to monitor the effectiveness of measures to reduce industrial pollution and detect problems at an early stage (i.e. before flora and fauna become affected).

Many qualitative scales provide nominal or ordinal levels of measurement. This restricts the range of statistics that can be applied to the data (Tourtellotte and Syndulko, 1989). For example, it may seem convenient to calculate a total score from individual test items in the Bayley Infant Scales or the IQ test. However, such an operation is statistically invalid. Summary statistics, such as sample means and standard deviations, are valid only with interval and ratio variables (Durward *et al.*, 1999). In other words, information obtained using measurement technology is generally more powerful because the magnitude of change can be reported.

Websites for Software and Hardware

www.adeptscience.co.uk	A commercial site for information on HpVee; data collection software and hardware and analysis software
http://www.ni.com/labview/what.htm	A commercial site for information on Labview; data collection software and hardware and analysis software
http://www.mathworks.com/products/matlab/	A commercial site for information on Matlab; data collection software and hardware and analysis
http://msdn.microsoft.com/visualc/default.asp	A commercial site for information on C++
www.amplicon.co.uk	A commercial site for information on instrumentation for measurement
www.analog.com	A commercial site for information on instrumentation and sensors for measurement
http://www.dspguide.com/pdfbook.htm	An online resource for digital signal processing
http://www.engineers4engineers.co.uk/metrics.htm	A resource site for information on units, measurement and instrumentation

Websites on Safety and Standards

http://gallery.uunet.be/esf/	European Safety Federation
www.mdss.com	Medical Device Safety Service
http://eur-op.eu.int	Office of the official publications of the EU
www.medical-devices.gov.uk	UK Medical Device Agency
www.fda.gov	US Food and Drug Administration
www.iecee.org	International Electrotechnical Commission (IEC) System for Conformity Testing and Certification of Electrical Equipment

References

Cohen, A. (2000). Biomedical signals: origin and dynamic characteristics; frequency-domain analysis. In Bronzino, J.B. (ed.), *The Biomedical Engineering Handbook*, 2nd ed., vol. 1. Boca Raton, FL: CRC Press & Springer, pp. 52-1–52-24.

Durward, B.R., Baer, G.D., and Rowe, P.J. (1999). *Functional Human Movement: Measurement and Analysis*. Oxford: Butterworth Heinemann.

Kondraske, G.V. (1989). Measurement science concepts and computerized methodology in the assessment of human performance. In Munsat, T.L. (ed.), *Quantification of Neurologic Deficit*. Stoneham: Butterworths, pp. 33–48.

Pandyan, A.D., *et al.* (1999). A review of the properties and limitations of the Ashworth and modified Ashworth Scales. *Clinical Rehabilitation* 3(5): 373–383.

Riddoch, G., Bristow, W.R., Cairns, H.W.B., Carmichael, E.A., Critchley, M., Greenfield, J.G., *et al.* (1943). *Aids to the Investigation of Peripheral Nerve Injuries*. Medical Research Council War Memorandum, No. 7, rev. 2nd ed. London: HMSO.

Tourtellotte, W.W., and Syndulko, K. (1989). Quantifying the neurologic examination: principles, constraints and opportunities. In Munsat, T.L. (ed.), *Quantification of Neurologic Deficit*. Stoneham: Butterworths, pp. 7–16.

Wade, D.T. (1992). *Measurement in Neurological Rehabilitation*. Oxford: Oxford University Press.

18

Randomised Trials

Douglas G. Altman

Introduction

A randomised trial is a planned experiment that is designed to compare two or more forms of treatment or behaviour, generically called *interventions*. Randomised trials were originally developed in agriculture as a means of getting valid comparisons between different ways of treating soil or plants, where it was known that there was inherent variability in the land being used (e.g. drainage and exposure to wind). The same principles extend to numerous other disciplines where a comparison of procedures or treatments is needed, in particular when humans are being studied.

The key idea of a controlled trial is that we compare groups of individuals who differ only with respect to their treatment. The study must be prospective because biases are easily incurred when comparing groups treated at different times and possibly under different conditions. It must be comparative (controlled) because we cannot assume what will happen to the patients in the absence of any therapy.

This chapter concentrates on randomised trials of healthcare interventions, better known as *clinical trials*. Studies of interventions in other fields, such as educational or policy interventions, share the general features. Clinical trials merit special attention because of their medical importance; some particular problems in design, analysis and interpretation; and certain ethical problems. The methodology was introduced into medical research over 60 years ago, the most famous early example being a trial of streptomycin in the treatment of pulmonary tuberculosis (MRC, 1948). This chapter will illustrate how the design and analysis of trials have to be tailored to individual circumstances.

Although clinical trials can be set up to compare more than two treatments, I will concentrate on the two-group case. Often one treatment is an experimental treatment, perhaps a new drug, and the other is a comparison or 'control' treatment which may be a standard treatment, an ineffective 'placebo' or even no treatment at all, depending on circumstances. Some trials compare different active treatments, or even different doses of the same treatment.

Research Methods for Postgraduates, Third Edition. Edited by Tony Greenfield with Sue Greener.
© 2016 John Wiley & Sons, Ltd. Published 2016 by John Wiley & Sons, Ltd.

This chapter provides a brief overview of a complex topic. There are many books devoted to clinical trials, of which those by Pocock (1983), Matthews (2000) and Wang and Bakhai (2006) are particularly recommended; see also Chapter 15 in Altman (1991).

Trial Design

Random Allocation

A vital issue in design is to ensure that, as far as is possible, the groups of patients receiving the different treatments are similar with regard to features that may affect how well they do. The usual way to avoid bias is to use *random allocation* to determine which treatment each patient gets. Indeed, randomisation is very widely considered to be essential. With randomisation, which is a chance process, variation among subjects that might affect their response to treatment will on average be the same in the different treatment groups. In other words, randomisation ensures that there is no bias in the way the treatments are allocated. There is no guarantee, however, that randomisation will lead to the groups being similar in a particular trial. Any differences that arise by chance can be at least inconvenient and may lead to doubts being cast on the interpretation of the trial results.

While the results can be adjusted to take account of differences between the groups at the start of the trial, major imbalance can be prevented at the design stage by using *stratified randomisation*. In essence, this involves randomisation of separate subgroups of individuals, such as patients with mild or severe disease. If you know in advance that a few key variables are strongly related to outcome, they can be incorporated into a stratified randomisation scheme. There may be other important variables that we cannot measure or have not identified, and we rely on the randomisation to balance them out. It is essential that stratified randomisation uses *blocking*; otherwise, there is no benefit over simple randomisation. With blocking, allocations to each treatment are balanced within each consecutive series, or block, of patients within each stratum. For example, blocks of size of six will each include three patients on each of two treatments ordered randomly within the block.

While randomisation is necessary, it is not a sufficient safeguard against bias (conscious or subconscious) when patients are recruited. The treatment allocation system should be set up so that the person entering patients does not know in advance which treatment the next person will get. A common way of concealing allocations is to use a series of sealed opaque envelopes, each containing a treatment specification. For stratified randomisation, two or more sets of envelopes are needed. For drug trials, the allocation is often done by the pharmacy, who produce numbered bottles that do not indicate the treatment contained. For non-pharmaceutical trials, bias is best prevented by using a telephone or web-based system.

Blinding

The key to a successful clinical trial is to avoid any biases in the comparison of the groups. Randomisation deals with possible bias at the treatment allocation, but bias can also creep in while the study is being run. Both patients and doctors may be affected in the way they respond and observe by knowledge of which treatment was given, especially if the outcome is subjective, such as reduction in pain. For this reason, ideally neither the patient nor the person who evaluates the patient should know which treatment was given. Such a trial is called *double blind*. If only the patient is unaware, the trial is called *single blind*.

When there is no standard effective treatment, it is reasonable *not* to give the control group any active treatment. However, it is often better to give the control group patients an inert or placebo treatment than nothing. Firstly, the act of taking a treatment may itself have some benefit to the patient, so that part of any benefit observed in the treatment group could be due to the knowledge or belief that they had taken a treatment. Secondly, for a study to be double blind, the two treatments should be indistinguishable. Placebo tablets should therefore be identical in appearance and taste to the active treatment, but pharmacologically inactive. More generally, when the control treatment is an alternative active treatment rather than a placebo, the different treatments should still be indistinguishable if possible. In some circumstances, such as in most surgical trials, it is impossible to disguise the treatment. In such cases, it helps to have the patients' outcome assessed by someone who is blind to which group they are in.

Many clinical trials show some apparent benefit of treatment in the placebo group, and there are often side effects too. Without a placebo, we cannot know how specific any benefit (or harm) is to the active treatment.

Sample Size

For parallel group studies, the calculation of sample size is based on comparing either means (*t*-test) or proportions (chi-square test) (see Chapters 24 and 25 for a more detailed discussion of sampling). Sample size calculations are based on the idea of having a high probability (called *power* and usually set at 80–90%) of getting a statistically significant difference if the true difference between the treatments is of a given size. Unfortunately, specifying this *effect size* is not easy.

For common diseases, such as heart disease, small benefits are worthwhile. Detecting small effects requires large trials. For example, a trial designed to have a high probability of detecting a reduction of mortality from 30% to 25% would need at least 1250 patients per group.

Ethical Issues

One of the main ethical issues is the provision of adequate information about the trial and the treatments to potential participants. In general, the patient should be invited to be in the trial and should be told what the alternative treatments are (although they will not know which they will get). The patient can decline, in which case they will receive the standard treatment. A patient agreeing to participate will have to sign a form stating that he or she understood the trial. In some cases, it is not possible to get informed consent, for example if the patients are very young, very old or unconscious.

No doctor should participate in a clinical trial if he or she believes that one of the treatments being investigated is superior, and the doctor should not enter any patient for whom he or she thinks that a particular treatment is indicated. In other words, the ideal medical state to be in is one of ignorance: the trial is done because *we do not know which treatment is better*. This is sometimes called the *uncertainty principle*.

The methodological quality of a trial is also an ethical issue. A trial that uses inadequate methods, such as failing to randomise, and thus fails to prevent bias may be seen as unethical. Likewise, having an adequate sample size is often considered to be an ethical matter.

Outcome Measures

In most clinical trials, information about the effect of treatment is gathered about many patient outcomes, sometimes on more than one occasion. There is the temptation to analyse each of these outcomes and see which differences between treatment groups are significant. This approach leads to misleading results, because multiple testing will invalidate the results of significance tests. In particular, just presenting the most significant results as if these were the only analyses performed is fraudulent.

It is preferable to decide in advance of the analysis which outcome measure is of major interest and focus attention on this variable when the data are analysed. Other data can and should be analysed too, but these variables should be considered to be of secondary importance. Any interesting findings among the secondary variables should be interpreted rather cautiously, more as ideas for further research than as definitive results. Side effects of treatment should usually be treated in this way.

Alternative Designs

The simplest and most frequently used design for a clinical trial is the *parallel group design*, in which two (or more) groups of patients receive different treatments concurrently. The most common alternative is the *crossover design*, in which all the patients are given both treatments in turn. Here, randomisation is used to determine the order in which the treatments are given. The crossover design has some attractive features, in particular that the treatment comparison is *within-subject* rather than *between-subject*. Because within-subject variation is usually much smaller than between-subject variation, the sample size needed is smaller than for a parallel group trial. There are some important disadvantages, however. In particular, patients may drop out after the first treatment, and so not receive the second treatment. Withdrawal may be related to side effects. Crossover studies should not be used for conditions that can be cured or may be fatal. Lastly, there may be a carryover of treatment effect from one period to the next, so that the results obtained during the second period are affected. In other words, the observed difference between the treatments will depend upon the order in which they were received. Crossover trials are discussed well by Senn (2002).

A variation of the parallel group trial is the *group sequential trial*, in which the data are analysed as the data accumulate, typically at 3, 4 or 5 time points. This design is suitable for trials that recruit and follow up patients over several years. It allows the trial to be stopped early if a clear treatment difference is seen, if side effects are unacceptable or if it is obvious that no difference will be found. Adjustment is made to significance tests to allow for multiple analyses of the data.

In some situations, groups of individuals (*clusters*) are randomised, usually for logistic reasons. Such studies are common in health education, for example. Clusters may be families, general practices, schools, hospitals, communities or geographical areas. Here, the cluster is the unit of randomisation rather than the individual. Special considerations are necessary in the design and analysis of such trials, as described by Donner and Klar (2000).

One further type of design is the *factorial design*, in which two active treatments, say A and B, are compared with each other and with a control. Patients are randomised into four groups, who receive either A only, B only, both A and B, or neither. This design is effectively two trials in one, for example comparing A versus placebo and, separately, B versus placebo. The design assumes that there is no interaction (or synergy) of the effects of A and B, but

Table 18.1   The main features of a trial protocol

Background and rationale	Outcomes
Objectives	Sample size
Trial design	Recruitment
Participants	Assignment of interventions
Interventions	Blinding (masking)
Outcomes	Data collection methods
Study setting	Data management
Eligibility criteria	Statistical methods
Interventions	Data monitoring

Chan *et al.* (2013).

occasionally the interaction is the main point of interest. Although rarely used in clinical trials, such designs (and more complex ones) are standard practice in agricultural field trials and in industrial research (see Chapter 19).

Protocol

As with all research, an important aspect of planning a clinical trial is to produce a protocol, which is a formal document outlining the proposed procedures for carrying out the trial. Recent guidelines detail what to include in a clinical trial protocol (Chan *et al.*, 2013). The main features of a trial protocol are shown in Table 18.1.

A detailed protocol must accompany an application for a grant for a trial, and the local research *ethics committee* will also require most of the above information. As well as aiding in the performance of a trial, a protocol makes the reporting of the results much easier as the introduction and methods section of the paper should be substantially the same as the methods sections in the protocol.

Many difficulties can be avoided by running a pilot study, which is valuable for assessing the quality of the data collection forms, and for checking the logistics of the trial, such as the expected time to examine each patient (which affects the number that can be seen in a session).

Analysis

The general methods for the statistical analysis of data, described in Chapters 28 and 29, apply to clinical trials, and you should read those chapters. Several specific problems arise in the analysis of clinical trials, some of which are considered here. A fuller discussion of bias in analysis is given by May *et al.* (1981) in addition to the textbooks referenced.

Comparison of Entry Characteristics

Randomisation does not guarantee that the characteristics of the different groups are similar. The first analysis should be a summary of the baseline characteristics of the patients in the two groups. It is important to show that the groups were similar with respect to variables that may affect the patient's response. For example, we would usually wish to be happy that the age distribution was similar in the different groups, as many outcomes are age related.

A common way to compare groups is with hypothesis tests, but a moment's thought should suffice to see that this is unhelpful (Altman and Doré, 1990). If the randomisation is performed fairly, we know that any differences between the two treatment groups *must* be due to chance. A hypothesis test thus makes no sense except as a test of whether the trial was indeed randomised. In any case, the question at issue is whether the groups differ in a way that might affect their response to treatment. That question is clearly one of clinical importance rather than statistical significance. If we suspect that the observed differences (imbalance) between the groups may have affected the outcome, we can take account of the imbalance in the analysis, as described further in this chapter.

Incomplete Data

Data may be incomplete for several reasons. For example, occasional laboratory measurements will be missing because of problems with the samples taken. It is important to use all the data available and to specify if any observations are missing. Also, some information may simply not have been recorded. It may seem reasonable to assume that a particular symptom was not present if it was not recorded, but such inferences are in general unsafe and should be made only after careful consideration of the circumstances.

The most important problem with missing information in trials relates to patients who do not complete the study and so their outcome is unknown. Some patients may be withdrawn by their doctor, perhaps because of side effects. Others may move to another area or just fail to return for assessment. Efforts should be made to obtain at least some information regarding the status of these patients at the end of the trial, but some data are still likely to be missing. If there are many more withdrawals in one treatment group, the results of the trial will be compromised, as it is likely that the withdrawals are treatment related.

A further difficulty is when some patients have not followed the protocol, either deliberately or accidentally. Included here are patients who actually receive the wrong treatment (in other words, not the one allocated) and patients who do not take their treatment, known as *non-compliers*. Also, sometimes it may be discovered after the trial has begun that a patient was not eligible for the trial after all.

The only safe way to deal with all of these situations is to keep all randomised patients in the trial. The analysis is thus based on the groups as randomised and is known as an *intention-to-treat analysis*. Any other policy towards protocol violations will depart from the randomisation which is the basis for the trial. Also, excluding some patients involves subjective decisions and thus creates an opportunity for bias. In practice, patients who are lost to follow-up and thus have no recorded outcomes will need to be excluded, but all other patients should be included. Missing outcomes for some patients cause major difficulties in both analysis and interpretation (Lachin, 2000). It is important to try to design and conduct a trial so as to minimise missing data; Little *et al.* (2012) discuss various strategies. Modern statistical methods, especially multiple imputation, are becoming more popular as a means of dealing with missing outcomes (Sterne *et al.*, 2009).

Adjusting for Other Variables

Most randomised trials are based on the simple idea of comparing two groups with respect to a main outcome variable of interest, for which the statistical analysis is simple. We may, however, wish to take one or more other variables into consideration in the analysis. One reason might be that the two groups were not similar with respect to baseline characteristics. We can thus perform the analysis with and without adjustment for those variables. If the results are similar, we can infer that the imbalance was not important and can report the simple comparison, but if the results are different we should use the adjusted analysis. Imbalance will only affect the results if the unbalanced variable is related to the outcome measure. It will not matter if one group is much shorter than the other if height is unrelated to response to treatment. The use of 'restricted' randomisation designed to give similar groups (e.g. stratified randomisation) is desirable as it reduces the risk of serious imbalance, but variables used in this way to balance allocations should always be adjusted for (although the impact of so doing tends to be minimal).

Subgroup Analyses

Even with a single outcome variable, there is often interest in identifying which patients do well on a treatment and which do badly. We can answer some questions like this by analysing the data separately for subsets of the data. We may, for example, redo the analysis including only younger patients and then only older patients. Although that approach is common, the correct way to address subgroups is to compare the outcome in one group with that in the other. Using a *test of interaction*, we would for example directly compare the estimates of treatment effect in younger and older patients rather than conduct two separate analyses (Altman and Bland, 2003).

Subgroup analyses pose serious problems of interpretation similar to those resulting from multiple outcome measures. It is reasonable to perform a few subgroup analyses as specified in the protocol. But if the data are analysed in numerous ways in the hope of discovering some 'significant' comparison, or because some aspect of the data suggests a difference, then they must be identified as exploratory.

Interpretation of Results

Single Trials

In most cases, the statistical analysis of a clinical trial will be simple, at least with respect to the main outcome measure, perhaps involving just a t-test or a chi-square test. Interpretation seems straightforward, therefore, but for one difficulty.

Inference from a sample to a population relies on the assumption that the trial participants represent all such patients. However, in most trials, participants are selected to conform to certain inclusion criteria, so extrapolation of results to other types of patient may not be warranted. For example, most trials of anti-hypertensive agents, such as beta-blocking drugs, are on middle-aged men. Is it reasonable to assume that the results apply to women too, or to younger or older men? In the absence of any information to the contrary, it is common to infer wider applicability of results, but the possibility that different groups would respond differently should be borne in mind. This issue was discussed by Rothwell (2005).

All Published Trials

In many fields, there have been several similar clinical trials, and it is natural to want to assess all the evidence together. It quickly becomes apparent when looking at the results of several clinical trials of the same treatment comparison that the results vary, sometimes markedly. We expect some variation in estimated treatment effects because of random variation, and should not be worried by it. The confidence interval for the treatment benefit observed in a single trial gives an idea of the range of treatment benefit likely to be observed in a series of identical trials of the same size.

A relatively recent development has been to move towards the identification, methodological appraisal and (optionally) a pooled statistical analysis of all published trials to get an overall assessment of treatment effectiveness. Such a study is known as a *systematic review*, and the analysis combining results from several studies as a *meta-analysis* (this latter term is sometimes used for the whole exercise) (Egger *et al.*, 2001). Systematic reviews can show a highly significant overall treatment benefit when most of the individual trials did not get a significant result. Again, this is to be expected, as many clinical trials are too small to detect anything other than an unrealistically huge treatment benefit. However, this new type of study has generated many new challenges for statisticians. One important aspect is the need to assess methodological 'quality', which is better thought of as considering whether the various trials might have some risk of bias, for example from failure to have blinded assessment of outcomes (Higgins *et al.*, 2011).

Reporting the Results of Clinical Trials

As hinted in this chapter, not all trials are done well. In addition, many studies have found that inadequate reporting of trials is widespread. The CONSORT Statement gives guidance for reporting clinical trial results, in the form of a checklist and diagram showing the flow of participants through the trials (Moher *et al.*, 2010; Schulz *et al.*, 2010). CONSORT is a widely accepted standard for reporting trials, and many medical journals expect authors to follow these recommendations. Full reporting of trials assists in the assessment of the quality of the methodology used in a trial, and thus helps the reader judge the reliability of the results.

While there are many aspects of the quality of a trial, the most important are those relating to steps taken to avoid bias, such as randomisation and blinding. Empirical evidence is accumulating that failure to use such methods does indeed lead to bias (Moher *et al.*, 2010).

References

Altman, D.G. (1991). *Practical Statistics for Medical Research*. London: Chapman and Hall.

Altman, D.G., and Bland, J.M. (2003). Interaction revisited: the difference between two estimates. *BMJ* 326: 219.

Altman, D.G., and Doré, C.J. (1990). Randomisation and baseline comparisons in clinical trials. *Lancet* 335: 149–153.

Chan, A.W., Tetzlaff, J., Gøtzsche, P.C., Altman, D.G., Mann, H., Berlin, J., *et al.* (2013). SPIRIT 2013 explanation and elaboration: guidance for protocols of clinical trials. *BMJ* 346: e7586.

Donner, A., and Klar, N.S. (2000). *Design and Analysis of Cluster Randomisation Trials*. London: Arnold.

Egger, M., Davey Smith, G., and Altman, D.G. (eds.). (2001). *Systematic Reviews in Health Care. Meta-analysis in Context*. London: BMJ Books.

Higgins, J.P.T., Altman, D.G., Gøtzsche, P.C., Jüni, P., Moher, D., Oxman, A.D., *et al.*, Cochrane Bias Methods Group, Cochrane Statistical Methods Group. (2011). The Cochrane Collaboration's tool for assessing risk of bias in randomised trials. *BMJ* 343: d5928.

Lachin, J.M. (2000). Statistical considerations in the intent-to-treat principle. *Controlled Clinical Trials* 21: 167–189.

Little, R.J., D'Agostino, R., Cohen, M.L., Dickersin, K., Emerson, S.S., Farrar, J.T., *et al.* (2012). The prevention and treatment of missing data in clinical trials. *New England Journal of Medicine* 367: 1355–1360.

Matthews, J.N.S. (2000). *An Introduction to Randomized Trials*. London: Arnold.

May, G.S., DeMets, D.L., Friedman, L., Furberg, C., and Passamani, E. (1981). The randomised clinical trial: bias in analysis. *Circulation* 64: 669–673.

Moher, D., Hopewell, S., Schulz, K.F., Montori, V., Gøtzsche, P.C., Devereaux, P.J., *et al.* (2010). CONSORT 2010 Explanation and Elaboration: updated guidelines for reporting parallel group randomised trials. *BMJ* 340: c869.

MRC. (1948). Streptomycin treatment of pulmonary tuberculosis. *BMJ* 2: 769–782.

Pocock, S. (1983). *Clinical Trials: A Practical Approach*. Chichester: Wiley.

Rothwell, P.M. (2005). External validity of randomised controlled trials: 'to whom do the results of this trial apply?' *Lancet* 365: 82–93.

Schulz, K.F., Altman, D.G., Moher, D., and the CONSORT Group. (2010). CONSORT 2010 Statement: updated guidelines for reporting parallel group randomised trials. *BMJ* 340: c332.

Senn, S. (2002). *Cross-over Trials in Clinical Research*, 2nd ed. Chichester: Wiley.

Sterne, J.A., White, I.R., Carlin, J.B., Spratt, M., Royston, P., Kenward, M.G., *et al.* (2009). Multiple imputation for missing data in epidemiological and clinical research: potential and pitfalls. *BMJ* 338: b2393.

Wang, D., and Bakhai, A. (eds.). (2006). *Clinical Trials: A Practical Guide to Design, Analysis, and Reporting*. London: Remedica.

19

Laboratory and Industrial Experiments

Tony Greenfield

Introduction

Experimental design and analysis comprise an essential part of the scientific method.

Every experiment should be well designed, planned and managed to ensure that the results can be analysed, interpreted and presented. If you do not do this, you will not understand properly what you are doing, and you will face the hazards of failing to reach your research goals, of wasting great effort, time, money and other resources in fruitless pursuits.

Other chapters in this book (Chapters 28 to 30) are about the analysis of experimental data. This chapter is about the design of experiments.

An experimental design is:

the specification of the conditions at which experimental data will be observed.

Experimental design is a major part of applied statistics, and there is an immense literature about it. In this chapter, I present only those aspects of experimental design that have most to contribute to the physical sciences, specifically to laboratory and industrial experiments. Such experiments are aimed primarily at improving products and processes.

Descriptions are necessarily brief and selective. Please read this chapter as an introduction and a reference. There are many good books on the subject, and a few of the best are listed at the end of this chapter.

There are many types of experimental design. In this chapter, we describe a range of designs selected for their usefulness to the physical sciences, including manufacturing and industries, and leading from the simplest to the more complex. These are summarized here.

Research Methods for Postgraduates, Third Edition. Edited by Tony Greenfield with Sue Greener.
© 2016 John Wiley & Sons, Ltd. Published 2016 by John Wiley & Sons, Ltd.

Descriptive

A sample of several test pieces, all from a standard material, is tested to determine the elementary statistics of a characteristic of that material. You may, for example, wish to report the mean and the standard deviation of the tensile strength of a standard material.

Comparative

Comparison against a Standard

You may wish to compare the characteristic of a new material against a specified industry standard. You would test a sample of several pieces and ask if there was sufficient evidence to conclude that the measured characteristic of this material was different from the standard specification.

Comparison of Two Materials with Independent Samples

You may have two materials, perhaps of different compositions or made by slightly different processes, or, even if they are claimed to be of the same composition and made by exactly the same process, they are made at different places. You wish to determine if they have the same or different properties, so you test a sample of several pieces from each material. These samples are independent of each other.

Comparison of Two Materials by Paired Samples

You may have two materials, and you wish to determine if they have the same or different properties. However, you wish to ensure, in the presence of uncontrollable outside influences, to make a fair comparison. For example, you may wish to expose samples of structural steel to the weather and measure their corrosion. One approach would be to expose test pieces in pairs, each pair comprising one item piece of each material. The data to be analysed would be the difference in corrosion measured between each pair by weighing them separately.

Response

Factorial Experiments

When new materials or manufacturing processes are being developed, there are usually several variables, or factors, that can influence a material property. Experiments to investigate the effects of several variables should be designed in which all of those variables are set at several levels.

> **Warning**: There is a widespread belief, still taught in schools and to undergraduates, that the best approach is to experiment with one variable at a time and to fix all the others. That approach is inefficient and uneconomic, and will not provide information about interactions between variables.

Two-level factorial experiments, in which each factor is set at two levels, high and low, are widely used during development studies.

Response Surface Exploration with Composite Designs

In the final stage of a development study, when you are seeking the conditions (e.g. the values of composition and process variables) that will yield the best value of a material property (e.g. the highest value of tensile strength), additional points must be added to factorial experiments so that curvature of the response can be estimated. These designs are known as *augmented or composite designs*.

Inter-laboratory Trials and Hierarchical Designs

Another class of experiment used in industry is the inter-laboratory trial for the purpose of estimating repeatability of test results within each of a set of laboratories and reproducibility of test results between laboratories. These are described more fully in several textbooks and in a British Standard (BS 5497).

Similar issues arise when comparing variability within and between batches of product.

Principles of Experimental Design

Statistical analysis of experimental results is necessary because of variation: *all test results vary*. This variation must therefore be considered when experiments are designed.

Descriptive Experiments

In a descriptive experiment, a characteristic of a standard material will be reported from the analysis of measurements on several test results. For example, the mean tensile strength of a sample of several test pieces will be calculated. This is unlikely to be the true value of the underlying population mean. If you calculated the mean tensile strength of another sample of several test pieces. it would be different. The calculated sample mean is therefore only an estimate, a *point* estimate, of the underlying population mean. In reporting it, you should report an interval in which you can confidently expect the population mean to lie: a confidence interval for the population mean.

This confidence interval depends on three things:

1. the variation of test results, expressed as the variance or standard deviation of the measured material property;
2. the number of pieces tested in the sample; and
3. the degree of confidence of the interval, loosely interpreted as the probability that the population mean is truly in that interval, usually as a percentage (for example: a 95% confidence interval).

The variation will have to be determined from the experiment. The number of test pieces must be specified before the experiment is performed. The degree of confidence is the choice of the experimenter and should be specified before the experiment is performed. Ideally, the experimenter should specify how large a confidence interval, and with what confidence, he would like. For example, he may specify: sample mean value ± 1.0 N/mm^2 with a confidence of 95%. The experiment would then proceed in four stages:

Stage 1: A preliminary experiment to estimate the underlying population variance and/or a review of similar experiments reported in the literature.

Stage 2: A calculation of the sample size N needed to estimate the specified confidence interval using the variance estimated in stage 1.

Stage 3: Test measurements on a sample of N pieces.

Stage 4: Calculation of the estimated mean, standard deviation and confidence interval.

These four stages are described fully in the textbooks given in the References.

Comparative Experiments

Statistical analysis of test results should never be regarded simply as a set of calculations leading to a clear-cut conclusion such as 'the effect is significant' or 'there is no significant effect'. The conclusion depends on the circumstances of the experiment and on the intentions of the experimenter which should be declared before the tests are performed. For example, the circumstances of an experiment may ordain how likely an effect is to be declared as statistically significant *if it exists*; the intentions of the experimenter will include a statement of what she considers to be a *technically* significant effect.

Four major steps must be taken before starting an experiment:

Step 1: State the alternative inferences that can be made from the experiment.

Step 2: Specify the acceptable risks for making the wrong inference.

Step 3: Specify the difference which must be demonstrated statistically so as to be of technical significance.

Step 4: Compute the necessary sample size.

These four steps are described more fully here, and they are described in detail in the textbooks in the References.

Step 1: State the alternative inferences that can be made from the experiment. These should be stated as alternative prior hypotheses.

When two materials are to be compared according to some property, the most usual comparison is between the mean values of that property. Even though the sample mean values $\bar{x}_1$ and $\bar{x}_2$ will differ, is there sufficient evidence to infer that the mean values of the underlying populations (μ_1 and μ_2) differ? If there is not sufficient evidence, you cannot refute the claim that the means of the underlying populations are the same. An assumption that they are the same is known as the *null hypothesis* (H_0). An assumption that they are different is known as the *alternative hypothesis* (H_a).

These may be stated symbolically as:

$$H_0: \mu_1 = \mu_2$$
$$H_a: \mu_1 <> \mu_2$$

In this case, the experimenter is concerned about *any* difference between the population means. This will lead to a two-sided test.

If the experimenter is interested in a new material only if it has a greater mean strength than the standard material, a one-sided test will be used and the alternative hypotheses will be:

$$H_0: \mu_1 = \mu_2$$
$$H_a: \mu_1 > \mu_2$$

The distinction must be made *before* the experiment is performed. The calculation of the sample size depends on the distinction.

> **Step 2**: Specify the acceptable risks for making the wrong inference. The wrong inferences are called the type 1 error and the type 2 error with probabilities α and β, respectively.

The possible inferences from a two-sided test may be understood from Table 19.1.

A *type 1 error* occurs when the experimenter accepts the alternative hypothesis (H_a) although the null hypothesis (H_0) is true. The probability of this occurring is α. Usually, α is specified as 0.05 (a 5% chance).

A *type 2 error* occurs when the experimenter accepts the null hypothesis (H_0) although the alternative hypothesis is true. The probability of this occurring is β which depends on the difference between the population means ($\mu_1 - \mu_2$). Usually, β is specified as 0.05 or 0.10 for a specified difference of technical importance. The probability of detecting this difference is ($1 - \beta$). Thus, if β is chosen to be 0.05 and the experiment is designed accordingly, there is a strong chance (a probability of 0.95) that the specified difference will be detected if it truly exists. More generally, a plot of the probability ($1 - \beta$) of correctly detecting a true difference (Δ) against Δ is known as the *power curve of the test*.

Table 19.1   Possible inferences from a two-sided test

		Inference	
		$\mu_1 = \mu_2$	$\mu_1 <> \mu_2$
Truth	$\mu_1 = \mu_2$	Correct probability = ($1 - \alpha$)	Type 1 error probability = α
	$\mu_1 <> \mu_2$ or $\mu_1 - \mu_2 = \delta$ where δ is not zero	Type 2 error probability = β which depends on δ	Correct probability = ($1 - \beta$)

Table 19.2 Burst strengths (kPa)
of filter membranes

Batch 1	Batch 2
267	284
262	279
261	274
261	271
259	268
258	265
258	263
258	262
257	260
256	259
251	246
250	241

Unfortunately, it is common for experiments to be performed without consideration of β or the power, and consequently important true effects may remain undetected. For example, consider the burst strengths (kPa) of two batches of filter membranes as shown in Table 19.2.

Suppose that a purpose of the experiment was to detect any difference exceeding 10 kPa. A power calculation ($\alpha = 0.05$) shows that with only 12 results for each batch, there is a probability of 0.75 of detecting a difference if it exists. Sample sizes of 23 would be needed to give a probability of 0.95 of detecting that difference.

Step 3: Specify the smallest difference (δ) of technical significance, which should have a specified probability $(1 - \beta)$ of being detected.

The purpose of many experiments is to discover an improvement in a material property, that which is being tested. In other experiments, the purpose may be to show that, under different circumstances, there is no substantial difference in the material property.

In either case, the experimenter should be able to state the smallest difference which he would regard as likely to have a practical or technical significance. For example, how much stronger, in terms of tensile strength, should one material be over another to make its selection preferable for a particular application? One percent, or 2%, or 0.5%? It may depend on the application.

The specification of this smallest difference (δ) is essential to the design of a comparative experiment.

Step 4: Compute the necessary sample size.

There are several formulae for calculating sample size. The correct choice of formula depends on the type of comparative experiment (comparison against a standard, comparison of two materials with independent samples and comparison of two materials by paired samples) and on whether the proposed test of comparison is one-tailed or two-tailed. The information needed in any of the calculations is the choice of α, β, δ and an estimate of the variance (σ^2) of the underlying population.

The choices of α, β and δ depend entirely on the opinions and purposes of the experimenter, as already explained. An estimate of the population variance (σ^2) may often be obtained from earlier experiments or from literature. Otherwise, a preliminary experiment of at least five test pieces must be done to obtain that estimate.

Response Experiments

Introduction to Response Experiments

Much research and development in the materials sciences are intended to establish relationships between materials properties and other factors which the technologist can control in the production of those materials. The properties are called *response variables* (also called *dependent variables*). The other factors which influence the response variables are called *control variables* (also called *independent variables*). The control variables are usually composition variables and process variables. All of these variables must be measurable.

- The variable we are most interested in studying is often called the *response* variable because it changes in response to changes in other variables. Also, it is usually a characteristic of a product by which we judge the usefulness of the product.
- For example, we may be performing an experiment about the production of an artificial silken thread. The process involves the stirring of a mixture, and the tensile strength depends, at least partly, on the stirring speed (Figure 19.1). In this case, tensile strength is an important property of artificial silk. It seems to respond to stirring speed, so it is a response variable. We show it along the vertical, or *y*, axis.
- The variable which we think is influencing, or controlling, the response variable is, in this case, stirring speed. We call it a *control* variable and show it along the horizontal, or *x*, axis.

All of the variables that may influence the response are collectively called *explanatory* or *predictor variables*. Sometimes there are other variables which may influence the response variables but which cannot be controlled, although they can be identified and measured. Common examples are the temperature and humidity of a factory workshop atmosphere. These variables are called *concomitant variables* (also called *covariates*).

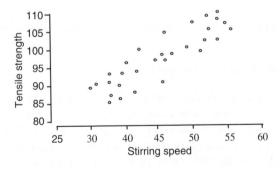

Figure 19.1   Graph showing tensile strength (response variable) against stirring speed (control variable).

The experimental design is the specification of the values of the control variables at which the response variables will be measured. The experiment should be designed according to the expected relationship between the response variable and the control variables. The expected relationship is a hypothesis. The hypothesis should be formulated as an algebraic model that can be represented in terms of the measurable variables.

The experimental design would be very simple, and few observations would be needed to fit the model if the expected relationship could predict the results exactly. However, there are several reasons why this cannot be achieved:

- the exact relationship cannot be known, because the model is only an approximation to reality;
- all measurements may be subject to time-dependent errors which we cannot identify but which show their presence by trends in the observed values of response variables; and
- all measurements are subject to random errors which represent other unidentified variables which, taken together, show no pattern or trend.

The experiment must therefore be designed so as to reduce the influence of these unknowns.

The statistical objectives of designed response experiments are to specify:

1. the number of observations;
2. the values of the control variables at every observation; and
3. the order of the observations.

with a view to:

1. ensuring that all effects in the model can be estimated from the observed data;
2. testing the reality of those effects by comparison with random variation;
3. ensuring that all effects can be estimated with greatest possible precision (reducing the influence of random variation);
4. ensuring that all effects can be estimated with the least possible bias, or greatest accuracy (reducing the effects of time-dependent errors);
5. suggesting improvements to the model; and
6. keeping within a budget of effort and cost.

Factorial Experiments

Two-Level Factorial Experiments
The two-level factorial design is fundamental to experimental design for the physical sciences.

When there is good reason to believe that, over the range of values of a control variable, the dependent variable is related to the control variable by a linear function of the form:

$$y = a + bx \tag{19.1}$$

where y is the response variable; x is the control variable; and a and b are coefficients to be estimated. Then, a and b can be estimated with the greatest precision if all observations are divided equally between the two ends of the range of x. A common fault among experimenters is to divide the range into $N - 1$ equal parts (where N is the number of planned observations) and to make one observation at each end and at each of the division points. A few intermediate points may be desirable as a check on the believed linearity if curvature is suspected.

In equation (19.1), the effect on y of a change in x of one unit is represented by the coefficient b, which is the slope of the line. Another way of representing the relationship between x and y is achieved by using a different notation. In this notation, the independent variables (the x's) are called *factors* and are represented by capital letters: A, B, C and so on.

The range of a factor is specified by the two ends of the range: the high and the low values of the factor. These are represented by lowercase letters with suffixes. For example, in the single-factor experiment, the high and low values of factor A would be a_1 and a_0, respectively. This lowercase notation is also used to represent the observed values of the dependent variable at the corresponding observation points.

The effect of factor A on the dependent variable (y) over the complete range of factor A is equal to:

$$\text{(mean value of } y \text{ at point } a_1) - (\text{mean value of } y \text{ at point } a_0)$$

which can be expressed as:

$$\text{effect of } A = \bar{y}(a_1) - \bar{y}(a_0)$$

or, more briefly,

$$A = a_1 - a_0 \tag{19.2}$$

Note a further abbreviation in that the capital letter A is used to denote the effect of factor A. Similarly, the mean value of y is simply

$$M = (a_1 + a_0)/2 \tag{19.3}$$

Now consider two factors, A and B, which can be represented as two variables in a plane with the dependent variable y along a third dimension perpendicular to the plane. The design is shown in Figure 19.2.

The high and low values of B are b_1 and b_0, respectively. If observations of y are made only at points defined by the extreme ranges of the two factors, there are four points which can be denoted by the combinations of letters as: a_0b_0, a_1b_0, a_0b_1, and a_1b_1. The notation can be abbreviated further to represent these four points as: (1), a, b, ab. The symbol (1) denotes the observation point at which all the factors are at their low levels. The point a is where factor A is at its high level but factor B is at its low level. The point ab is where both factors are at their high levels. The rule is that the high and low levels of factors are represented by the presence or absence, respectively, of lowercase letters.

Analysis is almost as easy as in the single-factor case. Using the combinations of lowercase letters to represent the values of y observed at the corresponding points, the average effect of factor A is:

$$A = \frac{a + ab}{2} - \frac{(1) + b}{2} \tag{19.4}$$

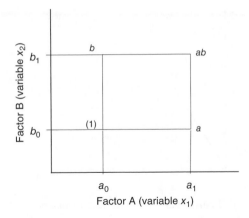

Figure 19.2   Graph showing a two-factor design.

That is, the effect of A is the difference between (the mean value of y observed at all the points where A was at its high level) and (the mean value of y observed at all the points where A was at its low level).

Similarly:

$$B = \frac{b + ab}{2} - \frac{(1) + a}{2} \tag{19.5}$$

The interaction of factors A and B may be defined as the difference between (the effect of A at the high level of B) and (the effect of A at the low level of B). It is denoted by AB. Thus:

$$AB = (ab - b) - (a - (1)) \tag{19.6}$$

This is exactly the same as: the difference between (the effect of B at the high level of A) and (the effect of B at the low level of A).

The estimation of these effects is equivalent to fitting the algebraic model:

$$y = \beta_0 + \beta_1 x_1 + \beta_2 x_2 + \beta_{12} x_1 x_2 \tag{19.7}$$

where y is the response variable; x_1 and x_2 are two control variables; and β_0, β_1, β_2 and β_{12} are algebraic coefficients.

Least-squares regression analysis is widely used for analysis of these and other experiments to be described. Computer software is available for this analysis which includes the estimation and testing of coefficients in equations such as equation (19.7). Chapter 29 discusses regression analyses in greater detail.

Two-Level Fractional Factorial Experiments

These principles of design and analysis of two-level factorial experiments can be extended to experiments involving any number of factors. See Figure 19.3 for an illustration of a

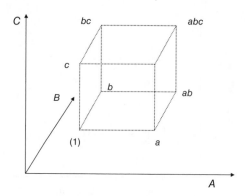

Figure 19.3   The three-factor situation.

three-factor situation. However, the number of observations in such an experiment increases exponentially with the number of factors. If there are n factors, the number of observations is 2^n. Thus:

Number of factors	Number of observations
1	2
2	4
3	8
4	16
5	32
6	64
7	128
8	256
9	512
10	1024

It is not unusual to have experiments with seven or more factors (control variables). Thrift demands an experiment with only a fraction of the experiments in a full design. If a suitable fraction can be found, the resulting experiment is called a *two-level fractional factorial*.

The theory and method of constructing these fractional experiments are described in textbooks. Also, software is available for the automatic design and analysis of these experiments (see the References).

Composite Designs

Whereas two-level factorial experiments, and their fractions, are suitable for fitting models that are linear in the main effects and including interactions, they are not suitable for estimating curvature of response if it exists. For example, if there is a single control variable, equation (19.1) may be suitable either if the relationship is genuinely linear for all values of x, or on the rising or decreasing slope of a quadratic response.

However, if the experiment is to be performed in a range of x that is close to the peak (or trough) of the quadratic response, curvature will have a major effect and must be estimated. This is particularly important if a purpose of the experiment is to estimate the value of x for which y is a maximum (or minimum).

Equation (19.1) must then be augmented as:

$$y = a + bx + cx^2 \tag{19.8}$$

Similarly, equation (19.7) must be augmented as:

$$y = \beta_0 + \beta_1 x_1 + \beta_2 x_2 + \beta_{12} x_1 x_2 + \beta_{11} x_1^2 + \beta_{22} x_2^2 \tag{19.9}$$

Designs for these augmented relationships are called *augmented* or *composite designs*. The theory and methodology of constructing them are described in several textbooks. Software is available for constructing and analysing them. Analysis is usually by least-squares regression.

An Example

Figure 19.4 shows how a catheter is fixed to a valve body. It enters through the A-channel and expands into the C-channel where it is held by a bush which is pressed into the end of the catheter. The purpose of the experiment was to discover the dimensions such that the catheter would be gripped with maximum security. The assembly is put into a tensile tester, and force is gradually increased until the catheter is pulled out. The response variable ($Y1$) is the disassembly force measure in newtons.

There are some constraints in the dimensions. The C-channel inner dimension (ID) must be greater than the bush outer dimension (OD). The C-channel ID must be greater than the A-channel ID. These constraints are avoided by using differences in variables $X3$ and $X4$.

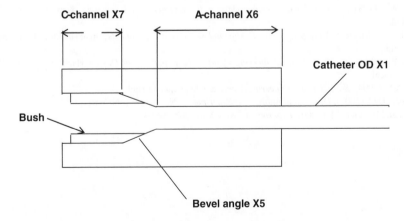

Figure 19.4   Catheter fixed to a valve body.

The control variables (measurements in millimetres, except for $X5$) are:

		Low	High	Increment
$X1$	Catheter OD	1.7	1.8	0.05
$X2$	Bush OD	1.6	2.0	0.1
$X3$	C-channel ID – bush OD	0.20	0.45	0.05
$X4$	C-channel ID - A channel ID	0.00	0.10	0.05
$X5$	Bevel angle	30°	90°	5°
$X6$	A-channel length	3.0	5.5	0.5
$X7$	C-channel length	0.5	4.0	0.5

The experiment was designed, using DEX (Greenfield, n.d.), to fit a model which included all quadratic effects and the first-order interactions: $X1.X2$, $X1.X3$, $X1.X4$, $X1.X6$, $X1.X7$, $X3.X5$ and $X4.X5$. The choice of interactions was based on experience, mechanical judgement and analysis of earlier trials.

The designed experiment has 47 observations, of which the first 32 comprise a quarter of a 2^7 factorial. The following 15 observations are axial and centre points that are added to estimate curvature as quadratic terms in the model. Thus, this is a composite experiment constructed by augmenting a fractional two-level experiment. It achieved its purpose of discovering the conditions that would yield a maximum disassembly force for the catheter valve. You can see a further explanation of the designed experiment of Metcalfe (1994, p. 420).

References

Atkinson, A.C., and Donev, A.N. (1992). *Optimum Experimental Designs*. Oxford: Oxford University Press.

Box, G.E.P., and Draper, N.R. (1987). *Empirical Model-Building and Response Surfaces*. New York: John Wiley.

Box, G.E.P., Hunter, W.G., and Hunter, J.S. (1978). *Statistics for Experimenters. An Introduction to Design*. New York: John Wiley.

Diamond, W.J. (1981). *Practical Experimental Designs for Engineers and Scientists*. New York: Van Nostrand Reinhold.

Greenfield, T. (N.d.). *DEX: A Program for the Design and Analysis of Experiments*. Great Hucklow, Derbyshire: Tony Greenfield.

Grove, D.M., and Davis, T.P. (1992). *Engineering Quality and Experimental Design*. Harlow: Longman Scientific and Technical.

Metcalfe, A.V. (1994). *Statistics in Engineering*. London: Chapman and Hall.

Montgomery, D.C. (2001). *Design and Analysis of Experiments*. New York: John Wiley.

Wu, C.F.J., and Hamada, M. (2000). *Experiments*. New York: John Wiley.

20

Experiments in Biological Sciences

Roger Payne

Experiments play a very important role in biological research. Indeed, much of the theory of design and analysis of experiments was originally developed by statisticians at agricultural or biological institutes. The methods are useful, however, in many other application areas. So this chapter should be of interest to all researchers even if their chosen area of study is not biology.

The aim of the chapter is to cover the main principles of the design and analysis of experiments. Further information can then be obtained either from the books listed at the end of the chapter, or from your local statistician.

Firstly, we consider the basic experimental *units*. In a field experiment, these may be different *plots* of land within a field. In animal experiments, they may be individual *animals* or perhaps *pens* each containing several animals. There is a great deal of practical expertise involved in setting out, managing, sampling and harvesting field plots. The accuracy of these aspects can be vital to the success of an investigation, and potential field experimenters are encouraged to read Dyke (1988, chaps. 2–6) for further advice.

Blocking

The units frequently have either an intrinsic or an imposed underlying structure. Often, this consists of a grouping of the units into sets known as *blocks*. For example, the animals in an experiment may be of several different breeds (this would be an intrinsic grouping). Alternatively, in a field experiment, you might decide to partition up the field into several different blocks of plots. The aim when allocating plots to blocks is to make the plots in each block as similar as possible, to try to ensure that pairs of plots in the same block are more similar than pairs in different blocks. Often, this can be achieved by taking contiguous areas of the field as blocks, chosen so that the blocks change along a suspected trend in fertility. However, blocks need not be contiguous. For example, in an experiment on fruit trees, you might wish to put trees of similar heights in the same block, irrespective of where they occur in the field. Blocks need not all have the same shape. In a glasshouse, you might find that the

Research Methods for Postgraduates, Third Edition. Edited by Tony Greenfield with Sue Greener.
© 2016 John Wiley & Sons, Ltd. Published 2016 by John Wiley & Sons, Ltd.

main variability is between the pots at the side and those in the middle. So you could have two L-shaped blocks around the edge, and other rectangular blocks down the middle. The key point is that you should think about the inherent variability of your experimental units, and then block them accordingly. Further advice can be found, for example, in Dyke (1988, chap. 2) or Mead (1988, chap. 2). An important final point is that you should remember the blocking when you are planning the husbandry of the trial. For example, if you cannot harvest all the units of the experiment in one day, you should aim to harvest one complete set of blocks on the first day, another set on the second and so on. Then any differences arising from the different days of harvesting will be removed in the statistical analysis along with the differences between blocks and will not bias your estimates of the treatments.

More complicated blocking arrangements can also occur, and are described later in the chapter. The alternative strategy of fitting spatial covariance models to the fertility trends is also discussed.

Treatment Structure

The purpose of the experiment will be to investigate how various treatments affect the experimental units. In the simplest situation, there is a single set of treatments: perhaps different varieties of a crop, or different amounts of a fertiliser or different dietary regimes for animals. Table 20.1 shows the field plan of a simple example in which fungicidal seed treatments to control the disease take-all were examined in a *randomised complete block design*.

In this design, the units, in this case field plots, are grouped into blocks (in columns) as described in the 'Blocking' section. Each treatment occurs on one plot in every block, and the allocation of treatments to plots is *randomised* independently within each block; so here, a random permutation of the numbers 1–3 was selected for each block and used to determine which treatment was applied to each of its plots. Designed experiments are usually analysed by analysis of variance. Table 20.2 shows output from GenStat for Windows (Payne, 2014a,b; Payne *et al.*, 2014a; or see genstat.co.uk). GenStat was originally developed by statisticians at a biological research institute, and so it has especially comprehensive facilities for the design and analysis of experiments.

Firstly, there is an *analysis of variance* table which allows the effects of the treatments to be assessed against the inherent variability of the units. The variability between the blocks is contained in the *Block stratum*, and the variability of the plots within the blocks is in the *Block.Plot stratum*. Within the *Block.Plot stratum*, the *Seed_treatments* line contains the variability that can be explained by the assumption that the seed treatments do have differing effects, and the residual line contains the variability between units that cannot be explained by either block or treatment differences. The *mean square* (m.s.) for blocks is over nine times that for the residual, indicating that the choice of blocks has been successful in

Table 20.1 An example of a randomised complete block design

150	150	100	0
0	100	150	150
100	0	0	100

Table 20.2   Analysis of variance

Analysis of Variance

Source of variation	d.f.	s.s.	m.s.	v.r.	F pr.
Block stratum	3	2.39990	0.79997	9.12	
Block.Plot stratum					
Seed_treatments	2	2.11449	1.05724	12.05	0.008
Residual	6	0.52630	0.08772		
Total	11	5.04069			

Tables of Means

Grand mean 7.149

Seed_treatments	0	100	150
	6.590	7.256	7.601

Standard Errors of Differences of Means

Table	Seed_treatments
rep.	4
d.f.	6
s.e.d.	0.2094

increasing the precision of the experiment. Similarly, the variance ratio (v.r.) of 12.05 for *Seed_treatments* shows that there is evidence that the treatments are different. The analysis will also provide tables showing the estimated mean of the units with each treatment, usually (as here) accompanied by a standard error to use in assessing differences between the means.

It is important to realise that the analysis is actually fitting a *linear model* in which the yield y_{ij} on the plot of block i to receive treatment j is represented as follows:

$$y_{ij} = \mu + \beta_i + t_j + \varepsilon_{ij}$$

where μ is the overall mean of the experimental units; β_i is the *effect* of block I; t_j is the *effect* of treatment j; and ε_{ij} is the *residual* for that plot (representing the unexplained variability after allowing for differences between treatments and between blocks).

In this analysis, the residuals are assumed to have independent normal distributions with zero means and equal variances. The variance ratio of 12.05 for treatments in the analysis of variance table can then be assumed to have an F distribution on 2 and 6 degrees of freedom (the degrees of freedom from the treatment and residual lines, respectively), leading to the probability value of 0.008 in the right-hand column, and the differences between the means divided by the standard error of difference can then be assumed to have a t distribution with six degrees of freedom (the degrees of freedom for the residual).

The effect t_j of treatment j represents the difference between the mean for treatment j and the overall mean. So when we are assessing whether the treatments are identical, we are actually seeing whether there is evidence that their effects are different from zero.

This way of representing the analysis becomes more useful in experiments when the treatments given to the units may differ in several different ways. For example, we may have several different fungicides to study and we may also want to try a range of different amounts; or we may want to investigate the effect of varying the amounts of several different types of fertiliser; or we may want to see how well different varieties of wheat are protected by different makes of fungicide. Each of these types of treatment is then represented by a different treatment *factor*, with *levels* defined to represent the various possibilities. One of the great advantages of a designed experiment is that it allows you to examine several different treatment factors at once. Suppose that we have two treatment factors N (nitrogen at levels 0, 180 and 230) and S (sulphur at levels 0, 10, 20 and 40) and we wish to examine all their combinations, again in a randomised complete block design. The factors N and S are said to have a *crossed* or *factorial* structure, and we can represent the yield y by the model

$$y_{ijk} = \mu + \beta_i + n_j + s_k + ns_{jk} + \varepsilon_{ijk}$$

We now have three *terms* to represent the effects of the treatments: the parameters n_j represent the *main effect* of nitrogen, s_k represent the *main effect* of sulphur and ns_{jk} represent the *interaction* between nitrogen and sulphur.

The analysis of variance table in Table 20.3 contains a line for each of these, to allow you to decide how complicated a model is required to describe the results of the experiment.

Table 20.3   Analysis of variance

Analysis of Variance

Source of variation	d.f.	s.s.	m.s.	v.r.	F pr.
Block stratum	2	0.30850	0.15425	3.44	
Block.Plot stratum					
N	2	4.59223	2.29611	51.22	<.001
S	3	0.97720	0.32573	7.27	0.001
N.S	6	0.64851	0.10808	2.41	0.061
Residual	22	0.98625	0.04483		
Total	35	7.51269			

Tables of Means

Grand mean 1.104

N		0	180	230	
		0.601	1.313	1.398	

S		0	10	20	40
		0.829	1.155	1.167	1.266

N	S	0	10	20	40
0		0.560	0.770	0.524	0.552
180		0.894	1.289	1.525	1.545
230		1.032	1.404	1.454	1.700

Standard Errors of Differences of Means

Table	N	S	N, S
rep.	12	9	3
d.f.	22	22	22
s.e.d.	0.0864	0.0998	0.1729

When analysing a factorial experiment, we would like to find a simple model to explain the situation. The full model above will estimate the means for the sulphur and nitrogen treatments as:

N × S means	N0	N180	N230
S0	0.560	0.894	1.032
S10	0.770	1.289	1.404
S20	0.524	1.525	1.454
S40	0.552	1.545	1.700

=

μ	+	n_j: N0	N180	N230	+	s_k		+	ns_{jk}	N0	N180	N230
1.104		−0.503	0.209	0.294		S0	−0.276		S0	0.234	−0.144	−0.090
						S10	0.051		S10	0.118	0.075	−0.044
						S20	0.063		S20	−0.141	0.148	−0.007
						S40	0.162		S40	−0.211	0.071	0.141

It will be much easier to describe what is happening if there is no interaction. The model will then be:

$$y_{ijk} = \mu + \beta_i + n_j + s_k + \varepsilon_{ijk}$$

leading to a table of means:

N × S means	N0	N180	N230	=	μ	+	n_j: N0	N180	N230	+	s_k	
S0	0.326	1.038	1.122		1.104		−0.503	0.209	0.294		S0	−0.276
S10	0.652	1.364	1.448								S10	0.051
S20	0.665	1.377	1.461								S20	0.063
S40	0.763	1.475	1.559								S40	0.162

and you will notice that we can decide on the best level of nitrogen without needing to consider how much sulphur is to be applied, and on the best level of sulphur without needing to think about the level of nitrogen on the plot. This is what we mean by saying that the two factors do not interact: the *interaction* assesses the way in which the changes in yield caused by the various levels of nitrogen differ according to the amount of sulphur or, equivalently, the way in which the response to amount of sulphur differs according to the level of nitrogen.

This idea can be extended similarly to three or more factors. These *factorial arrangements* thus have the advantage that we can examine several different types of treatment at once. If there are no interactions, we can present one-way tables of means, and these will each have the same replication (number of units for each level of the treatment) as they would have if we had performed individual experiments of this size for each treatment factor in turn; the only difference is that the factorial experiment has fewer residual degrees of freedom, but 22 (above) is ample! More importantly, by examining the interaction we can assess how valid our conclusions for each factor are over a range of values of other factors.

Treatment factors need not always have a crossed structure. They can also be *nested* one within another. For example, we may have several strains of two different species of aphid, Mp and Rp, and wish to examine their esterase levels. Suppose that we have strains Mp_1 ... Mp_4 and Rp_1 ... Rp_3 with several individuals of each strain. We will certainly be interested in assessing differences between Mp and Rp. We may also be interested in how much variation there is amongst $\{Mp_1, Mp_2, Mp_3 \text{ and } Mp_4\}$ and amongst $\{Rp_1, Rp_2 \text{ and } Rp_3\}$; that is whether there is variability of the strains beyond the variability of the individual aphids. The model of interest (assuming that there is no blocking) would then be:

$$y_{ijk} = \mu + s_i + st_{ij} + \varepsilon_{ijk}$$

where the parameters s_i represent the effects of the species ($i = 1, 2$), and st_{ij} represent the strain *within*-species effects ($j = 1 ... 4$ for $i = 1$; $j = 1 ... 3$ for $i = 2$).

Notice that the model does not contain a strain main effect. The actual number allocated to each strain is only a labelling; it does not imply any special similarity, for example between the strain numbered 2 for Mp and the strain numbered 2 for Rp.

Other Types of Blocking Structure

Sometimes more complicated blocking structures may be required. For example, there may be more than one way of forming the units into groups. Perhaps we need to cater for large fertility trends running both along and across the field; or we may have insufficient pens in an animal experiment to complete the experiment all at once and need to subdivide the animals, from several different breeds, into batches to be examined in successive weeks. One possibility, if we have two blocking factors each with the same number of levels as the number of treatments that we wish to examine, is to use a *Latin square*. An example, for five dietary treatments, is shown in Table 20.4.

Notice that each diet occurs once in each row and once in each column. So we have simultaneously blocked the units by rows (Week) and by columns (Breed). In the analysis, we will be able to estimate and eliminate both the row differences and the column differences, leading to a smaller residual mean square, and thus smaller standard errors for differences

Table 20.4 Five dietary treatments and their schedules

5 × 5 Latin square (before randomisation)	Breed 1	Breed 2	Breed 3	Breed 4	Breed 5
Week 1	Diet A	Diet B	Diet C	Diet D	Diet E
Week 2	Diet B	Diet C	Diet D	Diet E	Diet A
Week 3	Diet C	Diet D	Diet E	Diet A	Diet B
Week 4	Diet D	Diet E	Diet A	Diet B	Diet C
Week 5	Diet E	Diet A	Diet B	Diet C	Diet D

between the treatment means. Table 20.4 shows the plan before randomisation. To randomise, we need to select a random permutation for the rows and then another one for the columns.

We have now seen two types of blocking structure. The randomised complete block design has a *nested* structure with the individual units nested within the blocks, while the Latin square has a *crossed* structure of rows crossed with columns. These operations of nesting and crossing provide the basis for the more complicated arrangements that are sometimes needed for sophisticated trials. For example, the *split-plot design* extends the ideas of the randomised block design to have a further nesting of *sub-plots* within plots. Alternatively, we may have replicated Latin squares (rows crossed with columns, all nested within replicates).

All of these designs can be generated using menus in GenStat for Windows. Further examples can be found in Cochran and Cox (1957) and Payne (2014b, chap. 4).

Assumptions of the Analysis

Analysis of variance assumes firstly that the model is additive: that is, differences between treatment effects must remain the same however large or small the underlying size of the variable measured. So, for example, in a randomised block design, we are assuming that the theoretical value of the difference between two treatments remains the same within a block where the recorded values are generally low compared to one where the values are generally high. If your design has more than one replicate of each treatment within each block, you can check this by fitting block × treatment interactions, but usually this is not possible. An alternative method, which checks for the common form of non-additivity where treatment effects are proportional, is to fit Tukey's single degree of freedom for non-additivity. Non-additivity can also cause interactions between treatment factors, but of course these may also occur for genuine reasons, for example caused by one treatment modifying the mode of action of another.

Secondly, the variance must be homogeneous: the variability of the residuals should be the same at high values of the response variable as at low values. If this assumption does not hold, the standard errors presented will be too large for differences between treatments with low means and too small for differences between larger means, and this may cause incorrect conclusions to be drawn. Homogeneity of variance can easily be assessed by plotting the residuals against the fitted values: if the variance is homogeneous, the residuals should lie within a uniform band.

Thirdly, the residuals are assumed to have independent normal distributions. Non-normality can be assessed by plotting the residuals as a histogram or by plotting the residuals, sorted into ascending order, against values that would be expected from a normal distribution (a *normal plot*). Non-normality is usually also associated with non-homogeneity of variances.

Transformations

Failures of the assumptions can often be corrected by transforming the data. The transformations described in most textbooks are designed to stabilise the variance (assumption 2): for example, the square root transformation for counts or the angular transformation for percentages. A frequent mistake is to use the angular transformation blindly, without regard to the way in which the percentages have been obtained; it is important to realise that it is appropriate only when they are based on binomial data (e.g. a number r diseased out of n examined).

However, it is equally if not more important to consider the additivity of the model. Otherwise, as mentioned here, the resulting interactions can make the results difficult if not impossible to interpret. In some situations, a transformation can be chosen both to provide additivity and to stabilise the variance. With data where the treatments take the effect of a proportionate increase (or decrease), the standard errors will often be proportional to the means; a logarithmic transformation will then correct both aspects. With percentages representing proportions of diseased material, treatment effects are often found to be approximately proportional to the amount infected for low percentages, while for percentages near to 100% they tend to be proportional to the amount uninfected. If the percentages are obtained by visual assessment of areas such as infected parts of leaves, the standard errors tend to show the same pattern: for low percentages, the eye tends to examine the amount infected, while nearer to 100% it is the amount uninfected that is assessed. In this situation, a logit transformation, $\log(p/(100 - p))$, would be appropriate.

Further information about model assumptions and transformations can be found in Mead (1988, chap. 11), Mead and Curnow (1983, chap. 7) and John and Quenouille (1977, chap. 14).

Generalised Linear Models

If additivity and homogeneity of variance cannot both be corrected simultaneously by transformation, a generalised linear model should be used (McCullagh and Nelder, 1989). However, although these are readily available in statistical systems such as GenStat, they do require rather more statistical expertise to specify the models and to interpret the results than in ordinary analyses of variance.

Spatial Covariance Models

If you have many treatments to compare, as for example in a variety trial, it may by impossible to put one of each into a uniform block. Specialist designs, such as Alpha designs, are available for these situations and can be constructed by systems like GenStat and CycDesigN (see http://www.vsni.co.uk/software/cycdesign). The analysis uses a rather more complicated method known as REML, which stands for *residual* (or *restricted*) *maximum likelihood*. One advantage of this method, however, is that you can model the fertility trends in a more flexible way, by specifying models to describe the covariances between the plots (Gilmour *et al.*, 1995). This can result in substantial increases in precision. Further details are given by Galwey (2014), Payne (2014b, chap. 5) and Payne *et al.* (2014b, chap. 3).

Repeated Measurements

Special care is needed with experiments where the same units are observed at successive times. *Repeated measurements* like these can show complicated correlation patterns, and you may not be able to assume that the necessary *distributional assumptions* hold for analysis of variance (see the 'Assumptions of the Analysis' section). A statistician should be able to advise on alternative methods like the analysis of summary statistics, the multivariate analysis of variance, the use of anti-dependence structure or the modelling of inter-time correlation structure (see Payne, 2014b, chaps. 5 and 8; Payne *et al.* 2014c, chap. 4).

Conclusion

We have illustrated in this chapter the three main principles of experimentation (known as the three Rs): *replication*, the need to have more than one unit for each treatment so that you can ascertain the intrinsic level of variability of the units and so decide whether the differences between the treatments go beyond what we might expect to occur by chance; *randomisation*, the need to allocate units to treatments at random, to avoid any biases; and *blocking*, ways of grouping the units in order to improve the precision of the experiment. These concepts, together with the ideas of factorial treatment structure, are fundamental to a successful experiment, whether in agriculture or in any other research area.

References

Cochran, W.G., and Cox, G.M. (1957). *Experimental Designs*, 2nd ed. New York: Wiley.

Dyke, G.V. (1988). *Comparative Experiments with Field Crops*, 2nd ed. London: Griffin.

Galwey, N.W. (2014). *Introduction to Mixed Modelling: Beyond Regression and Analysis of Variance*, 2nd ed. Chichester: Wiley.

Gilmour, A.R., Thompson, R., and Cullis, B.R. (1995). Average Information REML, an efficient algorithm for variance parameter estimation in linear mixed models. *Biometrics* 51: 1440–1450.

John, J.A., and Quenouille, M.H. (1977). *Experiments: Design and Analysis*. London: Griffin.

McCullagh, P., and Nelder, J.A. (1989). *Generalized Linear Models*, 2nd ed. London: Chapman and Hall.

Mead, R. (1988). *The Design of Experiments: Statistical Principles for Practical Applications*. Cambridge: Cambridge University Press.

Mead, R., and Curnow, R.N. (1983). *Statistical Methods in Agriculture and Experimental Biology*. London: Chapman and Hall.

Payne, R.W. (ed.) (2014a). *The Guide to GenStat, Part 1: Syntax and Data Management*. Oxford: VSN International.

Payne, R.W. (ed.) (2014b). *The Guide to GenStat, Part 2: Statistics*. Oxford: VSN International.

Payne, R., Murray, D, Harding, S., Baird, D., and Soutar, D. (2014a). *GenStat for Windows*, 17th ed. Oxford: VSN International.

Payne, R., Welham, S., and Harding, S. (2014b). *A Guide to REML in GenStat*, 17th ed. Oxford: VSN International.

21

Survey Research

David de Vaus

What Is a Survey?

Survey research and questionnaire research are not the same thing. Although questionnaires are frequently used in surveys, there is no necessary link between surveys and questionnaires. There are two distinguishing characteristics of surveys: the form of data and the method of data analysis. Neither of these features requires questionnaire-based data collection – in-depth interviews, observation, content analysis and so forth can also be used in survey research.

Form of Data

Surveys are characterised by a structured or systematic set of data that I will call a *variable by case data grid*. This involves collecting information about the same variables from a sample of cases. This information is placed in a grid (see Table 21.1) in which each row represents a case (person) and each column represents the variables or information collected about each case.

Method of Analysis

While experimental studies rely on control groups, random allocation to groups and experimental interventions (see Chapter 20), surveys rely on existing variation in the sample rather than on creating it with interventions. Surveys also control for the influence of external factors by using statistical controls in data analysis, while experiments control external factors by random allocation to control groups (de Vaus, 2001).

Dimensions of Surveys

Units of Analysis

Not only are surveys not restricted to questionnaires, but also they are not restricted to studies of individual people. The units about which information is collected (the unit of analysis) may

Research Methods for Postgraduates, Third Edition. Edited by Tony Greenfield with Sue Greener.
© 2016 John Wiley & Sons, Ltd. Published 2016 by John Wiley & Sons, Ltd.

Table 21.1   A variable by case data grid

			Variables		
		Sex	Age (years)	Political orientation	Social class
Cases	Person 1	Male	36	Progressive	Working
	Person 2	Male	19	Moderate	Lower middle
	Person 3	Female	30	Progressive	Upper working
	Person 4	Male	55	Traditionalist	Upper middle
	Person 5	Female	42	Traditionalist	Middle

be countries, years, organisations, events or some other unit. If a unit of analysis other than a person is used (e.g. countries), the names of countries would be placed on the side of the data grid and we would collect information pertaining to countries (e.g. population size, area, density, unemployment rate, literacy rate, life expectancy etc.).

Types of Surveys

Descriptive

Social researchers can try to answer two fundamental questions about society. *What* is going on (descriptive research), and *why* is it going on (explanatory research)? Surveys can be an effective way of describing a phenomenon and have been used widely for this. Most governments conduct surveys to obtain an accurate description of population size and characteristics (censuses), unemployment levels, expenditure patterns, experience of crime, housing needs, health levels and health behaviours. Repeated descriptive surveys taken over time can describe changes such as changing patterns of family formation and changes in employment levels.

Explanatory

Explanatory surveys try to account for the phenomena they describe. Why are some people poorer than others? Why are people having fewer children? Why is the level of home ownership declining? Our initial, non-empirical attempts to answer these questions are our theories. These speculative theories require testing, and these theories guide the design of the survey, the data we collect and the way we analyse the data.

Content

When constructing a survey, we need to identify what sort of information we need to collect. In descriptive surveys, we must collect information about the phenomenon we are seeking to describe. This requires being clear about precisely what this phenomenon is. This requires clarification of the concepts implied in the research question. For example, if we want to describe unemployment levels, what do we mean by *unemployment*? Are women who are not in the paid workforce because they are looking after young children unemployed? Are students without paid work unemployed? Is a retired person unemployed? What about someone who works only one hour a week?

As well as defining our terms, the following questions help focus the description further:

1. What is the *time frame* for the description: the present time, sometime in the past or change over time?
2. What is the *location* of the interest: a particular region, a country or a comparison of countries?
3. How *specific* is the interest? Do we want to look at a population overall, or do we want to break down our analysis into subgroups (e.g. men vs. women, young vs. old or urban vs. rural)?

An explanatory survey will need to be clear about:

1. What is to be explained (dependent variable)?
2. What are the possible causes (independent variables)?
3. On which possible causes will the survey focus?
4. What are the possible mechanisms (intervening variables) by which the causal factors might produce their effect?

When specifying the content of a survey, try to be very clear about the type of information you need. For example, in a survey of individuals, the information you require will normally fall into one of the following categories:

Attributes: For example, gender, age, class and occupation.

Behaviour: What people *do* (e.g. hours worked, participation in voluntary groups and alcohol consumption).

Knowledge: What does the respondent *know* about the topic?

Beliefs: What a person thinks is *true* or *false* (e.g. do they believe that capital punishment reduces crime?).

Attitudes: What a person thinks is *desirable* (e.g. *should* capital punishment be introduced?).

Questionnaire Design

Although survey data can be obtained with other methods, the structured questionnaire remains the most common method of obtaining a structured set of survey data. It is therefore the method of data collection on which I will focus for the remainder of this chapter.

Concepts to Questions

In designing a questionnaire, the first step is to develop specific questionnaire items for the concepts that are employed in the research question. This involves translating what are often vague and abstract concepts into specific and concrete indicators of the concepts. This process involves 'descending the ladder of abstraction'. This involves:

- defining the concept;
- identifying different dimensions of the concept; and
- developing indicators for these dimensions and sub-dimensions (see de Vaus, 2014).

Principles in Designing Questions

There are at least eight principles to guide the design of questionnaire items (for more detail, see de Vaus, 2014).

1. **Reliability**: Assuming that the person does not actually change, a reliable question is one to which respondents give the same response on different occasions. Ambiguous or vague wording must be avoided to ensure that respondents would 'read' the question consistently on different occasions.
2. **Validity**: We must be sure that a question actually measures what we say it does. For example, if we ask about church attendance to measure religiousness, we must be sure that it actually does measure this concept.
3. **Discrimination**: Explanatory survey analysis relies on variation in the sample on key variables. Good measures should be sensitive to measuring real and meaningful differences in a sample. For example, if you want to look at the effect of heavy work demands on family well-being, then the measures of these concepts must be able to tap meaningful differences in work demands and family well-being.
4. **Response rate**: Non-response to questions needs to be minimised because of both the loss of information and the data analysis difficulties it introduces. Intrusive, insensitive, irrelevant or repetitive questions as well as those that are difficult to understand and answer or have insufficient response categories can produce frustration and non-response.
5. **Same meaning for all respondents**: If different respondents interpret a question in different ways, they effectively answer different questions, thus making the survey analysis largely meaningless. For example, if I use the term *old people* in a question, then respondents will define 'old' in quite different ways and, in effect, be answering different questions.
6. **Relevance**: Each question must earn its place in your survey. For each question, ask yourself whether it really is necessary.
7. **Exhaustiveness (inclusiveness)**: There must be sufficient response alternatives so that all respondents can answer the question.
8. **Exclusiveness**: The alternate responses should be mutually exclusive so that only one response for each variable is applicable to any respondent.

Question Wording

Table 21.2 provides a checklist for evaluating questions. An answer of 'YES' to any of these questions indicates that the question requires further revision. These ideas are expanded in de Vaus (2014), Foddy (1993), Sudman and Bradburn (1982), Bradburn and Sudman (1979), Bradburn *et al.* (2004), Alreck and Settle (1995), Sudman *et al.* (1996), Krosnick and Presser (2010) and Martin (2006).

Table 21.2   Question wording checklist

1. Is the language complex?	10. Is the question leading?
2. Is the question double-barrelled?	11. Is the respondent unlikely to have the
3. Is the question negative?	necessary knowledge?
4. Will the words have different meanings for	12. Is there a prestige bias?
different people?	13. Is the question too precise?
5. Is the question ambiguous?	14. Does the question artificially create
6. Is the frame of reference for the question	opinions?
unclear?	15. Is the question wording unnecessarily detailed
7. Does the question have dangling alternatives?	or objectionable?
8. Is the question a 'dead giveaway'?	16. Does the question contain gratuitous
9. Can the question be shortened?	qualifiers?

Response Formats

There is a wide variety of response formats in questionnaires. When framing a question, you must first decide whether it will be a closed question in which respondents select from a number of pre-set responses or whether it will be an open question to which respondents frame their own responses. There are advantages and disadvantages to both sorts of questions, but in general you are best advised to minimise the number of open questions you use (Bradburn and Sudman, 1979; Sudman and Bradburn, 1982; Foddy, 1993).

With each closed question, you will need to decide on a response format. There is insufficient space here to go into each type, but see de Vaus (2014) and Alreck and Settle (1995) for fuller discussion.

1. **Rating scales**: These require respondents to *select* one alternative from a set of ordered categories. This format comes in different styles, including:
   - Likert scales
   - Numerical rating scales
   - Feeling thermometers
   - Score out of 10
   - Semantic differential.
2. **Rankings**: These require respondents to *rank* a set of alternatives.
3. **Checklists**: These involve providing a list of items and asking respondents to select all that apply to them.
4. **Selecting between alternative attitudes statements**: This format involves describing alternative attitudes towards some matter and asking which of the attitudes is closest to their own.

Which Response Categories

Number of categories: There is no agreement about how many response alternatives should be provided (Schwarz *et al.*, 1985).

No opinion and don't know responses: Although some people do not offer these alternatives, it is generally desirable to do so since there are many issues to which

Table 21.3 Sample sets of response alternatives

Response alternatives	http://www.au.af.mil/au/awc/awcgate/edref/smpl-srv.pdf Appendix H p.83
The intensity of words	http://www.au.af.mil/au/awc/awcgate/edref/smpl-srv.pdf Appendix G p.65

people genuinely have no opinion. Forcing them to express an opinion where they really do not have one is to create false and unreliable answers (Bishop *et al.*, 1980).

Including a middle alternative: There is disagreement about whether a middle alternative such as *neither agree nor disagree* should be provided. While excluding this option stops people 'sitting on the fence', this faces the danger of artificially creating opinions (Presser and Schuman, 1980).

Response Alternatives

For questions where the response categories can be ranked from high to low in some respect, the websites in Tables 21.3 and 21.4 provide some excellent sets of well-evaluated sets of response alternatives.

Layout and Structure of Questionnaire

There are six areas to which attention needs to be given when assembling questions into a questionnaire. The particular ways of dealing with these matters will vary somewhat depending on the way in which the questionnaire is administered (see later discussion in this chapter).

Answering procedures: Ensure that the way in which questions are to be answered is clear (e.g. circle responses, tick boxes etc.).

Instructions: The questionnaire should include (a) general instructions such as when to complete, how to return the questionnaire, do not consult and so on; (b) section introductions; (c) question-answering instructions; and (d) navigational

Table 21.4 Sources of questions

Rather than unnecessarily developing new questions, it makes sense to use well-developed and tested questions that have been used in reputable surveys. The sites here are excellent sources of questions and provide ideas about question format and structure:

1. The Question Bank   https://www.ukdataservice.ac.uk
2. Wide range of national social surveys
    http://www.icpsr.umich.edu/icpsrweb/ICPSR/studies?q=general+social+survey&searchIn=ALL
    http://www.icpsr.umich.edu/icpsrweb/ICPSR/ssvd/index.jsp

In addition, many of the online survey design packages and questionnaire design packages include libraries of questions (see Table 21.5). There are also a number of excellent handbooks of collections of questions on a range of topics.

instructions which instruct respondents which question to answer next when they are required to skip questions.

Filter questions: Since some questions may only apply to subsets of respondents, it is best to use questions to filter the sample so that respondents are directed to questions that apply to them but skip those that do not apply.

Use of space: In self-administered paper questionnaires, avoid cluttering the questionnaire. Make use of margins, list question responses down the page and leave sufficient room for open-ended questions.

Question order: Question order can affect response rates and the way in which people answer particular questions. There are a number of general principles in ordering questions:
- Ensure that the flow of the questionnaire makes sense.
- Commence with questions that respondents will enjoy answering (easy, factual and concrete and relevant).
- Group related questions together.
- Leave open-ended questions to near the end, if possible.
- Introduce a variety of question formats to make the questionnaire interesting.

Software for Producing Questionnaires

The task of questionnaire layout has been made easier by the power of widely available word processors. Specialised software that has been developed for electronic surveys has made the process even simpler. Some of these packages are listed in Table 21.5. This software, for which evaluation versions are available for downloading, can produce both electronic and professional-looking paper questionnaires.

Modes of Questionnaire Administration

Questionnaires can be self-administered or administered by trained interviewers. With the development of computer technologies, we now have both 'low-tech' questionnaires that

Table 21.5 Internet survey software: sites containing free evaluation copies of these packages or access to free trials

Infopoll	http://infopoll.com/live/surveys.dll/web
Key Survey	https://www.keysurvey.com
Survey Writer	www.surveywriter.com
Survey Gold 8	www.surveygold.com
Sawtooth Survey System	www.sawtoothsoftware.com
Survey System	www.surveysystem.com
SNAP Survey	www.snapsurveys.com
Allegience	www.allegiance.com/products/surveys
SurveyMonkey	www.surveymonkey.com
Sphinx survey	www.sphinxsurvey.com

Table 21.6   The technology of administering questionnaires

PAPI	Paper-and-pencil interview
CAPI	Computer-assisted personal interview
CATI	Computer-assisted telephone interview
CASI	Computer-assisted self-interview
Email with executable questionnaire	CASI/disk by mail (DBM)-style questionnaire distributed by email
Dynamic online survey	Questionnaire written with a survey-writing software package that has similar interactive features to a CAPI, CATI or CASI questionnaire.

rely on traditional pencil-and-paper approaches and sophisticated computer-based 'high-tech' questionnaires. Table 21.6 summarises the main ways in which questionnaires are administered.

The use of particular low-tech and high-tech questionnaires will depend on the methodology for administering questionnaires. There are four main ways in which questionnaires are administered:

1. **Face-to-face administration**: This is performed by a trained interviewer. These may be PAPI or CAPI interviews.
2. **Telephone**: A trained interviewer using CATI techniques and equipment usually conducts these interviews.
3. **Postal**: These self-administered questionnaires are usually PAPI style, but some are disk by mail (DBM) or CASI-style interviews.
4. **Internet based**: There are two broad ways in which internet surveys are administered:
   (a) **Email**: Used to distribute and collect questionnaires. These email questionnaires come in different formats, including:
      • a formatted questionnaire sent as an email attachment; and
      • an interactive questionnaire in the form of an executable file, which also can be sent as an email attachment.
   (b) **Web pages**: Questionnaires are increasingly administered over the internet. Interactive questionnaires with similar features to the interactive features of CAPI and CASI methods are available with internet questionnaires. This form of survey administration has become increasingly widespread as access to the internet has exploded.

Web-based questionnaires can be designed easily with the aid of special software designed for this purpose (see Table 21.5). Further information about the range of software with which to conduct computer-assisted surveys can be gained by searching the sites http://www.meaning.uk.com/your-resources/software-database/search-the-database/ or http://www.websm.org/c/1283/Software/?preid=0. Web surveys can also be designed and administered online. There are a number of sites where this can be done at no charge. Examples are shown in Table 21.7.

Ethics

Survey researchers should *adhere* to ethical principles when conducting surveys. The researcher has responsibilities to at least four categories of people:

Table 21.7 Online questionnaire design and administration sites

SurveyWriter	www.surveywriter.com
Response-o-matic	www.response-o-matic.com
Instant Survey	www.instantsurvey.com
Free Online Surveys	http://freeonlinesurveys.com
Zoomerang	www.zoomerang.com
Kwiksurveys	www.kwiksurveys.com
Survey Monkey	www.surveymonkey.com
Create Survey	www.createsurvey.com

The public: The responsibility here is to ensure that the survey results are presented fairly in public and to correct distortions in the way others use your survey results in public.

Clients and sponsors: Researchers should respect the confidentiality and rights of sponsors and not make unwarranted claims about one's expertise or about the value of techniques that will be employed.

The profession: It is important not to act in ways that will discredit the profession or make it difficult for other researchers in the future to conduct research because of your actions.

Respondents: The responsibilities to respondents involve conducting the survey in such a way as to ensure that:
• Informed consent is given by the respondent.
• Participation is voluntary.
• Confidentiality is protected.
• No harm comes to participants.

The full codes of ethics of a number of different professional organisations provide a much fuller outline of the types of ethical issues that must be taken account of in social science research, and survey research in particular (Table 21.8).

As telephone and web-based surveys have become increasingly common, ethical issues regarding the practice of what can be quite intrusive survey approaches have come to the fore. This has led to the formulation of codes of ethics that relate specifically to this form of surveying. Examples of these codes can be viewed on the website of the European Society

Table 21.8 Codes of ethics for survey and social science research

American Association of Public Opinion Research	http://www.aapor.org/Standards-Ethics/AAPOR-Code-of-Ethics.aspx
Council of American Survey Research Organizations	http://www.casro.org/?page=TheCASROCode2014
American Sociological Association	http://www.asanet.org/about/ethics.cfm
British Sociological Association	http://www.britsoc.co.uk/media/27107/StatementofEthicalPractice.pdf?1463851665645

for Opinion and Marketing Research (ESOMAR) (http://www.esomar.org/uploads/public/knowledge-and-standards/codes-and-guidelines/ESOMAR_Guideline-for-conducting-Research-via-Mobile-Phone.pdf).

Practicalities

When conducting a survey, there are many practical issues that need to be built into any proposal or plan. These matters will affect the type of survey you end up conducting. The practical considerations will often mean that the survey you *do* is not necessarily the survey you *want*. The real survey may be different from your ideal survey. To assist with covering the practical side of the survey design, ensure that you have answered the following types of questions:

1. Will you realistically have access to the type of sample that your research problem assumes? Will the sample produce the variability on the key variables that you require?
2. What is your sample size? Will the sample size be large enough to test your propositions, to divide it up to analyse sub-groups and to generalise?
3. What are the difficulties in contacting the sample?
4. How important is it to be able to generalise beyond the sample?
5. What are your timelines? How realistic are these? Be serious!
6. Have you planned for something to go wrong and for most things to take considerably longer than you anticipate?
7. What is your Plan B if your response rate is too low?
8. Have you obtained ethics clearance?
9. Have you established ways in which respondents can contact you?
10. What is your budget?
11. Is the survey a solo or a team effort?
12. Do you have the expertise to conduct the survey?
13. Do you have experts who can give advice and provide feedback?
14. Have you arranged to conduct a pilot test?
15. How will coding be conducted?
16. How will data be checked and cleaned?
17. Who will enter data? By what means will data be entered (automatic, manual, optical scan etc.)?
18. Have you thought through how you will analyse the data? Do you have all the questions you require to do the analysis you plan to conduct? Are variables at the right level of measurement for the analysis you plan? Is the sample type appropriate for your data analysis plans (e.g. probability sample)?

Pros and Cons of Each Method of Administration

It is not possible to recommend one method of questionnaire administration over the others. Each has its strengths and weaknesses, and these strengths and weaknesses will be context dependent. They will depend on your topic, your own skills, resources and time, the characteristics of your sample and so forth.

Rather than urge a particular method of data collection, it is more useful to provide a checklist to guide your choice in the context of your own survey. When comparing the methods for your survey, consider how well each is likely to perform in relation to these matters. (Dillman (2007) provides a useful assessment of some administration methods against some of these criteria, and internet sites for various internet survey software packages provide further guidance in regard to internet-based methods; see Table 21.9.)

Over the last few decades, survey response rates have been declining for a number of reasons. One response to this is to enable potential respondents to choose from the survey mode that is most convenient. The strategy is that such multi-mode surveys will increase response rates by making it easier for diverse samples to respond. However, the potential problem with this approach is that the method by which questionnaires are administered can affect responses, especially for sensitive questions where the degree of anonymity can influence responses to these questions. The jury is still out both on the effectiveness of multi-mode surveys in improving response rates and on the effect of these mixed-mode approaches on data quality (Sakshaug *et al.*, 2010).

Table 21.9   Checklist for determining the best method of data collection (Dillman, 2007)

Response rates	*Ability to avoid distortion due to:*
• General samples	• Interviewer characteristics
• Special-purpose samples	• Interviewer opinions
• Representative samples	• Influence of other people
• Availability of unbiased sampling frame	• Allows opportunities to consult
• Nature of likely sample bias and relevance of this to study	• Avoids subversion
• Avoidance of refusal bias	• Automatic coding (ease and reduction of error)
• Control over who completes the questionnaire	• Error checking
• Gaining access to the selected person	• Item non-response controls
• Locating the selected person	• Enables anonymity
Effects on questionnaire design	Implementing the survey
Ability to handle:	• Ease of finding suitable staff
• Long questionnaires	• Speed of data collection and returns
• Complex questions	• Speed of deployment
• Boring questions	• Cost
• Item non-response	• User friendliness
• Filter questions	• Ease of feedback to respondents
• Control over order in which questions are read and answered	• Ability to follow up non-responders
• Appropriateness in obtaining full-text responses to open-ended questions	• Ease of use
• Special question modules possible	• Confidentiality and privacy protection
• Enjoyment of completing questionnaire	• Accuracy (opportunity for human error):
• Ability to have follow-up questions	○ respondent
• Range of 'stimulus' materials that can be used (e.g. graphics, audio, lists and video)	○ coder
• Appropriateness for sensitive questions	• Level of automation
	• Respondent access to help to clarify questions or answering procedures
Quality of answers	• Longitudinal studies:
• Minimise social desirable responses	○ Respondent tracking
	○ Piping (also within survey)

References

Alreck, P., and Settle, R. (1995). *The Survey Research Handbook*. New York: McGraw-Hill.

Bishop, G., Oldendick, R., Tuchfarber, A., and Bennett, S. (1980). Pseudo-opinions on public affairs. *Public Opinion Quarterly* 44(2): 198–209.

Bradburn, N., and Sudman, S. (1979). *Improving Interview Method and Questionnaire Design*. San Francisco: Jossey-Bass.

Bradburn, N., Sudman, S., and Wansink, B. (2004). *Asking Questions: The Definitive Guide to Questionnaire Design for Market Research, Political Polls, and Social and Health Questionnaires*. San Francisco: Jossey-Bass.

de Vaus, D. (2001). *Research Design in Social Research*. London: Sage.

de Vaus, D. (2014). *Surveys in Social Research*, 6th ed. London: Routledge.

Dillman, D. (2007). *Mail and Internet Surveys: The Tailored Design Method*, 3rd ed. New York: Wiley.

Foddy, W. (1993). *Constructing Questions for Interviews and Questionnaires: Theory and Practice in Social Research*. New York: Cambridge University Press.

Krosnick, J., and Presser, S. (2010). Question and questionnaire design. In Marsden, P. and Wright, J. (eds.), *Handbook of Survey Research*, 2nd ed. Bingley: Emerald Group Publishing.

Marsden, P., and Wright, J. (eds.) (2010). *Handbook of Survey Research*, 2nd ed. Bingley: Emerald Group Publishing.

Miller, D. (1991). *Handbook of Research Design and Social Measurement*. Newbury Park, CA: Sage.

Presser, S., and Schuman, H. (1980). The measurement of the middle position in attitude surveys. *Public Opinion Quarterly* 44(1): 70–85.

Sakshaug, J., Yan, T., and Tourangeau, R. (2010). Nonresponse error, measurement error and mode of data collection: tradeoffs in a multi-mode survey of sensitive and non-sensitive items. *Public Opinion Quarterly* 74: 907–933.

Schwarz, N., Hippler, H., Deutsch, B., and Strack, F. (1985). Response scales: effects of category range on reported behaviour and comparative judgements. *Public Opinion Quarterly* 49(3): 388–395.

Sudman, S., and Bradburn, N. (1982). *Asking Questions: A Practical Guide to Questionnaire Design*. San Francisco: Jossey-Bass.

Sudman, S., Bradburn, N., and Schwarz, N. (1996). *Thinking about Answers: The Application of Cognitive Processes to Survey Methodology*. San Francisco: Jossey-Bass.

Further Reading

Bethlehem, J., and Biffignandi, S. (2011). *Handbook of Web Surveys*. New York: Wiley.

Biemer, P., and Lyberg, L. (2003). *Introduction to Survey Quality*. New York: Wiley.

Couper, M. (2008). *Designing Effective Web Surveys*. New York: Cambridge University Press.

de Vaus, D.A. (2013). Survey methods. *Oxford Bibliographies Online*. www.oxfordbibliographies.com (select 'Sociology', and then select 'Survey Methods').

Fink, A. (1995). *How to Sample in Surveys*. Thousand Oaks, CA: Sage.

Groves, R., Fowler, F., Couper, M., Lepkowski, J., Singer, E. and Tourangeau, R. (2011). *Survey Methodology*. New York: Wiley.

Kalton, G. (1983). *Introduction to Survey Sampling*. Beverly Hills, CA: Sage.

Marsh, C. (1982). *The Survey Method: The Contribution of Surveys to Sociological Explanation*. London: George Allen & Unwin.

Martin, E. (2006). *Survey Questionnaire Construction*. Research Report Series, Survey Methodology #2006-13. http://www.census.gov/srd/papers/pdf/rsm2006-13.pdf

Rosenberg, M. (1968). *The Logic of Survey Analysis*. New York: Basic Books.

22

Theory and Practice of Qualitative Research

Irena Ograjenšek

What Is Qualitative Research, and Why Use It?

The use of qualitative research builds on the core assumption that there are important phenomena or processes of a social or behavioural nature that cannot be measured quantitatively. In other words, researchers who are using qualitative methods attempt to understand complex phenomena by getting to know the behaviours or cognitions of persons or organizations involved, as well as their values, rituals, symbols, beliefs and emotions (Frankfort-Nachmias and Nachmias, 2000). This can be done via four key data channels:

- talk and speech;
- documents and texts;
- observations; and
- visual objects (e.g. photographs, drawings and video).

Gal and Ograjenšek (2010) state that, broadly viewed, by default any exploratory research has a qualitative core that helps to shape initial ideas and sharpen the sense of *what is important* to study or measure, *what can* be studied or measured and, if so, *how it can* be systematically studied or measured.

Going beyond the use of qualitative methods as part of exploratory processes, Strauss and Corbin (1998) claim that qualitative research provides researchers with a creative yet systematic process for analysing complex data and either identifying or generating multiple meanings from these data.

In the process, researchers not only describe phenomena but also, when identifying underlying influential factors (e.g. attitudes and motivations), create a theoretical framework for further (either qualitative or quantitative) verification.

Research Methods for Postgraduates, Third Edition. Edited by Tony Greenfield with Sue Greener.
© 2016 John Wiley & Sons, Ltd. Published 2016 by John Wiley & Sons, Ltd.

Selected practical examples are given in Box 22.1.

Box 22.1   When does it make sense to use qualitative research?

You want to …

- study poorly known phenomena and want to *develop hypotheses* for further testing and *questions* to be included in the questionnaire survey.
- better understand the *motives, feelings, values, attitudes* and *perceptions* that underline and influence behaviour (e.g. how customers perceive a certain service; how satisfied they are with it, and why; how employees perceive certain process changes etc.).
- capture the *language* and *imagery* people use to describe and/or relate to a product, service, brand, organisation, nation, country, political idea or the like.
- better understand a target audience's *perceptions* of communication *messages*.
- better understand the *context* and *meaning* of the *quantitative data* obtained in a quantitative research process.

Qualitative versus Quantitative Research

How does qualitative research differ from quantitative research? The core differences are summarized in Table 22.1.

Qualitative research is unavoidable in the exploratory phase of research, specifically in cases which call for an inductive research approach; in other words, when we study poorly known phenomena for which no or very loose theoretical framework(s) exists. In such cases, when we need to start creating a theoretical framework from scratch on the basis of our initial empirical observations, the value of qualitative research cannot be emphasised enough. It helps us develop hypotheses for further testing and questions to be included in the questionnaire survey used in the confirmatory phase of research.

A fairly recent example of a phenomenon for which a theoretical framework had to be created from scratch is electronic word of mouth (e-WOM). When it first started spreading through social networks, researchers had nothing to lean on in order to characterize it. The theory behind the classic word of mouth was useless because of the completely different modes, communication channels and impact e-WOM has in both the intimate and business spheres of an individual's activity. Ultimately, a specific qualitative observation-driven technique (netnography) had to be applied in order to capture and understand e-WOM.

On the other hand, if the phenomena we are dealing with are well known, and have already been operationalized in multiple different ways, it might make sense to skip the qualitative exploratory phase and head straight for quantitative confirmatory research. Concepts such as employee motivation, service quality and customer loyalty are only a few from a long list of illustrative examples.

The fact that the researcher is used as the measurement instrument in the qualitative research process prevents him or her from remaining objectively separated from the subject matter. The richness of data makes the research process very time-consuming. And a small(er) sample size means that generalization (if attempted at all) will be based on a subjective assessment.

Table 22.1 Core differences between qualitative and quantitative research

Research	Qualitative	Quantitative
Aim	• Provision of very detailed descriptions • Understanding of underlying reasons and motivations	• Classification of features • Counting and quantification • Construction of statistical models
Focus	Complex and broad (researcher may not know or may only know roughly what (s)he is looking for)	Concise and narrow (researcher knows what (s)he is looking for)
Applicability in the research process	• Unavoidable in exploratory phase • Possible in confirmatory phase	• Possible in exploratory phase • Unavoidable in confirmatory phase
Sample size	Usually smaller than 30 units	Has to enable statistical testing, therefore usually from 30 units upwards
Tools of measurement	Researcher as the measurement instrument	• Questionnaires in social sciences • Gauges in natural and technical sciences
Predominant data form	Words, pictures, objects and nonverbal information	Numbers
Character	Subjective because respondent-led and/or interpretive	Objective because adhering to pre-identified measurement standards

That is a risky undertaking because the studied sample might be atypical of the population. As shown in Box 22.2, in the framework of qualitative research, we usually aim to include cases on the basis of their perceived information richness and not necessarily with the goal of facilitating sample representativeness. Furthermore, providing a measure of the precision of estimates is not generally supported through qualitative research unless a very Bayesian perspective is taken.

Box 22.2   Typical cases in a qualitative sample.

Purposive sampling is a typical sampling strategy in qualitative research. We are systematically on the lookout for information-rich cases. Examples include, but are not limited to:

- **Extreme or deviant cases** – For example, best-in-class versus dropouts or outstanding entrepreneurial successes versus dramatic failures
- **Intense but not extreme cases** – For example, good versus bad students or above-average versus below-average students
- **Typical cases** – For example, an average student, an employee with an average number of monthly absences or a researcher with an average number of research publications

> - **Critical cases** – For example, opinion leaders whose acts and behaviours influence other members of the group they belong to
> - **Entry cases** – People who help us start rolling the snowball – they represent our entry point into a closely knit community which is reluctant to open up to a researcher.

So why bother with qualitative research? Because time and again, quantitative research manages to miss important contextual details. This, in turn, leads to poor outcomes of decision-making processes despite the rigour of the scientific method.

Furthermore, qualitative research is inevitable when dealing with:

- **hard-to-reach subjects**: This includes especially members of hidden populations, such as people living with HIV, drug abusers, rape victims and people with eating disorders, as well as people lacking time – busy managers, stressed mums and the like.
- **difficult (because sensitive) topics**: For example, an individual's sexual orientation or terminal disease, and its effects on family life.

Therefore, a pragmatic researcher will embrace qualitative *and* quantitative methodologies, apply them either simultaneously or consecutively (depending on the needs of the research project at hand) with the goal of obtaining the best possible research outcome(s) and not fail to document their weaknesses and limitations.

Typologies of Qualitative Research

Two core typologies of qualitative research can be proposed.

The first typology differentiates between *desktop* and *field research*. Desktop research is set in the virtual space. The researcher can draw knowledge and conclusions from databases of bibliographical information, official statistics databases as well as available web (including social media) content. Field research can take place in either an artificial (e.g. in a test market) or a real-life (e.g. in home or at work) setting.

The second typology differentiates between *questioning-driven* and *observation-driven* qualitative research.

Questioning-driven qualitative research is focussed on discovering respondents' attitudes, motivations, expectations, intentions, preferences and so forth. These cannot be accounted for in any other way than in a dialogue.

Observation-driven qualitative research is a systematic process of recording the behavioural patterns of people, objects and events as they are witnessed by a human observer or recorded in the process of a mechanical observation.

The second typology is used as a starting point for the subsequent discussion of available qualitative research methods.

Questioning-Driven Qualitative Research Methods

Overview of Methods

Questioning-driven qualitative research methods include:

- personal interviews;

- dyads;
- triads;
- mini groups;
- focus groups; and
- bulletin boards.

The most important differences among these methods concern:

- **number of participants** – ranging from two to indefinite;
- **level of structure** – ranging from broadly defined topic areas to be covered to very specific questions;
- **level of flexibility** – ranging from total flexibility which allows for respondent-led flow of conversation to total inflexibility which only allows responses to pre-specified questions;
- **length of the time lag** – between asking the question and receiving the responses (minimal if responses are provided in real time; lengthy if questions are posted on bulletin boards and responses are not actively sought);
- **use of support materials** – ranging from none to the whole array of pictorial, audio and video records or even the actual products; and
- **use of modern information and telecommunication technology** – which can only facilitate collection of evidence (e.g. a recording device) or can also provide a virtual discussion platform.

Let us take a quick look at each method in turn.

> **Personal interview** – This is an interaction between two persons: an interviewer and a respondent (sometimes referred to as an *interviewee*). In its classic form, this interaction takes place face-to-face in real time. Although best supporting the development of intimacy between an interviewer and a respondent, which is a necessary prerequisite to obtain information of the highest possible quality, this form is neither cost nor time efficient. Fortunately, modern information and telecommunication technology supports more cost- and time-efficient derivations of a personal interview in form of telephone interview, Skype interview, email-facilitated interview or web interview.

> **Dyad** – This interaction is among three persons: an interviewer and two respondents involved in the same process (e.g. research paper preparation, home purchase, process improvement or medical treatment). Depending on the nature of the qualitative research at hand, the respondents can be a couple, a pair of co-workers or a study pair. Theoretically, a dyad can also include people with significantly different levels of knowledge (e.g. a teacher and a pupil), insight (e.g. a doctor and a patient) or power (e.g. an employer and an employee), although that further complicates the research process and might lead to intentional or unintentional bias in research results.

> **Triad** – This is an interaction among four persons: an interviewer and three respondents involved in the same process. These could be members of the same

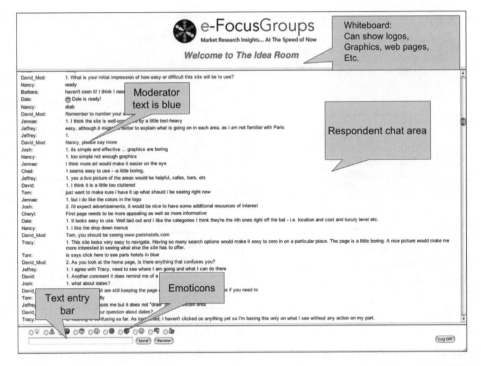

Figure 22.1   Typical elements of an online focus group – the respondent view. www.e-FocusGroups. com. Reproduced with permission.

study or quality circle, regular users of a given product or service, patients receiving the same treatment and so on.

Mini group – This group consists of four to six members. The discussion is facilitated by a trained moderator.

Focus group – A group of up to 12 individuals. It is led by a trained moderator who facilitates the discussion among members. In its classic form, a focus group usually lasts from one and a half to two hours. However, if any elements of creative thinking are introduced (such a focus group might also be referred to as a *creative group*), from three hours upwards can be spent dealing with a given topic. An illustration of an online focus group is given in Figures 22.1 (the respondent view) and 22.2 (the client view). In comparison, the client view is much richer and makes it possible for the clients to interact with the moderator in real time (during the group discussion), thus saving time and money.

Bulletin board – A questioning-driven qualitative research technique which makes it possible to include a large number of respondents from widely dispersed geographical areas. The responses are therefore not sought in real time. This makes the method useful only if the researcher is not pressed for time as it might take several weeks or even months to accumulate a large enough number of responses to make the search for underlying patterns feasible.

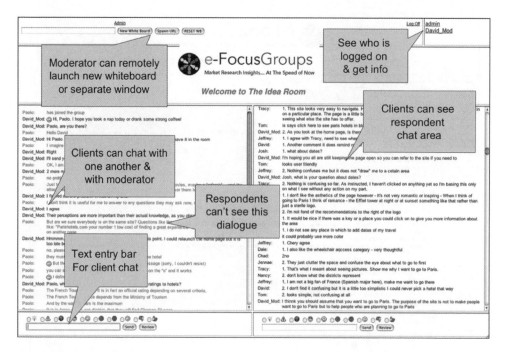

Figure 22.2   Typical elements of an online focus group – the client view. www.e-FocusGroups.com. Reproduced with permission.

Types of Questions and Questioning Techniques

The traditional differentiation between open and closed questions is not good enough for the needs of a qualitative researcher. The typology of questions and questioning techniques needs to be based on the content and research goals. Table 22.2 shows the most useful types of questions to be deployed in the qualitative research process.

Further to the typology of questions in Table 22.2, some other questioning techniques might prove useful in the qualitative research process, particularly the so-called *projective ones*. Not only are they very useful as icebreakers, but also they unleash respondents' creativity and imagination, often providing us with insights that are deeper than initially hoped for.

Some of the most useful projective techniques are described in Table 22.3.

Figure 22.3 shows an example of a collage. How would you interpret it – does it stand for a product used by an older, family-oriented lover of traditional cooking recipes or a young, hip, urban health food fanatic?

Active Listening

In order to facilitate active listening in the qualitative research process, it is advisable to get familiar with the most relevant guidelines. They are summarized in Box 22.3.

Table 22.2   Types of questions

Type of question	Aims and characteristics	Illustrative examples
Direct and factual questions	Provide background information	• Which PhD study programmes were you considering before deciding for this one? • Why did you decide for this one?
Grand tour questions	Enable a reconstruction of a routine, procedure, activity or event that took place at a particular time in a respondent's life. It is non-threatening, as there is no single correct answer; allows the respondent to demonstrate expertise and experience.	• You mentioned that you enrolled in a PhD programme. What happened that you decided to do so? How long did it take to decide? Which persons or factors influenced your decision to enrol into this particular PhD study programme?
Structural questions	Help understand how a respondent organized his or her feelings and knowledge within a particular area. Might be neutral or leading.	• Neutral question: 'Please describe the ideal PhD supervisor'. • Leading question: 'Here are some characteristics of a person who might be considered an ideal PhD supervisor: an ideal PhD supervisor is knowledgeable, reliable and quick to provide feedback. How does this description match with your image of an ideal PhD supervisor?'
Contrast questions	Help uncover differences in attitudes and perceptions by setting and comparing an item or an object to another. Each mentioned attribute is probed to determine the extent to which that attribute is important and meaningful. Each of these new reasons is then probed.	• How does a PhD study programme at this school differ from the PhD study programme at a competitive school? (If, for example, the tutorial support is mentioned) What makes the tutorial support so special?
Hypothetical questions	Help understand responses. A plausible situation is presented to a respondent, who is asked to verbalize how he or she would react.	• Imagine you were approached by the school with the leading PhD study programme in your field and offered a full scholarship. How would you feel? What would you be thinking? What types of questions would you ask? What type of responses would you be given?

Table 22.2   (*Continued*)

Type of question	Aims and characteristics	Illustrative examples
Third-person questions	Ask for elaboration within the context of an anonymous, non-present person which makes the process non-threatening.	• You said that, in your opinion, availability of a full scholarship is the most important factor in the selection of a PhD programme. Others whom I have talked to have said the same thing. But 'in my last group', several people said that full-scholarship availability was really less important than the possibility to interact with the leading experts in the field. What do you think about this point of view?

Figure 22.3   A sample collage. Adapted after a GfK GRAL-ITEO tržne raziskave d.o.o. collage.

Table 22.3   Typical projective techniques

Technique	Description
Word association	Word association asks a respondent to quickly react to the presentation of words or phrases with the first thing or things that come to mind.
Sentence and story completion	Sentence and story completions require a respondent to draw on his or her own attitudes and beliefs to quickly complete an incomplete sentence or story with the first phrase that comes to mind or anything else that makes sense.
Personification	Personification is a technique that seeks to establish the image and character of a product, brand, organisation, country and so on, by relating it to some well-known person, theatrical character or even animal.
Shopping list	The respondents are asked to speculate about characteristics of different types of individuals based on their shopping lists (types of products they purchase).
Picture projection	Picture projection techniques use visual data as the basis for respondents to construct stories or descriptions. A respondent is shown a picture and then asked to provide the dialogue, thoughts or feelings of depicted individuals.
Picture and product sorting or mapping	Picture and product sorting or mapping is used to identify different ways that respondents view other individuals, brands, products, places, activities and so on – which are comparable (i.e. grouped together) and which differ in the eyes of the respondent. When the process of mapping is finished, respondents have to provide explanations for their decisions regarding why they did or did not group the given pictures or products together.
Collage	Collage is a visual image formed from the selection and arrangement of many smaller images that express respondents' underlying feelings and attitudes towards themselves, a particular brand, an entire product category, an organisation, a political idea and so on. When the collage is finished, respondents are asked to provide explanations of the images.

Box 22.3   Guidelines facilitating active listening.

Active listeners ...

- take interest in what respondents say by relating what they are saying to how it affects them.
- focus on *what* participants say and not on *how* they say it.
- do not let their attitude towards the respondent or their beliefs distract or bias them.
- wait to judge responses until after they clearly understand what the respondents are saying.
- listen for issues, ideas and themes, not just focus on facts.

- observe nonverbal signals and use them to help interpret what is said.
- do not allow the note taking to become a distraction.
- concentrate on listening and showing complete attention by using eyes, ears and body position to demonstrate attentiveness.
- resist distractions not related to the research problem at hand.
- keep an open mind and do not react to emotional words or controversial statements.
- ask additional questions for clarification.
- paraphrase respondents' comments to confirm accurate interpretation (but do not summarize!).
- keep asking, 'Why?'

Questioning-Driven Qualitative Research in Practice

This section uses four examples from research practice to demonstrate how a specific 'need to know' leads to the application of questioning-driven qualitative research, and how conclusions then go beyond what would have been possible through the exclusive use of quantitative research.

Example 22.1

Zeithaml *et al.* (1990) initiated a systematic effort to understand how consumers and managers perceive the key attributes of service quality, and what discrepancies might exist between the perceptions of consumers and of service marketers and managers that may cause service quality shortfalls. Towards that end, they started an exploratory phase in 1983 involving in-depth interviews with executives as well as 12 focus groups, led by trained facilitators, with customers drawn from four different service segments: retail banking, credit card institutes, securities brokerages as well as product repair and maintenance. The sampling process intended to ensure that the conceptual insights would be applicable to a broad cross-section of industries.

Findings from the focus groups and executive interviews provided the researchers with the empirical foundation for creating the now-famous SERVQUAL model which defines *service quality* as the gap between expected and perceived service attributes measured on the Likert scale (Ograjenšek, 2007). More specifically, content analysis of focus group discussions led to identification of 10 initial dimensions of service quality as perceived by customers: reliability, responsiveness, competence, access, courtesy, communication, credibility, security, understanding the customer, and tangibles (tangible proofs of service, e.g. a train ticket or an invoice).

Further work that combined factor-analytic studies and qualitative refinements enabled the researchers to reduce the 10 themes to five key ones: tangibles, reliability, responsiveness, assurance and empathy. These five themes serve as the basis for many initiatives to measure customer satisfaction, and to date they have been used worldwide in thousands of academic and business studies on customer perceptions and expectations of services.

Example 22.2

More than 80 semi-structured interviews with key acquisition decision makers in processes of mergers and acquisitions were the key data collection technique in a qualitative study accounting for the seller's perspective carried out by Graebner and Eisenhardt (2004). The findings were amended by tracking of real-life acquisition processes and sources of secondary data (e.g. company reports and company websites). This resulted in an emergent theoretical framework explaining how company acquisition occurs from the seller's perspective and what are the roles of price, timing, strategic and emotional factors in the process.

Example 22.3

Focus groups and in-depth interview techniques have been among the qualitative methods used by researchers attempting to understand the needs and work practices of users of official statistics agencies' services. Marchionini and Levi (2003) report using interviews and focus groups with service providers and non-specialist end users, along with transaction log analysis and email content analysis, to develop a user-task taxonomy for government statistical data and identify desired features in a web-based statistical tables browser.

Example 22.4

The Critical Incident Technique (CIT; Gremler, 2004) uses stories about 'critical incidents' as data. The CIT technique has been at the core of Bitner *et al.*'s (1990) research on customers' views of the nature of service failures (as distinguished from the service providers' point of view on what constitutes service failures). They collected 700 'incidents' (stories of very good or very bad service encounters) from customers of airlines, hotels and restaurants.

The primary results of CIT studies are the themes and categories that emerge through a process of content analysis and classification. Some statistical analysis of the codes generated in the process is also possible when the number of incidents is large. Accordingly, Bitner *et al.* (1990) categorized the collected incidents to isolate the particular events and related behaviours of contact employees that cause customers to experience service encounters as highly satisfactory or dissatisfactory.

Their analysis identified three key groups of perceived service failures (employees' response to service delivery problems, employees' response to customer needs and requests as well as unprompted and unsolicited employee actions). A total of 12 specific and detailed subthemes that sketch a wide terrain of possible service problems was developed and constitutes the heart of the findings. Bitner *et al.* (1990) pointed to possible uses of the resulting classifications both in further research and in managerial practice, such as in developing customer satisfaction monitoring programs, designing and improving service procedures and policies and training contact personnel to reduce service failures.

Observation-Driven Qualitative Research Methods

Theory of Observation-Driven Qualitative Research

Perhaps you remember the following quote:

> In conversation one day Sherlock Holmes asked Watson how many steps there were to the Baker Street apartment. Watson responded that he did not know. Holmes replied, 'Ah, Watson, you *see* but you do not *observe*'.

Table 22.4 What can be observed?

Category	Illustrative example
Human actions	Workers' movement patterns in factory
Verbal behaviour	Statements made by airline travellers waiting in line
Expressive behaviour	Facial expressions and other forms of body language
Spatial relations	Proximity of middle managers' offices to the president's office
Temporal patterns	Time needed to perform a given task on the factory floor or behind the service counter
Physical objects	Modes of use of a given device or facility
Verbal records	Hidden messages in the address by the Prime Minister
Pictorial records	Pictorial evidence of a given person's network of acquaintances

The difference between seeing and observing is further emphasized by the question "When is an observation scientific?" The answer seems simple. An observation is scientific if it:

- serves a formulated research purpose;
- is planned systematically;
- is recorded systematically; and
- is subject to checks on reliability and validity.

Table 22.4 summarizes categories which can be observed along with illustrative examples.
    Observation processes differ in the following basic characteristics:

- **Visible versus hidden** – it has to be pointed out that the known presence of an observer may alter the behaviour of the observed persons.
- **Structured versus unstructured** – in the first case the observer is told what to specifically focus on, and in the second he or she is free to decide what to record and report.
- **In a natural versus in an artificial setting** – for example, in a real versus test market.
- **Human versus mechanical observation** – human observers introduce their particular form of bias into the research process, but it also has to be pointed out that people will always find ways of cheating a machine – be it a POS terminal, a telemetric device or something else.

When carried out during a longer period of time within a specific community, a simple observation study turns into an *ethnographic study* (if carried out in a real-life setting, e.g. in the home, at school, at work or on the go) or a *netnographic study* (if carried out in a virtual setting, observing what has been going on discussion forums, in blogs, in chat rooms, in different types of social media, in newsrooms etc.).

    Both types of studies can lead to various forms of results, such as taxonomies, mappings or descriptions of themes and meanings ascribed to events, objects or social processes. They both revolve around the idea that the events which happen in people's lives cannot be understood without taking into account the context in which people live and function and the cultural meaning they ascribe to social interactions, symbols and the like. An introduction to ethnographic research is a standard fixture in books on social research methods (e.g. Somekh and

Lewin, 2011; Bryman, 2012). Kozinets (2002) is a useful starting point for those interested in carrying out a netnographic study of their own.

Observation-Driven Qualitative Research in Practice

This section uses four examples from research practice to demonstrate how a specific information need leads to the application of observation-driven qualitative research, and how conclusions then go beyond what would have been possible through the exclusive use of quantitative research.

Example 22.1

Ben-Zvi (2003) was interested in understanding how learners construct meaning of visual distributions and of variability within and between distributions, and to that effect studied junior high school students working with their teacher and peers in a computerized environment. Pairs of students were observed within their classroom during an extended period of engagement with a curriculum-based data investigation. The whole process was videotaped, and a qualitative analysis took into account all actions, discussions, computer interactions and gestures within the context in which they occurred. The research goal was to trace the emergence of beginners' reasoning about variation, including the development of cognitive structures and the sociocultural processes of understanding and learning. The analysis showed that most of the learning took place through dialogues between the students themselves. The video analysis also pointed out that the teacher's interventions at the students' request had important catalytic effects.

Example 22.2

Treisman (1992) wanted to understand the causes for a massive failure of minority students, in particular African-American students, taking mathematics and science at the University of California, Berkeley. His team initially conducted a quantitative survey of several thousand Berkeley students regarding possible causes for failures of such students in calculus classes. Responses pointed to four key possibilities: that black students fail due to having lower motivation, inadequate preparation, lack of family support and lower family incomes, compared to other groups. However, analysis of interviews with minority students and their families refuted all hypotheses, and left the researchers without a good explanation for the observed phenomena of class failure and dropout.

Treisman, a trained mathematician, then initiated an ethnographic study. He and his team moved in to live with students at their apartments, and observed differences in how groups of 20 Chinese and 20 African-American calculus students studied and solved class assignments outside the classroom. The analysis of videos and interviews showed, to the researchers' amazement, that the Chinese students tended to learn from each other in group sessions that they organized for mutual support and development of shared understanding, while the African-American students frequently worked by themselves. The differences in learning styles and social habits could not be predicted based on quantitative or survey data regarding students' individual backgrounds. The findings led Treisman to re-evaluate existing notions about learning processes of graduate students and to design a group-oriented workshop course at Berkeley that produced dramatic positive results in improving the performance of black students.

Example 22.3

Perčič (2009) carried out an ethnographic study of the effects of so-called *shopping diets* (an anti-consumerism trend whose adherents aim to reduce shopping by (re)discovering the joys of their own food production as well as barter) based on a netnographic analysis of shopping dieters' web posts dealing with their positive (e.g. extended real-life social network, raised environmental consciousness, raised appreciation for the small joys of life as well as huge monetary savings) and negative experiences (e.g. huge time consumption). Many of the participants ultimately embraced the self-imposed experiment as a sustainable way of living; those who did not go all the way still felt better about themselves because they positively contributed towards the preservation of the planet.

Example 22.4

Attempting to characterize workplace skills, Noss *et al.* (1999) studied how nurses use knowledge of 'average' in their work. They showed that mathematical and statistical meanings of average are not class-based mathematical abstractions, but are integrated with contextual meanings within the particular purposes for which averages are used by nurses, such as when monitoring the blood pressure of critically ill patients. These findings could never have been discovered in a quantitative research setting.

Software Support of Qualitative Research

Qualitative research is supported by several software packages. The best known among them is NVivo, produced by QSR International (http://www.qsrinternational.com/product). In addition to data formats such as audio files, videos, digital photos, Word, PDFs, spreadsheets, rich text, plain text and web content, NVivo also supports the capture of social media data (e.g. Facebook, Twitter and YouTube posts).

NVivo is based on the grounded theory method which builds on an inductive as opposed to deductive research approach. As we know, it makes sense to apply it when dealing with poorly researched phenomena. Instead of beginning with hypotheses formulation and then proceeding with data collection, we begin with data collection. Regardless of the form and format, the collected data are then stored in the NVivo environment.

In the next step, audio files, videos, digital photos, Word, PDFs, spreadsheets, rich text, plain text, web content and social media data are carefully checked and marked with a series of codes which the researcher creates 'on the go', adding, deleting and changing the codes when repeatedly studying a certain NVivo entry.

The codes are subsequently grouped into similar concepts, and from the concepts the categories are formed. They represent the basis for theory creation. In other words, the theory is *induced* from the empirical evidence.

The latest versions of NVivo come with an in-built comprehensive sample project which facilitates quick development of user skills and is further supported by an excellent series of tutorials (see http://www.qsrinternational.com/support_tutorials.aspx).

Reliability and Validity of Qualitative Research

The literature on qualitative research (e.g., Strauss and Corbin, 1998; Klein and Myers, 1999) emphasizes that researchers should use various processes to establish or improve the reliability and validity of the analyses and interpretations of the information collected through qualitative channels.

For example, whenever possible and justified by the research question, two or more separate coders should conduct systematic content analysis, so that inter-rater agreement could be measured and reported.

Similarly, when carrying out observation studies, two or more observers should be deployed simultaneously in order to account for gender- and age-specific points of view. All other observer characteristics which might play an important role in the observation process (e.g. education, occupation, nationality and religion) should also be acknowledged and the observer contingent shaped accordingly.

Various triangulation techniques may also be used to improve the soundness of the emerging conclusions. Noss *et al.* (2007) were involved in a large project on techno-mathematical literacies in the workplace in which a mix of ethnographic methods, observations, worker interviews and other qualitative methods were used. In describing the methods used in their study, which examined workers' use and interpretations of production graphs in a manufacturing setting, they noted:

> Our general method ... is to organise feedback and validation meetings for the purposes of triangulation. In feedback meetings, which are company specific, we present our findings to a group of managers in the company and discuss how far they resonate with their own experience. In broader validation meetings, we present our conclusions to managers from different companies within the sector in order to draw out similarities and differences. (p. 370)

This quote illustrates the need for, and importance of, synthesizing the results of qualitative research. Gal and Ograjenšek (2010) point out that a lot of qualitative research is being conducted as part of small-scale studies by different researchers who are disconnected from each other. Collaboration and partnership of researchers in the qualitative arena – for example, by comparing and jointly evaluating their research results (the idea of meta-analysis) – should be given more emphasis in the future. This in turn could broaden the scope of inference for the results of small qualitative studies.

References

Ben-Zvi, D. (2003). Reasoning about variability in comparing distributions. *Statistics Education Research Journal* 3(2): 42–63. http://www.stat.auckland.ac.nz/~iase/serj/SERJ3(2)_BenZvi.pdf

Bitner, M.J., Booms, B.H., and Tetreault, M.S. (1990). The service encounter: diagnosing favorable and unfavorable incidents. *Journal of Marketing* 54: 71–84.

Bryman, A. (2012). *Social Research Methods*, 4th ed. Oxford: Oxford University Press.

Frankfort-Nachmias, C., and Nachmias, D. (2000). *Research Methods in the Social Sciences*. New York: Worth.

Gal, I., and Ograjenšek, I. (2010). Qualitative research in the service of understanding learners and users of statistics. *International Statistical Review* 78(2): 287–296.

Graebner, M.E., and Eisenhardt, K.M. (2004). The seller's side of the story: acquisition as courtship and governance as syndicate in entrepreneurial firms. *Administrative Science Quarterly* 49: 366–403.

Gremler, D.D. (2004). The critical incident technique in service research. *Journal of Service Research* 7: 65–89.

Klein, H.K., and Myers, M.D. (1999). A set of principles for conducting and evaluating interpretive field studies in information systems. *MIS Quarterly* 23: 67–88.

Kozinets, R.V. (2002). The field behind the screen: using netnography for marketing research in online communities. *Journal of Marketing Research* 39: 61–72.

Marchionini, G., and Levi, M.D. (2003). Digital government information services: the Bureau of Labor statistics case. *Interactions* 10(4): 18–27.

Noss, R., Bakker, A., Hoyles, C., and Kent, P. (2007). Situating graphs as workplace knowledge. *Educational Studies in Mathematics* 65: 367–384.

Noss, R., Pozzi, S., and Hoyles, C. (1999). Touching epistemologies: meanings of average and variation in nursing practice. *Educational Studies in Mathematics* 40: 25–51.

Ograjenšek, I. (2007). SERVQUAL surveys. In Ruggeri, F., Kenett, R.S., and Faltin, F.W. (ed.), *Encyclopedia of Statistics in Quality and Reliability*. Chichester: Wiley, pp. 1797–1801.

Perčič, E. (2009). *Time as Value: Possible Advantages of Consumption Reduction and Change*. Master's thesis, University of Ljubljana, Faculty of Social Sciences. (Published in Slovenian under the title 'Čas kot vrednost: možne prednosti zmanjševanja in spreminjanja potrošnje')

Somekh, B., and Lewin, C. (2011). *Theory and Methods in Social Research*, 2nd ed. London: Sage.

Strauss, A., and Corbin, J. (1998). *Basics of Qualitative Research: Techniques and Procedures for Developing Grounded Theory*. Thousand Oaks, CA: Sage.

Treisman, U. (1992). Studying students study calculus: A look at the lives of minority mathematics students in college. *College Math Journal* 23: 362-372.

Zeithaml, V.A., Parasuraman, A., and Berry, L.L. (1990). *Delivering service quality*. New York: Free Press.

23

Kansei Engineering

Lluis Marco-Almagro

An Introduction to Kansei Engineering

Users of products and services – all of us – are becoming more and more demanding. In this beginning of the twenty-first century, we not only want products that work well and satisfy our needs, but also want products that we like. When customers are questioned on what they want, a list of needs normally referring to *functionality* is obtained. Designers and engineers can translate this voice of the customer into technical parameters, so that the product fulfils those needs (this is often done using *quality function deployment* (QFD); a simple online search gives many results with information about this tool). However, customers do not usually explain their emotional needs, probably because they are not aware of having them or are unable to tell which they are. Even when those emotional needs are discovered, it is not obvious which technical properties of the product will elicit those desired emotions.

For many years, designers were not very interested in emotions (Norman, 2004). The focus was more on making usable products. But making products that work well and fulfil user expectations is not enough. When analysing the products we normally use, we realize that we love some gadgets that are far from perfect but that we just like. Or we have a deep appreciation for a product because of the person who gave it to us as a present, or because it reminds us of the good times we had using it. These emotional aspects attached to products cannot be disregarded. The massive success of some 'emotional products' in recent decades, such as Apple's iPod (first launched in 2001), confirms this tendency.

How do designers incorporate this 'emotional touch' when creating a new product? They usually rely on their intuition, creativity and experience. But they also use different qualitative and quantitative methods to collect information on how products are perceived and used. Several of these methods can be grouped under the umbrella term *emotional design*. As emotions elicited by products are difficult to measure directly, emotional design relies basically on self-reported techniques (users explain their first impression on a product or what they feel when using it). On the qualitative side, focus groups and interviews are quite common, not only

Research Methods for Postgraduates, Third Edition. Edited by Tony Greenfield with Sue Greener.
© 2016 John Wiley & Sons, Ltd. Published 2016 by John Wiley & Sons, Ltd.

in emotional design but also in market research in general. Some other qualitative methods are more specific to industrial design: experience diaries (issuing notebooks to users so that they can make notes of their experiences with a product over a period of time) and field observations (watching people in the environment in which they would normally experience a product). Jordan (2002) and Martin and Hanington (2012) contain descriptions of many of these qualitative methods.

Although qualitative methods can give rich information on the emotions elicited by a product, they have several difficulties:

- Results from qualitative approaches depend a lot on the person leading the focus group or performing the interview.
- Qualitative approaches, especially interviews, require a lot of time. So usually only a few interviews are performed and conclusions are derived from asking a small amount of people.
- It can be difficult to obtain product design guidelines due to the fact that users are not typically thinking in a designer's paradigm.

Quantitative approaches such as questionnaires overcome these difficulties (although they pose some others, the main one probably being the fact that they are necessarily reductionist).

Kansei engineering (KE) is a quantitative method used in emotional design and usually based on the use of questionnaires. The Japanese word *kansei* means sensitivity or sensibility. Simon Schütte (2005: p. 36) proposes the following explanation of kansei: 'Kansei is an individual's subjective impression from a certain artifact, environment or situation using all the senses of sight, hearing, feeling, smell, taste and the sense of balance as well as recognition'. For example, a beautiful photograph of a landscape full of snow can elicit the kansei 'calmness'. In a KE study, the main purpose is discovering which technical parameters of a product convey the chosen emotions. Imagine we are studying watches. We want to discover, say, that if the face of a watch is rectangular, the watch is perceived as being elegant. Or that analogue displays are perceived as classical, whereas digital displays are perceived as modern. In this way, we not only get information about the specific watches used in the study, but also discover general rules that are valid even for watches not yet created. Therefore, it is a method for incorporating emotions in the product development phase.

In KE, there is an attempt to describe the whole range of emotions a product can convey. We are not modelling a unique response (such as the elements that make people prefer one watch over the others) but several responses (such as the elements that provoke people to perceive a watch as being modern, elegant, reliable and so on). There are as many responses as necessary concepts to cover the whole range of expected emotions.

The method was first proposed by researcher Mitsuo Nagamachi in the 1970s. He had a background in psychology and medicine and was working at that time in Hiroshima University's Faculty of Engineering (Childs *et al.*, 2003). The term *kansei engineering* was first used in 1986 by Kenichi Yamamoto, then the president of Mazda Motor (Schütte, 2005). Professor Nagamachi soon adopted this term. Since then, KE has been successfully used in several sectors: automotive, apparel, electronic home products and packaging, among others, initially in Japan and Korea. A paper by Mitsuo Nagamachi (1995) published in the journal *Applied Ergonomics* can be considered as a seminal paper where Professor Nagamachi presents his proposal to the world scientific community.

A Model for Kansei Engineering Studies

KE is clearly a multidisciplinary field, where scholars with different backgrounds can provide ideas and improve methods: engineers, designers, psychologists, sociologists, marketing experts and so on. But perhaps one of the most distinctive characteristics of KE is that it is based on collecting quantitative data (usually ratings made by users), as opposed to other qualitative methods used in emotional design. Once data are collected, statistical methods are commonly used to link the physical properties to the elicited perceptions.

I consider the paper 'Concepts, methods and tools in kansei engineering' (Schütte *et al.*, 2004) a key paper in the effort to translate the ideas and procedures of KE to the Western world.

This chapter describes the steps needed for conducting a KE study. This proposed model systematizes the phases that comprise a KE study, so it brings clarity to the procedure. Figure 23.1 shows a slightly modified version of the original model, taken from the proposal by Marco-Almagro (2011).

I will briefly describe each step of the model here.

Choice of Domain

Choosing the domain obviously includes deciding which product is the protagonist of the study. But this is not the only task. In addition to deciding on the product, the choice of

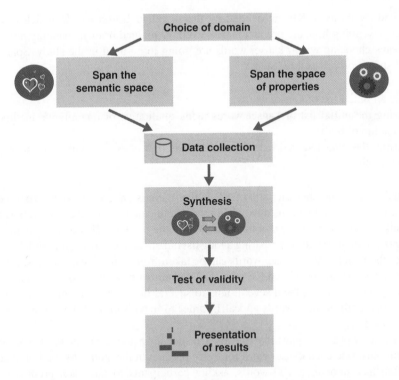

Figure 23.1   A model for developing kansei engineering studies.

domain also requires:

- Defining the target group to which the product is addressed (Schütte *et al.*, 2004). Participants in the study have to match the characteristics of the target group (so they are basically a sample from the whole population represented by the target group). Usually, a fairly homogeneous group of people (from the point of view of personal and socioeconomic characteristics) is chosen.
- Defining the kind of presentation for the product. There are basically two possibilities: showing real physical products (either prototypes or fully functional products) or showing representations of the product (2D representations, like photographs, or 3D representations, perhaps using a virtual reality environment).
- Defining the context for presentation. The atmosphere of the place where the experiment is conducted can have an effect on the emotions elicited by the product. The same product can convey different emotions depending on the environment. The usual recommendation is conducting the data collection in an atmosphere as neutral as possible.

For example, the product domain could be analogue watches to be used by middle-aged women.

Spanning the Semantic Space

The elicited emotions in KE are defined by the so-called *kansei words*, which usually are adjectives describing how the product is perceived. Explained briefly, spanning the semantic space means choosing which kansei words are going to be used in the study. Spanning the semantic space comprises three steps:

1. Setting an initial list of kansei words.
2. Reducing the initial list of kansei words using qualitative or quantitative methods (or a combination of both).
3. Proposing the final reduced list of kansei words (the ones that will be used in the data collection phase).

As I stated before, one characteristic of KE is that there is an attempt to describe the whole range of emotions a product can convey. This endeavour can only be successful if the semantic space really covers all possible emotions elicited by the product. This is why the first step in spanning the semantic space is preparing a long list of kansei words using all available sources: the longer the better. These kansei words can be taken from magazines, manuals, catalogues, websites, users, experts or related KE studies. The number of kansei words on this initial list can be very large, varying from several tens to several hundreds. This first word collection must continue until no new words appear in order to make sure that the complete semantic space is considered.

Each kansei word will play the role of a response in the experiment. Therefore, each participant must rate each kansei word for every product in the study. As the initial semantic space often has more than 100 words, asking participants to rate each product on every word from the initial semantic space is totally unfeasible. So there must be a selection of kansei words using data reduction methods. Any reduction implies a loss of information,

Figure 23.2   Affinity diagrams are often used for spanning the semantic space in kansei engineering studies.

but the idea is removing words that express very similar emotions and are thus essentially redundant.

This reduction of kansei words can be performed using an affinity diagram: kansei words with similar meanings are grouped together (Figure 23.2). For each group, one of the kansei words will be chosen as representative of that group. As the number of words in the initial semantic space can be huge, this grouping can be done in several steps. Imagine we have 300 words in the beginning. An affinity diagram can collapse these words into 40 groups, and a word is selected representing each of these 40 groups. In a second affinity diagram, the 40 words are grouped again, and 16 final groups are created, thus producing a final set of 16 words.

Multivariate methods can also help in this word reduction process, but obviously a preliminary data collection is then needed. The common procedure is choosing a small set of products, and asking some people to rate the kansei words using the same scale planned for the data collection phase. Once we have the data, they are summarized by calculating averages among all raters, so we finally have a number for each product and kansei word. Words are then grouped using a cluster analysis: words that are close to other words – in the sense that they have a similar profile of ratings through all products, and thus are perceived as similar – will be part of the same group.

Kansei words for the watches example could be *modern*, *elegant* and *comfortable*.

Spanning the Space of Properties

The space of properties lists the physical properties of the product that can have an effect on the elicited kansei. Spanning the space of properties comprises the following steps:

1. Making a list of all possible physical product properties, and selecting the ones that apparently have the largest impact on the users' kansei.
2. Preparing the design matrix, which is a matrix saying how many products will be used for the experiment and establishing the value of each property for each one of the products.
3. Selecting the products (or producing product prototypes) according to the design matrix.

Spanning the space of properties in KE is similar to choosing factors in an industrial experiment (see Chapter 19 of this book). If the domain of interest is watches, physical properties that are under producers' control are, for example, the shape of the watch or colour of the face. These two properties could be factors in a KE study. The factor shape of the watch could have two levels: round shape and rectangular shape. The factor colour of the face could have more than two levels: golden colour, white colour, light blue colour and so on.

How many factors must be selected? The answer should be as many factors as necessary to describe the whole space of properties of the product. In practice, the number of factors we can choose is somehow determined by the number of products that will be rated by participants in the KE study. A good recommendation is defining the design matrix using factorial experiments (Chapter 19 has an introduction to two-level factorial experiments). Doing this, we ensure that the set of stimuli to be evaluated is suitable to be analysed with a regression analysis.

Data Collection

The data collection phase is one of the most important ones (if not the most important) in a KE study. The principle of 'garbage in, garbage out' applies here: if our raw data are not reliable, the conclusions from our study will be poor or even erroneous, no matter the sophistication of the statistical analysis performed later.

The common procedure for collecting data for a KE study is the following: each participant in the study is presented with a product, and he or she then rates this product on all the kansei words. Ideally, both products and kansei words are shown randomly to each participant. Ratings can be done either on semantic differential scales or Likert scales. Semantic differential scales (Osgood *et al.*, 1957) show a statement with two opposite words, one on the left and another on the right (Figure 23.3, top). The commonly used Likert scales show a statement that the respondent is asked to evaluate according to the level of agreement or disagreement (Figure 23.3, bottom). Both Likert scales and semantic differential scales are used in KE studies, although Likert scales are more common.

KE studies usually employ 5-point or 7-point scales. There are many papers comparing the behaviour of different scales (e.g. Dawes, 2008; Lawless *et al.*, 2010), but all of them agree on the following main conclusions:

- Reliability and validity are good in scales that have at least five response categories (3-point scales allow too little differentiation).

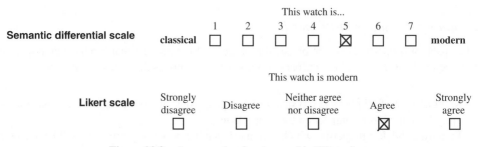

Figure 23.3   An example of scales used in KE studies.

- Results are basically the same regardless of the number of response categories used when moving from 5-point scales to 9-point scales.
- Scales with more than nine points do not seem to be needed, as they do not add more discernibility.

Keep in mind that the principles in designing questions for questionnaires stated in Chapter 21 of this book on survey research also apply here.

The location where the data collection takes place is important, as explained in the choice of domain phase. The time needed to complete the study must also be taken into account. The rule here is simple: the shorter the better. However, this of course depends on the number of kansei words and stimuli used in the study.

At the end of the data collection phase, we have a rating for each stimuli and each kansei word given by each participant in the study. Usually, these data are collapsed and the average for all participants on each stimulus and kansei word is later used.

Synthesis

Once the data collection is completed, data must be analysed. The synthesis phase is the core of KE, as the link between emotions and physical properties of products is made here. For every kansei word, product properties are found that affect the kansei word. Using again the watches example, one conclusion could be that watches with a rectangular face are perceived as elegant.

In statistics, regression analysis – in its broader sense – can be used to model the relationship between a dependent variable and one or more independent variables. In a KE study, each kansei word from the semantic space acts as a response (dependent variable), whereas each factor from the space of properties is an independent variable. Chapter 19 of this book already stated that a least-squares regression analysis is widely used for analysis of factorial experiments. This is also the case here: the aim is estimating (at least) the main effects and (if possible) second-order interactions.

Two facts almost always characterize KE datasets:

1. Independent variables are categorical factors, having two or more levels.
2. The response is discrete, usually integers from an ordinal scale (ranging from 1 to 7 or from 1 to 5).

As the response is discrete, a multiple linear regression cannot be directly used for modelling the data, as a continuous response is needed. The most common way of solving this issue is working with the mean of all participants' ratings (another possibility is directly using the ordinal response with an ordinal logistic regression). As independent variables are all categorical, dummy variables must be used. A commonly used approach in KE studies is analysing the data using a method called *quantification theory type I* (QT1). The method was first proposed by Chikio Hayashi (1952). QT1 is simply a multiple regression model where all independent variables are categorical. The method allows the inclusion of all levels of each categorical variable in the model, avoiding the need to have reference levels (which converts the model into something more difficult to interpret by non-statisticians). The procedure, although without naming it QT1, is explained in detail in Sweeney (1972).

You can find more information on linear regression in Chapter 29 of this book. All statistical packages – such as Minitab, JMP or SPSS – are able to perform linear regression (even Excel can be used for that). These statistical applications usually have a good help facility, with explanations that go straight to the point and examples that can be reproduced with the software. I think these help facilities in statistical software are a highly recommended starting point for learning about statistical methods.

Test of Validity and Presentation of Results

Once the synthesis phase is completed, significant factors are revealed and their effect on each of the kansei words is discovered. Some confirmatory experiments must be run in the test of validity phase to confirm the results from the synthesis phase. In these confirmatory experiments, we do not try to discover which factors affect each kansei word (this has already been done), but to explicitly validate our discoveries. If all confirmatory experiments are successful, conclusions must be compiled to prepare the presentation of results step. If some confirmatory experiments do not provide the expected result, previous phases of the KE study must be revisited.

The last phase in the model for conducting KE studies is the presentation of results. As users of KE are often industrial designers, marketing experts and technicians not necessarily used to statistical terms, clear explanations coming from the statistical analysis and visual outputs are very convenient. Descriptive graphs that are easy to understand for everybody with a bit of effort, and that directly translate numerical data into useful information, are especially appealing. For example, radar plots and profiles for kansei words can be used to check similarities among words.

Many KE studies perform a principal component analysis (PCA) of the kansei words, so that the kansei words are represented in the first two principal components, giving a visual representation of the semantic space. PCA is a commonly used multivariate technique that helps to understand the underlying data structure. The objective is to take p variables $X_1, X_2, \ldots,$ X_p and find combinations of these to produce new variables (called *principal components*) $Z_1, Z_2, \ldots, Z_p$. These principal components are uncorrelated (meaning that each one captures different dimensions of the data) and have decreasing variances (so that probably only the first principal components are enough to capture an important amount of the variability in the dataset). The usefulness of PCA in KE will hopefully become apparent in the example explained in the 'An Example Using Printed Paper Sheets' section.

A good introduction to multivariate methods such as PCA (mentioned in the presentation of results phase) and cluster analysis (mentioned in the spanning the semantic space phase) can be found in Manly (2004). This book briefly describes the mathematical procedures to perform these techniques, but focuses on the concepts and explains how to interpret results using well-selected examples. As I stated before, I also recommend reading the explanations about multivariate techniques that can be found in the Help function of popular statistical packages.

An important output from a KE study is the list of rules that explain how each property of the product (each factor) affects each kansei word. This can help in deciding the most appropriate property settings for eliciting the desired perception.

An Example Using Printed Paper Sheets

I want to conclude this short overview of KE with an example. The aim of the example is twofold: clarify the procedure and encourage the reader to formulate his or her own simple KE study. Although KE is commonly used with industrial products, this is obviously not compulsory. In fact, I think many research opportunities come from trying to apply this procedure to non-manufacturing products or services. So the choice of domain for this example is printed paper sheets.

The purpose is discovering how several typographic properties affect the perception the reader has when faced with a printed paper sheet. The conclusions may be interesting for those producing academic material or industrial reports. I conducted the study for the sake of curiosity. Participants were university teachers, students and industry technicians used to reading technical reports.

Two students helped in the process of spanning the semantic space. A long list of kansei words (a bit more than 100) related with typography was produced. The list came from searching for adjectives in specialized websites (basically blogs devoted to typography such as ilovetypography.com or typographica.org). An affinity diagram was used to reduce the initial list of kansei words to the following eight final words: *elegant, formal, modern, clear, traditional, dense, serious* and *funny*.

Three factors were used to define the space of properties: font used, spacing between lines and having justified paragraphs or not. Spacing between lines was set to single and 1.5 lines. Font used was set to Times New Roman 12 (a *serif* font, and the default one in Microsoft Office until version 2003) or to Calibri 11 (a *sans serif* font, and the default one in Microsoft Office from version 2007). As each factor had two levels, it was possible to arrange a set of prototypes according to a 2^3 factorial design, so having all possible combinations (Table 23.1). It was easy to prepare prototypes of paper sheets according to the matrix shown in Table 23.1. On the right, there is an excerpt of each of the stimuli used.

Fourteen people rated each of the eight paper sheets on each of the eight kansei words using 7-point ordinal scales. Stimuli (paper sheets) were randomized for each participant. The results (averaged by participants) are shown in Table 23.2.

As a 2^3 factorial design was used for this experiment, common techniques for calculating main effects and interactions in factorial experiments can be used here (see Chapter 19 for more information). In fact, simple descriptive graphs such as the radar plot in Figure 23.4 (drawn with Excel) can be used to extract conclusions. Looking at this graph, one can see that stimuli 2, 4, 6 and 8 get high ratings (around 5), whereas stimuli 1, 3, 5 and 7 have much lower ratings (around 3.5). Checking Table 23.1 with the whole list of stimuli, we realize that stimuli 2, 4, 6 and 8 had the text justified, contrary to stimuli 1, 3, 5 and 7. So one conclusion is that when paragraphs are justified, paper sheets are perceived as more elegant, formal and serious.

Other conclusions that can be extracted after analysing the data are:

- Paragraphs written with Calibri 11 are perceived as more modern than paragraphs written with Times New Roman 12.
- The perception of density is affected by line spacing: single is considered denser.
- The perception of clarity is also affected by line spacing: 1.5 lines is considered clearer. But it is also affected by justification: justified paragraphs increase the perception of clarity.

Table 23.1 The design matrix for the typography example

	Justified	Spacing	Font	
1	No	Single	Calibri 11	Lorem ipsum dolor sit amet, consectetuer adipiscing elit. Phasellus eget mi at mi vehicula condimentum. Phasellus eros justo, rutrum ut, viverra eu, eleifend vitae, ligula. Morbi interdum. Vestibulum odio velit, ullamcorper ut, dapibus nec, egestas nec, augue. Nam luctus tortor vel massa venenatis nonummy. Nunc purus eros, fermentum ac, convallis eu,
2	Yes	Single	Calibri 11	Lorem ipsum dolor sit amet, consectetuer adipiscing elit. Phasellus eget mi at mi vehicula condimentum. Phasellus eros justo, rutrum ut, viverra eu, eleifend vitae, ligula. Morbi interdum. Vestibulum odio velit, ullamcorper ut, dapibus nec, egestas nec, augue. Nam luctus tortor vel massa venenatis nonummy. Nunc purus eros, fermentum ac, convallis eu,
3	No	1.5 lines	Calibri 11	Lorem ipsum dolor sit amet, consectetuer adipiscing elit. Phasellus eget mi at mi vehicula condimentum. Phasellus eros justo, rutrum ut, viverra eu, eleifend vitae, ligula. Morbi interdum. Vestibulum odio velit, ullamcorper ut, dapibus nec, egestas nec, augue. Nam luctus tortor vel massa venenatis nonummy. Nunc purus eros, fermentum ac, convallis eu,
4	Yes	1.5 lines	Calibri 11	Lorem ipsum dolor sit amet, consectetuer adipiscing elit. Phasellus eget mi at mi vehicula condimentum. Phasellus eros justo, rutrum ut, viverra eu, eleifend vitae, ligula. Morbi interdum. Vestibulum odio velit, ullamcorper ut, dapibus nec, egestas nec, augue. Nam luctus tortor vel massa venenatis nonummy. Nunc purus eros, fermentum ac, convallis eu,
5	No	Single	Times New Roman 12	Lorem ipsum dolor sit amet, consectetuer adipiscing elit. Phasellus eget mi at mi vehicula condimentum. Phasellus eros justo, rutrum ut, viverra eu, eleifend vitae, ligula. Morbi interdum. Vestibulum odio velit, ullamcorper ut, dapibus nec, egestas nec, augue. Nam luctus tortor vel massa venenatis nonummy. Nunc purus eros, fermentum ac,
6	Yes	Single	Times New Roman 12	Lorem ipsum dolor sit amet, consectetuer adipiscing elit. Phasellus eget mi at mi vehicula condimentum. Phasellus eros justo, rutrum ut, viverra eu, eleifend vitae, ligula. Morbi interdum. Vestibulum odio velit, ullamcorper ut, dapibus nec, egestas nec, augue. Nam luctus tortor vel massa venenatis nonummy. Nunc purus eros, fermentum ac,
7	No	1.5 lines	Times New Roman 12	Lorem ipsum dolor sit amet, consectetuer adipiscing elit. Phasellus eget mi at mi vehicula condimentum. Phasellus eros justo, rutrum ut, viverra eu, eleifend vitae, ligula. Morbi interdum. Vestibulum odio velit, ullamcorper ut, dapibus nec, egestas nec, augue. Nam luctus tortor vel massa venenatis nonummy. Nunc purus eros, fermentum ac,
8	Yes	1.5 lines	Times New Roman 12	Lorem ipsum dolor sit amet, consectetuer adipiscing elit. Phasellus eget mi at mi vehicula condimentum. Phasellus eros justo, rutrum ut, viverra eu, eleifend vitae, ligula. Morbi interdum. Vestibulum odio velit, ullamcorper ut, dapibus nec, egestas nec, augue. Nam luctus tortor vel massa venenatis nonummy. Nunc purus eros, fermentum ac,

Table 23.2   Results for the typography example (averaged by participants)

Stimuli	Elegant	Formal	Modern	Clear	Traditional	Dense	Serious	Funny
1	2.857	3.643	3.286	3.786	4.357	5.000	4.429	2.786
2	4.500	4.714	3.615	4.286	4.571	4.357	5.429	2.286
3	2.857	3.286	3.500	4.071	4.143	4.286	3.643	2.571
4	5.143	4.857	4.214	5.429	4.143	3.214	4.500	2.786
5	2.571	3.143	2.643	3.500	4.571	5.143	3.714	2.429
6	4.786	5.214	3.143	4.214	5.143	5.214	4.929	2.143
7	3.357	3.429	3.143	4.714	4.071	3.857	3.929	2.429
8	5.071	5.071	3.143	5.214	4.786	3.357	5.071	2.143

Figure 23.5 shows the kansei words in the two first principal components of a PCA (computed and graphically represented with the statistical software Minitab). This is sometimes called the representation of the semantic space. Recall that serious, formal and elegant are very close one to the other (meaning that when a paper sheet is perceived as elegant, it is also perceived as serious and formal). On the contrary, dense and clear are opposed (meaning that when a paper sheet is perceived as dense, it is not perceived as clear).

You can use these conclusions for deciding the presentation of your own reports! And, of course, you can conduct similar studies to discover users' perceptions on issues of your interest.

Data from KE studies usually allow rich analysis. In our example, we could also analyse the results stratifying in different groups (e.g. university teachers and industrial technicians).

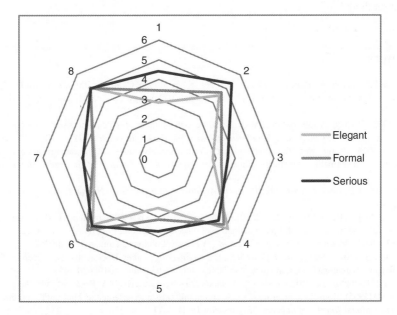

Figure 23.4   Radar plot for kansei words *elegant*, *formal* and *serious*.

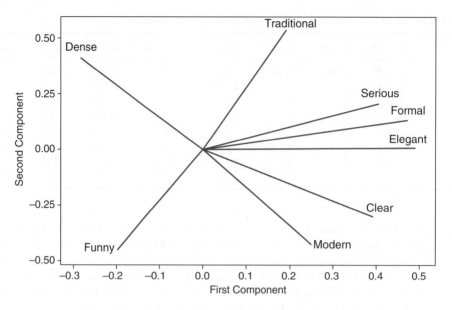

Figure 23.5   The semantic space in the typography example.

Many enhancements can be suggested for KE data analysis: using more complex models for explaining each Kansei word, detecting outliers in multidimensional datasets such as those coming from KE studies, performing a participants' segmentation according to their ratings and so on (Marco-Almagro, 2011). Kansei engineering is a nice area to test and apply many statistical methods!

References

Childs, T., De Pennington, A., Rait, J., Robbins, T., Jones, K., Workman, C., *et al.* (2003). *Affective Design (Kansei Engineering) in Japan.* Report from a DTI International Technology Service Mission. Leeds: University of Leeds.

Dawes, J. (2008). Do data characteristics change according to the number of scale points used? An experiment using 5-point, 7-point and 10-point scales. *International Journal of Market Research* 50(1): 61–77.

Hayashi, C. (1952). On the prediction of phenomena from qualitative data and the quantification of qualitative data from the mathematico-statistical point of view. *Annals of the Institute of Statistical Mathematics* 3(2): 69–98.

Jordan, P.W. (2002). *Designing Pleasurable Products: An Introduction to the New Human Factors.* London: Chapman & Hall/CRC.

Lawless, H.T., Popper, R., and Kroll, B.J. (2010). A comparison of the labeled magnitude (LAM) scale, an 11-point category scale and the traditional 9-point hedonic scale. *Food Quality and Preference* 21(1): 4–12.

Manly, B.F.J. (2004). *Multivariate Statistical Methods: A Primer.* London: Chapman & Hall/CRC.

Marco-Almagro, L. (2011). *Statistical Methods in Kansei Engineering Studies.* Doctoral thesis, Universitat Politecnica de Catalunya – BarcelonaTech, Barcelona. Available at http://hdl.handle.net/10803/85059

Martin, B., and Hanington, B. (2012). *Universal Methods of Design.* Beverly, MA: Rockport Publishers.

Nagamachi, M. (1995). Kansei engineering – a new ergonomic consumer-oriented technology for product development. *International Journal of Industrial Ergonomics* 15(1): 3–11.

Norman, D.A. (2004). *Emotional Design: Why We Love (or Hate) Everyday Things.* New York: Basic Books.

Osgood, C.E., Suci, G.J., and Tannenbaum, P.H. (1957). *The Measurement of Meaning*. Urbana: University of Illinois Press.

Schütte, S. (2005). *Engineering Emotional Values in Product Design: Kansei Engineering in Development*. Doctoral thesis, Linköping University, Linköping.

Schütte, S., Eklund, J., Axelsson, J., and Nagamachi, M. (2004). Concepts, methods and tools in kansei engineering. *Theoretical Issues in Ergonomics Science* 5(3): 214–231.

Sweeney, R.E. (1972). A transformation for simplifying the interpretation of coefficients of binary variables in regression analysis. *The American Statistician* 26(5): 30–32.

24

Principles of Sampling

Peter Lynn

Introduction

In much research, it is necessary to *sample* units for study. Examples are:

- A geographer who wishes to estimate the prevalence of a certain plant in a field will not study the entire field but will divide the field into small areas and sample some of those areas.
- A sociologist who wants to ascertain the proportion of the population who have experienced a certain event will not interview the whole population but will select a sample.
- In everyday life, you may:
  - taste your cooking to see if a dish is ready; and
  - sample cheese before you buy it.

In all of these situations, it is important *how* the sample is selected. If the sample is not typical of the total set of units in which you are interested, it will fail to serve its purpose.

Why Sample?

It is usually possible, at least in principle, to study all of the units which form the population of interest to the study. Reasons why this is rarely done are cost, feasibility (including time) and quality.

Cost

There is often a real marginal cost associated with the inclusion of each unit of study: the cost of the time of the researcher, experimenter, interviewer and field worker, and the cost of equipment and materials. So the budget may constrain the sample size. Even without that

Research Methods for Postgraduates, Third Edition. Edited by Tony Greenfield with Sue Greener.
© 2016 John Wiley & Sons, Ltd. Published 2016 by John Wiley & Sons, Ltd.

constraint, a smaller sample may leave more money for other stages of the project or for other projects.

Feasibility

If a study of the quality of an industrial output involves destruction of the samples, there is no point in studying the entire output. If results are needed by a particular deadline, there may be insufficient time to study all units.

Quality

Concentration of effort on a sample can increase the quality of the research which may then lead to more accurate results (see the 'Bias, Variance and Accuracy' section). For example, in an interview survey with a modest sample size, it should be possible to recruit a team of highly capable interviewers and provide them with personal training and briefing. With a much larger sample, it may be more difficult to find enough high-quality interviewers and infeasible to brief them personally. This could adversely affect the quality of the data.

Sample Design

The procedures and mechanisms that collectively constitute the method of sample selection are known as the *sample design*. At its simplest, a sample design is a *sampling frame* (the list of units from which the sample will be selected), with a specification of the procedures to sample from the frame. In scientific study, the procedures are usually *objective*, with units chosen by a chance mechanism rather than by subjective selection. Thus, if the same sample design is applied repeatedly, it is likely that different units will be sampled each time due to the play of chance.

Inference

Scientific sampling is to provide a means of making *inferences* about the population of interest using observations made on the sample. The observations are collectively called *sample statistics*, whereas the unknown descriptors of the population are called *population parameters*. So, sample statistics are used to make inferences about population parameters. For example, if we observe a sample mean, $\bar{x}$, we might then make an inference about the probability of the corresponding population mean, μ, being within a certain distance of $\bar{x}$. We might decide on the appropriate probability and distance, using information about how the sample was selected and information about the composition of the sample. A particular sample statistic may be called an *estimator* of a particular population parameter. For example, the sample might be an estimator of the population mean. The particular value of that sample mean statistic observed by the study will be called the study *estimate*. So, if the sample mean is 12.6, then 12.6 is the estimate of the population mean.

There are two basic approaches to inference. One is the *design-based approach*, and the other is the *model-based approach*. Chapter 10 of Thompson (2002) provides an introduction to similarities and differences between the two approaches, which are described and discussed in Cassel *et al.* (1977) and Särndal (1978). Increasingly, sampling methods combine elements of both approaches (Brewer, 2002; Kalton, 2002).

The design-based approach relies on the randomisation present in the design to provide the theoretical basis for estimation. This approach produces estimation methods which are unbiased, and the precision of which can be estimated, without requiring knowledge of the population structure.

With the model-based approach, the researcher must develop a statistical model that adequately describes the distribution of variables. With this method, the sampling mechanism is irrelevant, and the accuracy of estimators depends on the adequacy of the specification of the model. Great care must be taken with this approach. An adequately specified model can prove an impossible goal.

In the remainder of this chapter, we assume application of the design-based approach to inference.

Bias, Variance and Accuracy

Any method of selecting a sample for scientific study should be objective and should result in a *representative sample* (Kiaer, 1895). *Objectivity* is usually interpreted as meaning that the selection method should not permit any subjective influence, and it should be *unbiased*. If a sample design is unbiased, then the average value of a sample statistic, across a large number of repetitions of the study, will equal the corresponding population parameter (Moser and Kalton, 1971: sects. 4.3 and 5.1). This does not mean that the statistic based on the one sample actually selected will necessarily equal the population value. There could be a lot of variation across the different samples that could be selected. This variation is measured by the *sampling variance* (Moser and Kalton, 1971: sect. 4.3; Mohr, 1990: sect. 3). The smaller the sampling variance, the greater the chance that the sample statistic based on the one sample actually selected will be close to the corresponding population parameter.

Bias and variance are properties of the sample design, not of the particular sample selected. The complete set of all estimates that could be obtained, corresponding to all the samples that could be selected using the chosen sample design, is known as the *sampling distribution*. A sampling distribution may be summarised in tabular form, or as a histogram or graph. For example, Figure 24.1 shows three sampling distributions produced by three different sampling designs, all proposed to measure the same thing. Design A produces a symmetrical distribution, centred on the value of population parameter. So design A is unbiased. Design B has an identically shaped sampling distribution to design A but is not centred on the population parameter. So design B is biased and is inferior to design A if other factors, such as cost and ease of implementation, are equal. Design C also has a symmetrical sampling distribution, centred on the same value as design B, so it is equally biased. But the distribution is more compact. C has smaller sampling variance than B. So C is superior to B, other things being equal.

One question is left unanswered by Figure 24.1: is design A superior to C? If we consider bias, A is preferable, but if we consider variance, C performs better. The concept of error provides a means of answering the question. Error is a property of a particular selected sample, not of the sample design. It is simply the difference between the sample estimate and the population parameter. A sensible objective is to minimise the expected magnitude of error.

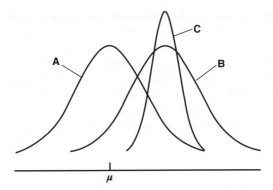

Figure 24.1   Three different sampling distributions produced by three different sample designs A, B and C, where μ is the population parameter.

This can be achieved by using the concept of *statistical accuracy*, a performance criterion that encompasses bias and variance simultaneously. A design can never guarantee an error-free estimate unless it is unbiased and has zero variance.

Accuracy can be measured by a quantity known as *root mean square error* (RMSE). The RMSE of an estimator can be thought of as a measure of the average magnitude, across the sampling distribution, of the difference between the estimate and the population value: the average error (for a precise definition, see Kish, 1965: pp. 13, 60). RMSE might not always be the most appropriate criterion for comparing the desirability of different research designs. But in general, it is a very useful concept and has the advantage of ease of interpretation. A biased design with low variance can produce a more accurate estimator than an unbiased design with higher variance.

Another term sometimes used when discussing estimators is *precision*. Precision is simply the converse of variance. Precision is the converse of variance. A precise estimator is one with low sampling variance, and vice versa. If unbiased designs are being considered, then precision is synonymous with accuracy. In general, precision is only once component of accuracy; bias is the other.

Precision is usually measured by a quantity known as the *standard error* (for a definition and discussion, see Moser and Kalton, 1971: sect. 4.3, 4.4). For many sorts of unbiased estimators, the standard error has a known and fixed relationship to the area under the curve of the sampling distribution. For example, in most situations the population parameter plus or minus two standard errors will encompass approximately 95% of the area under the curve. In other words, 95% of the possible samples under the specified design will produce an estimate within plus or minus two standard errors of the true population value. It is this relationship which allows the computation of confidence intervals and the testing of hypotheses (Mohr, 1990).

Precision (sampling variance) is determined by three factors: the inherent variation among the population, the sample size and other aspects of the sample design such as clustering and stratification. The sample size, n, is an important determinant of precision (larger samples yield greater precision), but it is not the only one. In most situations, variance is approximately inversely proportional to n, so doubling the sample size will halve the sampling variance. The number of units in the population, N, is virtually irrelevant. But it is possible to obtain precise estimators with small n. And it is equally possible to obtain imprecise estimators with large n, if the sample is not appropriately designed. It is the *design* of a sample that should give us faith in a study's results. Sample size alone is a fairly meaningless indicator.

Cost is important in the design of a study. It has a valid influence on study design. A main aim of the researcher, though not often stated explicitly, should be to maximise accuracy per unit cost. It is nearly always possible to improve the accuracy of a procedure, but this usually incurs a cost. The task of the research designer is to effect an appropriate balance between cost and accuracy.

Non-probability Sampling

There are sometimes compelling reasons for a researcher to consider sample designs which involve selecting the more easily accessible units in the population. *Accessibility sampling* is the term for designs where this is the main consideration. Such designs can have cost and administrative benefits and are common in some fields of research. But there is a drawback. There is a risk of bias, and the design provides no means of assessing this bias. For example, if the health of fish in a lake was assessed by measuring the first 10 fish caught from the most convenient place on the shoreline, the design would be biased if the propensity to be in that part of the lake and to be caught was in any way related to health.

To reduce the risk of such bias, the research may turn to *purposive sampling*. With purposive sampling, the researcher recognises that there may be inherent variation in the population of interest. He attempts to control this by using subjective judgement to select a sample which he believes to be 'representative' of the population. Purposive sampling *can* lead to very good samples, but there is no guarantee that it will. Its success depends on two assumptions:

- The researcher can identify in advance the characteristics which collectively capture all variation.
- The chosen sample will correctly reflect the distributions of these characteristics.

Two factors likely to cause contravention of these assumptions are imperfect knowledge of the population structure, and prejudiced selection. Yates (1935) showed that even experts are generally unable to purposively select an error-free sample.

Accessibility and purposive sampling are often combined, as with quota sampling methods used in some interview surveys (see Chapter 25) and in many web surveys (Bethlehem, 2010). However, any design that involves an element of purposive selection is open to criticism. Such designs to not necessarily lead to 'unrepresentative' samples, but there is no way to measure the likely quality of the samples. Probability sampling provides the means to do this.

Probability Sampling

Probability sampling or *random sampling* (they mean the same thing) is often thought the only defensible selection method for serious scientific study unless it is simply not feasible. This is a natural corollary of the design-based approach to inference (see the 'Inference' section).

Probability sampling refers to sample designs where units are selected by some probability mechanism, allowing no scope for the influence of subjectivity. Every unit in the population must have a known and non-zero selection probability. Note that the probabilities need not be equal. The advantages of probability sampling are that it enables the avoidance of selection biases and it permits the precision of estimators to be assessed, using only information that is collected from the selected sample. Furthermore, because precision can be estimated, this provides a tool for making informed estimates of the likely effect of changes to aspects of the design. Thus, you can choose between competing potential designs, according to precision and cost.

The theory of probability sampling is well developed. Thompson (2002) provides a good introduction with references to more specialised texts. There are practical strategies for implementing probability sampling designs in various fields. For example, see Cochran (1977) or Kish (1965) on surveys, Cormack (1979) on biological studies, Hohn (1988) on geological studies; Keith (1988) on environmental sampling, Ralph and Scott (1981) on sampling birds, Seber (1982, 1986) on sampling animals, Deming (1960) on business research and Metcalfe (1994) on industrial and engineering research. See also Chapter 3 of this book.

The simplest form of *probability sampling*, and the one with which other forms can be compared, is *simple random sampling*. This is the design under which every possible combination of n units from the population of N units is equally likely to be selected. Note that this implies that each individual unit has an equal selection probability. Simple random sampling is therefore *design unbiased*.

Systematic Sampling

Systematic sampling is another design that gives each unit an equal selection probability. The population units are listed, and a sample is taken by selecting units at fixed intervals down the list. For example, to sample 100 from a list of 2000, you would generate a random start between 1 and 20 and select that unit and every 20th thereafter until the end of the list was reached. Each unit has an equal chance of selection because each of the 20 possible samples, corresponding to the 20 possible random start points, has an equal probability of being selected and each unit belongs to one, and only one, of those 20 samples. But it is not simple random sampling because there are only 20 samples that could be drawn. Any other combination of units cannot be selected.

There are two main reasons for sampling systematically. One is administrative convenience. It may be easier for the sampler to count through a list or a set of files than to sample randomly. The other is that systematic sampling incorporates an element of stratification if the list is ordered in a way that is related to the subject of the study.

Proportionate Stratified Sampling

Proportionate stratified sampling involves dividing the study population into a number of groups – known as *strata* – and ensuring that the proportion of sample units selected from each stratum equals the proportion of population units in the stratum. The aim of *proportionate stratification* is to guarantee that the sample reflects the structure of the population, at least in terms of one or more important variables. For example, if the sample design for a medical study is to select patients systematically from a list ordered, or stratified, by age, the sample will be certain to reflect the age distribution of the population. If age is related to the variable(s) of interest, this is beneficial. Simple random sampling would not guarantee reflection of the age structure. You might, just by chance, select all the older patients. Stratification can be incorporated within a probability sample design. It does not violate the principle of random sampling.

Stratification can be either implicit or explicit. *Explicit stratification* requires the creation of distinct strata, the determination of a sample size and the selection of a sample from each stratum. For example, the geographical region of interest to a study may be divided into grid squares each of which would be assigned to one of three strata: densely inhabited, sparsely inhabited or uninhabited. Within each stratum, a simple random sample of grid squares could be selected for study.

With *implicit stratification*, the sample frame is sorted before it is sampled systematically. The sample size in each stratum is not fixed in advance, although it will not vary much.

Stratification usually brings modest gains in precision (see again 'Bias, Variance and Accuracy') and is worth doing unless it is prohibitively time consuming and expensive.

Disproportionate Stratified Sampling

Sometimes, different proportions of units are required in different strata, requiring the use of unequal sampling fractions. Explicit stratification is necessary. One reason for unequal sampling fractions is to provide sufficiently large samples for separate analysis in each stratum. For example, in a study of the population of England and Wales a sample of 2000 people might provide sufficient precision for many purposes. But if separate estimates were required for each of the two countries, the sample in Wales may be too small for estimates to be useful. Simple random sampling may select 1900 in England and only 100 in Wales. An unequal sampling fraction could be imposed to ensure a sample of 1725 for England and 275 for Wales, say.

Unequal sampling fractions may be used if it is impractical to give all units an equal selection probability. For example, in line–intercept sampling of vegetation cover, the size of a patch of vegetation is measured whenever a randomly selected line intersects it. This results in large patches having higher probabilities of inclusion in the sample.

Unequal fractions may be used to increase the precision of estimates based on the total sample. This can be achieved by using larger sampling fractions in strata which are inherently more variable (Moser and Kalton, 1971: pp. 93–99; Thompson, 2002: pp. 122–124). That this is sensible can be seen by considering an extreme situation where the population falls into two strata within one of which there is no variation at all. There would be no point in selecting more than one unit from that stratum because knowledge of any one unit confers knowledge of the complete stratum. So, the sampling fraction should be much higher in the other stratum.

In the examples in this section, unequal sampling fractions were obtained by specifying an exact sample size for each stratum. Another method is *weighted sampling* in which units in one stratum would have a different probability of selection than units in another. In the study of England and Wales, each person in Wales may have a selection probability three times that of each person in England. To produce unbiased estimates for England and Wales as a whole, it is necessary to *weight* the data to restore the correct distribution across the strata and this can cause a loss in precision.

Multi-stage Sampling

It is sometimes desirable to select the sample in two or more stages, to reduce cost or effort. For example, in a survey where people are interviewed in their own homes, a sample of small geographical areas such as postcode sectors or electoral wards may be selected first followed by a sample of addresses drawn from each sampled area. This provides efficient workloads for interviewers, and is much more cost-effective than a sample spread thinly over the whole country. Similarly, a survey collecting data from patients' medical records might first select a sample of GP practices, and then a sample of patients registered at each practice. This reduces the number of practices that need to be contacted, thus easing study administration.

Multi-stage sample designs are efficient solutions in many contexts, but they may result in less precise estimators than single-stage samples of the same size. This is because study units within each first-stage entity are often less variable than units in the whole population. This increases sampling variance. For example, two people living in the same postal sector, or registered at the same GP practice, may be more similar than two people sampled at random from the whole population, in terms of many variables. The homogeneity of first-stage entities can be measured by the *intra-class correlation coefficient*, ρ (Kish, 1965: sect. 5.4). The higher the intraclass correlation, the more detrimental it will be, in precision, to cluster the sample.

In multi-stage sampling, the selection probability of each unit will be the product of the probability of selecting the first-stage entity to which the unit belongs and the conditional probability of selecting the unit given that the first-stage entity has been selected. Thus, the relationship between these probabilities needs to be considered carefully when the sample is designed, and the probabilities must be recorded carefully to allow appropriate weighting (see the 'Weighting' section).

Capture–Recapture Sampling

The classic use of *capture–recapture sampling* is to estimate the total number of units in a population. There are variants of the technique, but the basic idea is to select a sample, attach an identifying mark to each sample unit, return these units to the population, select another independent sample and observe what proportion of the second sample has already been included in the first sample. If both samples are truly independent, this proportion should be an unbiased estimator of the population proportion included in the first sample. Thus population size can be estimated as the size of the first sample divided by this proportion.

Capture–recapture sampling has been used on many animal populations and on elusive human populations such as the homeless. See Cormack (1979) and Chapter 18 of Thompson (2002) for further discussion.

Adaptive Sampling

Adaptive sampling is frequently used to sample rare or elusive populations. It requires an initial selection and study of a probability sample. Then, whenever a unit exhibits a high or interesting value of the variable of interest, a further sample of units is taken close, usually geographically, to the interesting unit. For example, in a survey of a rare mineral resource, a probability sample of small areas might be studied. Most areas would reveal a zero occurrence of the mineral, but where the mineral is found there might be an increased probability that adjacent areas will also host the mineral. This intuitive idea of increasing the study coverage often brings increases in precision.

Adaptive sampling is used in the study of rare animal and vegetation populations, as well as in geology, and in epidemiological studies of human populations. Thompson (1988, 1990, and 2002: chaps. 23–26) provides further discussion.

Sample Size

An important element of sample design is the determination of the sample size. As described in the 'Bias, Variance and Accuracy' section, sample size affects the precision of estimators, although it is not the only element of sample design to do so. This means that if the research can specify in advance the required level of precision, it is possible to determine, at least approximately, the sample size that would be required to deliver that precision. See Barnett (1991, sects. 2.5 and 2.9).

Weighting

Weighting is when a numerical coefficient is attached to an observed value, usually by multiplication, so as to give that value a desired degree of importance. There are several reasons why it may be desirable to *weight* data before analysis (Lynn, 2005). The most common are given in this section.

To Correct for Unequal Selection Probabilities

If different units in the population had different selection probabilities, sample estimates would be biased unless each sampled unit is given a weight proportion to the reciprocal of its selection probability. For example, in the unequal sampling fractions discussed in the 'Probability Sampling' section, units sampled in Wales should be given a weight one-third that of units in England.

To Adjust for Non-response

If there are some selected units for which no data could be collected (this is common in interview surveys and in medical studies), then sample estimates will be biased if the propensity to respond is related to any of the variables of interest. Non-response bias may be compensated by weighting. This requires some knowledge about the relationship between responding and non-responding units, or between responding units and the total population in terms of some characteristics which may be related to propensity to respond and to variables of interest. There are various ways to develop such weighting (see e.g. Elliott, 1991, 1999).

To Correct for the Effects of Sampling Variance

The proportion of units in a random sample that have a certain characteristic may happen, by chance, to differ greatly from the corresponding population proportion. If the characteristic is believed to be related to variables of interest to the study, the data could be weighted to match the population profile. This will tend to reduce the variance of estimates (whereas the other two motivations for weighting, described in this chapter, will tend to reduce or eliminate bias).

Calibration weighting is a term used to describe a class of methods which involve ensuring that the weights used to determine the sample distribution meet certain constraints. A subclass of calibration methods is known as *post-stratification*. This involves weighting a sample to match the known population profile in terms of certain strata and thus has similarities with stratified sampling. Post-stratification and its likely effects are discussed in Holt and Smith (1979). The development of general calibration methods is outlined in Deville and Särndal (1992). In practice, calibration methods will tend to counteract multiple sources of error simultaneously, not just sampling variance. Lundström and Särndal (1999) propose that calibration is appropriate for non-response adjustment, for example.

Summary

Sampling is a complex discipline, yet it is of primary importance in many studies. It is the foundation on which much study is built. For many purposes, it is not necessary to be closely familiar with even the few sampling techniques mentioned in this chapter, but it is always important to consider how a sample is to be drawn and what effect that sampling method might have on the data. Sampling methods for scientific study should be objective and should maximise accuracy of estimation, per unit cost, as far as possible. This will require the strict application of appropriate probability selection methods which will then allow estimation of the accuracy obtained.

The particular sample design issues that are of primary importance vary between disciplines and are discussed in other chapters of this book.

References

Barnett, V. (1991). *Sample Survey Principles and Methods*, 2nd ed. London: Edward Arnold.

Bethlehem, J. (2010). Selection bias in web surveys. *International Statistical Review* 78: 161–188.

Brewer, K. (2002). *Combined Survey Sampling Inference: Weighing of Basu's Elephants*. London: Edward Arnold.

Cassel, C.M., Särndal, C.E., and Wretman, J.H. (1977). *Foundations of Inference in Survey Sampling*. New York: John Wiley & Sons.

Cochran. W.G. (1977). *Sampling Techniques*, 3rd ed. New York: John Wiley & Sons.

Cormack, R.M. (1979). Models for capture-recapture. In Cormack, R.M., Patil, G.P., and Robson, D.S. (eds.), *Sampling Biological Populations*. Fairland, MD: International Co-operative Publishing House.

Deming, W.E. (1960). *Sample Design in Business Research*. New York: Wiley.

Deville, J.C., and Särndal, C.E. (1992). Calibration estimators in survey sampling. *Journal of the American Statistical Association* 87: 376–382.

Elliot, D. (1991). *Weighting for Non-response*. London: OPCS.

Elliot. D. (1999). *Report of the Task Force on Weighting and Estimation*. National Statistics Methodology Series Report No. 16. London: Office for National Statistics.

Hohn, M.E. (1988). *Geostatistics and Petroleum Geology*. New York: Van Nostrand Reinhold.

Holt, D., and Smith, T.M.F. (1979). Post stratification. *Journal of the Royal Statistical Society Series A* 142: 33–46.

Kalton, G. (2002). Models in the practice of survey sampling (revisited). *Journal of Official Statistics* 18: 129–154.

Keith, L. (ed.) (1988). *Principles of Environmental Sampling*. New York: American Chemical Society.

Kiaer, A.N. (1895). Observations et experiences concernment des denombrements représentifs. *Bulletin of the International Statistical Institute* 8(2): 176–183.

Kish, L. (1965). *Survey Sampling*. New York: Wiley.

Lundström, S., and Särndal, C.E. (1999). Calibration as a standard method for treatment of nonresponse. *Journal of Official Statistics* 15: 305–327.

Lynn, P. (2005). Weighting. In Kempf-Leonard, K. (ed.), *Encyclopedia of Social Measurement*. London: Academic Press, pp. 967–973.

Metcalfe, A.V. (1994). *Statistics in Engineering*. London: Chapman and Hall.

Mohr, L.B. (1990). *Understanding Significance Testing. Quantitative Applications in the Social Sciences No. 6*. Newbury Park, CA: Sage.

Moser, C.A., and Kalton, G. (1971). *Survey Methods in Social Investigation*, 2nd ed. Aldershot, UK: Gower.

Ralph, C.J., and Scott, J.M. (eds.) (1981). *Estimating Numbers of Terrestrial Birds*. Studies in Avian Biology No. 6. Oxford: Pergamon.

Särndal, C.E. (1978). Design-based and model-based inference in survey sampling. *Scandinavian Journal of Statistics* 5: 27–52.

Seber, G.A.F. (1982). *The Estimation of Animal Abundance*, 2nd ed. London: Griffin.

Seber, G.A.F. (1986). A review of estimating animal abundance. *Biometrics* 42: 267–292.

Thompson, S.K. (1988). Adaptive sampling. *Proceedings of the Section on Survey Research Methods of the American Statistical Association* 784–786.

Thompson, S.K. (1990). Adaptive cluster sampling. *Journal of the American Statistical Association* 85: 1050–1059.

Thompson, S.K. (2002). *Sampling*, 2nd ed. New York: John Wiley & Sons, Inc.

Yates, F. (1935). Some examples of biased sampling. *Annals of Eugenics* 6: 102–213.

25

Sampling in Human Studies

Peter Lynn

Introduction

This chapter describes the main issues to be considered when designing a sample for a study of a human population. Sample units may be individual persons or may be groups of people such as households or families. A common research method used with human subjects is the interview survey, and much of the discussion is about sampling for interview surveys. The central issues are similar for any research that takes the collection of data directly from the selected sample. Important sampling issues for other research methods will be mentioned. These have much in common with interview surveys.

Much of this chapter refers specifically to studies in the United Kingdom, particularly the section on 'Sampling General Populations'. Some of the issues may be similar for research in other countries, but you should seek advice of a local expert.

Important Considerations

Target Population

Before you embark on the design of a sample, think carefully about the definition of the population about which you intend to make inferences. This is the *target population* (Chapter 24 describes the concept of inference). For example, consider a study about the attitudes of parents towards schools. Is the study concerned only with parents who currently have a child in school? Or are the views of those whose children will soon be starting school also relevant? And what about those whose children have recently left school? How should the study treat foster parents and other carers? Are you interested only in schools that cater for a particular age range? Is the study about all schools? Or only state schools? What about grant-maintained schools? Are you attempting to represent the whole of the United Kingdom? Or Great Britain? Or England and Wales?

Research Methods for Postgraduates, Third Edition. Edited by Tony Greenfield with Sue Greener.
© 2016 John Wiley & Sons, Ltd. Published 2016 by John Wiley & Sons, Ltd.

Only when you have defined the population of interest *precisely* will you be able to work out the best way to sample it. Then, if you realise that it is not possible to give all members of the population a chance of selection, you must restrict the *survey population* more than the target population. You should make this distinction explicit and acknowledge it in your thesis. Consideration of the nature of the excluded part of the target population may lead you to conclude that you cannot reasonably make inferences about the target population but that instead you should make statements only about the survey population. See Section 3.1 of Moser and Kalton (1971) for further discussion of the problem of defining the population.

Efficient Fieldwork

Much research involves visits to members of the sample to collect information from them. For example, you might interview people, or measure them, or observe certain behaviour. You therefore need to consider, and perhaps control, the geographical location of the sample.

For example, for large-scale interview surveys, it is common to select a sample of small areas, such as postcode sectors, and then a sample of addresses within each. It is possible to sample the areas so that, by taking an equal number of addresses in each area, you will have an equal-probability sample (see the 'Clustering' section). Then the number to sample within each area can be set to provide an efficient workload for one interviewer, bearing in mind the amount of work that one interviewer may want to do.

Even if you are doing all the fieldwork yourself, you will want as large a geographical spread of the population of interest as possible. This may be achieved by sampling some small areas and then concentrating your sample of people within those areas. It is better to use probability sampling, rather than purposive selection, for the reasons given in Chapter 24.

Fieldwork efficiency for some research methods is not related to the geographical distribution of the sample. An example is where the data are to be collected at a central point using a self-administered postal questionnaire, a web questionnaire or the telephone.

Clustering

The samples from sample designs where the study subjects are concentrated in several small areas are called *clustered samples*. Samples may be clustered in ways other than geographical, though geographical clustering is the most common form. Some degree of geographical clustering can be achieved indirectly, by sampling first-stage entities which provide some implicit clustering of study subjects, even though the geographical boundaries of the entities may overlap.

For example, in a study of hospital outpatients, it is likely that most outpatients will live close to the hospital where they were treated. If several hospitals are sampled at the first stage, the resultant sample will be clustered even though it might contain a few people who live a long distance from the hospital they attended.

In the rest of this chapter, for simplicity, first-stage entities in a sample design will be called *areas*, though the arguments apply equally to other sorts of first-stage entities.

Multi-stage sampling (see Chapter 24) is a necessary, but not sufficient, pre-requisite for clustered sampling. There is usually a trade-off between the reduced costs achieved with a clustered design and reduced precision.

Multi-stage designs should result in equal probability samples. The two simplest ways to achieve this are:

1. select areas with equal probabilities, and then select individuals with equal probabilities; and
2. select areas with probability proportional to size (*PPS sampling*; see Lynn and Lievesley, 1991: pp. 16–17), and then select individuals with probability proportional to the reciprocal of the selection probability of the area to which they belong.

The first method uses the same sampling fraction in each area, so the sample size will vary across areas in proportion to the population sizes. In the second method, the same number of individuals is selected in each area, so the fraction is inversely proportional to the area population size. The second method is preferable, for fieldwork planning and for statistical efficiency, but it is not always possible, as a suitable measure of size for each area must be known in advance.

Samples of the general population in the United Kingdom are typically clustered within administrative areas such as electoral wards or polling districts, or areas defined by postcodes such as postal sectors (Lynn and Lievesley, 1991; Bond and Lievesley, 1993). The choice of clustering unit is often influenced by the choice of sampling frame (see the 'Sampling Frames' section).

Stratification

If you need to over-sample some sub-groups of the population so that you can analyse them with reasonable precision, then the sub-groups must be identified before sample selection. There may be other reasons to stratify (see Chapter 24), so you must consider carefully what stratification factors are available and how they should be used.

Sampling Frames

The list of population members from which a sample is drawn is known as the *sampling frame*. The frame may have a physical existence, such as a printed list or a computer file. Alternatively, it may result from the application of a sampling method. In this section, we discuss the important properties of a sampling frame or sampling method. These are:

- No omissions (up-to-date)
- No ineligibles listed
- No duplicates
- Frame units correspond to study units
- Units are uniquely and fully identified
- Frame permits stratification and clustering
- Easy and inexpensive access
- Familiarity.

These properties are discussed in more detail by Lynn and Lievesley (1991: sect. 2.3). These characteristics of a frame will not always be attainable, as lists and records used in frames may have been compiled for purposes different from the purpose of the study.

The frame should completely cover the target population. There may be omissions because the frame was not designed to include all of your study or because some people have been excluded for other reasons. Reasons for omissions should be investigated, and careful consideration given to the extent that they might bias the results of your study.

The frame should not include individuals who are not part of your target population. Otherwise, your results may be distorted or, at best, you may have to expend some time identifying and excluding ineligible sampled cases. Similarly, duplicate entries may introduce bias unless the number of entries for each individual is known.

Ideally, the units listed on the frame should correspond exactly with the study units. If this is not the case, there should be a known linkage to enable calculation of selection probabilities. For example, to sample households, addresses are usually selected from a frame such as the Post Office's file of addresses (see the 'Sampling General Populations' section). There is not always a unique one-to-one link between an address and a household, but field workers can establish the number of households at each sampled address and whether each sampled household could have been selected via any other address.

The information on the frame should be sufficient to unambiguously identify each sampled individual and allow easy location on the ground. If this is not the case, it may be worth looking for supplementary information from other sources, such as better quality address details, before starting fieldwork.

For multi-stage sampling, the frame must identify suitable areas. Preferably, these should be areas for which there is also information available for stratification. The frame should be inexpensive and easy to use, and there are advantages in using a frame which has already been used as a sampling frame on previous occasions. You can learn from previous experiences and minimise the possibility of unexpected nasty surprises.

Sampling General Populations

Populations that might be of interest to particular studies can broadly be classified as either *general* or *special*. There is no precise definition of the difference between the two, but a general population can be thought of as one that includes a large proportion of the total population in the geographical area of interest to the study, perhaps a quarter or more. This would include 'all adults', 'all households with a telephone' and 'all married women'. A special population would be a smaller proportion of the total. It may be called a *minority population*.

Methods commonly used to sample general populations are quite distinct from those used to sample special populations. In this section, we discuss the former; in the 'Sampling Special Populations' section, the latter.

Address-based Sampling

General population sample designs in the United Kingdom usually involve the selection of addresses followed by the identification of relevant individuals associated with each selected address, using some clear definition of association. This is because there are no comprehensive lists of individuals available from which to sample directly in this country. The one exception is the electoral register (ER) which was previously used to sample electors (Foster, 1993), but now has very biased coverage due to the large number of electors who choose to opt out of the edited version of the register, which is the only version available for survey sampling.

The list most commonly used as a sampling frame of addresses in the United Kingdom is the Postcode Address File (PAF). An outline follows of this frame and how it might be used. See Lynn and Lievesley (1991: chap. 3) for further details. The council tax lists are also used sometimes but generally for local studies only as they can be accessed only through local authorities. Area sampling (Kish, 1965: pp. 301–358) is another way to sample residential addresses and is commonly used in North America. It has no advantages in Great Britain, where good lists of addresses exist.

The PAF is a computerised list of every address to which the Post Office delivers mail in the United Kingdom. It is split into two files: *large users* and *small users*. The latter is the one used to sample residential addresses, though it includes some non-residential property too. The PAF contains no information about the occupants at each address, so the only way to sample addresses is with equal probability. The main advantage of the PAF over the ER is greater coverage of residential addresses: only 1 or 2% are missing. It is also generally more up-to-date than the ER and, because it is computerised, complicated and large samples can be drawn with relative ease. The PAF is available free of charge from the UK Data Archive at the University of Essex for academic research purposes. Alternatively, samples can be commissioned from computer agencies, though their charges may be high. ER samples can be drawn by hand from public documents at either the National Statistics Library, in Titchfield, Hampshire, or the British Library in London. Alternatively, one computer agency, CACI Ltd, markets a computerised version of the ER, though this omits a small percentage of ER entries.

Telephone Number Sampling

If you intend to survey by telephone, it is not advisable to start by sampling addresses as described above. PAF does not include telephone numbers, so these would have to be obtained via directory enquiries or some equivalent commercial look-up service. It is unlikely that you would succeed in finding telephone numbers for more than half of the sampled addresses, leading to severe coverage problems.

Instead, telephone numbers should be sampled directly. Published phone books should not be used as a sampling frame, as the high incidence of unlisted (ex-directory) numbers in the United Kingdom can introduce considerable bias. Instead, a sampling method that gives a known chance of selection to all working numbers should be preferred. This can effectively be done by generating random numbers within ranges that are in service: a procedure known as *random-digit dialling*. This approach became feasible in the United Kingdom in the late 1990s (Nicolaas and Lynn, 2002). Alternative ways to sample telephone numbers include *plus-digit sampling* and *list-assisted sampling* methods (Collins, 1999).

You should note, however, that sampling landline phone numbers provides a rather biased coverage of the general population, as subgroups such as households with no one aged over 35 and households in urban areas are under-represented. If you wish to obtain a representative sample of the general population via a sample of phone numbers, you will need to use a dual-frame approach, selecting separate samples of landline numbers and mobile numbers, and then combining the two samples. Such an approach has both practical and statistical complications (Brick *et al.*, 2006; Keeter *et al.*, 2007; Gabler *et al.*, 2012).

Other Methods

Other methods attempt to directly sample people, bypassing the need to go via addresses or telephone numbers. However, in the United Kingdom, these methods do not constitute probability sampling methods (see Chapter 24), as there is no available sampling frame of members of the general population.

Many forms of quota sampling, often used in conjunction with face-to-face interviewing, are of this type. Typically, quota sampling may involve interviewers stopping people on the street or in some other public place until they have interviewed a pre-determined number of people (the *quota*) within each of a number of different categories. The categories may be defined by attributes such as age, gender and working status.

Another example of non-probability sampling of people is the use of opt-in web panels. Many companies in the United Kingdom offer to administer your web survey to a representative national sample. However, these samples are drawn from a large panel which has been accumulated through voluntary opt-in, often using methods such as pop-up adverts on websites. However carefully constructed, a sample selected from such a panel cannot really said to be representative. Though there are a few examples of web panels based on probability sampling, there are as yet none in the United Kingdom (Baker *et al.*, 2010; Nicolaas *et al.*, 2014).

Sampling Special Populations

Broadly, there are three methods to sample a special, or minority, population:

1. Screening a general population sample;
2. Sampling from an existing administrative list; and
3. Constructing a sampling frame.

A brief description of each follows. There is fuller treatment in Hedges (1978) and Kalton and Anderson (1986).

Screening a General Population Sample

Screening involves the selection of a sample of addresses, as you would for a general population sample, and then checking each sampled address (called a *screen interview*) to identify members of the population of interest. Those identified can then be interviewed. This process can involve a large amount of work. Its viability depends largely on the incidence of the special population in the general population. For example, if each of only 1% of addresses contains a member of the special population, then to find 300, say, you would need to screen 30,000 addresses. If the incidence were 10%, you would need to screen only 3000.

The amount of detail that needs to be collected in the screen interview is important. In some situations, it may be possible to screen by methods other than personal interviewing, such as with a postal questionnaire or by telephone. *Screening errors* could also be important, particularly if the definition of membership of the special population is complicated. If the screening questions are designed to be conservative, in the sense that some people identified as eligible turn out to be ineligible when the detailed data are collected by the main study instrument, then some effort will have been wasted in interviewing these *false positives*. On the other hand, if the questions are tightened to minimise the risk of false positives, some eligibles might be missed. These are *false negatives*. This could introduce bias. A balance has to be struck.

Sampling from an Existing Administrative List

It is sometimes possible to identify an existing list which is adequate as a frame. But there may be problems of access which have to be negotiated carefully. The main problems to confront when you use an existing list are those of coverage (omissions and ineligibles) and selection probabilities (duplicates). A clustered sample for fieldwork efficiency may be difficult to design as administrative files are often inflexible. Sometimes, a choice must be made between the use of an imperfect existing list and the screening of a general population sample. For example, it might be possible to use the DVLA files of registered keepers of motorcycles to sample motorcycle riders. But a keeper is not the same as a rider. Some people will keep more than one motorcycle each and will be listed more than once. On the other hand, there are around one million motorcycles in Britain, so it might be feasible to screen a general population sample, though it would be expensive.

Constructing a Sampling Frame

It may be possible to construct a frame where none exists. Perhaps several existing lists, each of which covers part of the population of interest, could be combined. *Snowballing* techniques (Goodman, 1952) may be used to expand the frame. The main problem could be to assess the completeness of the constructed frame. When you combine lists, you should pay special attention to overlap, with duplicate entries on the combined list. A recent refinement of snowball sampling is referred to as *respondent-driven sampling* (RDS). RDS provides greater credibility to the approach, as it includes estimation of selection probabilities and associated measures of statistical precision (Heckathorn, 1997, 2002; Tyldum and Johnston, 2014).

It may not be necessary to construct a frame of the individuals of interest if a frame can be constructed of first-stage units that can be sampled before enumeration of individuals within each sampled unit. For example, it would be an enormous task to identify all families who live in bed and breakfast accommodation in a particular area and to list them. It might be possible to identify and list all bed and breakfast *establishments*, a sample of which could be drawn, and to identify eligible families in each.

Weighting

As described in Chapter 24, study data should usually be weighted for several reasons (Lynn, 2005). With general population samples, different units will usually have different selection probabilities. This will certainly be true if the design is to select addresses and then to sample one person at each address. In this situation, people who live alone will have a greater chance of selection than those in multi-person households. The data should be weighted to correct the imbalance (see Lynn and Lievesley, 1991: chap. 4).

In studies of human subjects, there will almost always be some non-response. If the sampling frame provides some information about each individual, this information could be used to develop weighting to adjust for non-response. Even if the frame does not contain any useful information, as with the general population frames of addresses, it may be possible to obtain some basic details about non-respondents as well as respondents, by observation in the field. For example, variables related to housing type and area characteristics could be observed. If variables observed for all sample members are related to propensity to respond, and also to survey measures, then the information can be used to reduce non-response bias (Elliott, 1991; Lynn, 1996).

General population samples can also be compared with external population data, such as the decennial census of population. If discrepancies are apparent, weighting can be used to adjust the sample profile. If discrepancies can be assumed to be caused by non-response bias or by sampling variance, then weighting is an appropriate remedy. However, you should be careful when you compare your data with external data. Differences, even quite subtle ones, in the definitions used, the question wording and the data collection mode, can produce an artificial outcome to your analysis. In this situation, weighting might do more harm than good by introducing biases.

References

Baker, R., Blumberg, S.J., Brick, J.M., Couper, M.P., Courtright, M., Dennis, J.M., *et al.* (2010). AAPOR Report On Online Panels. *Public Opinion Quarterly* 74: 711–781.

Bond, D., and Lievesley, D. (1993). Address-based sampling in Northern Ireland. *Journal of the Royal Statistical Society Series D: The Statistician* 42: 297–304.

Brick, J.M., Dipko, S., Presser, S., Tucker, C., and Yuan, Y. (2006). Nonresponse bias in a dual frame sample of cell and landline numbers. *Public Opinion Quarterly* 70: 780–793.

Collins, M. (1999). Editorial: sampling for UK telephone surveys. *Journal of the Royal Statistical Society Series A* 162(1): 1–4.

Elliot, D. (1991). *Weighting for Non-Response*. London: OPCS.

Foster, K. (1993). The electoral register as a sampling frame. *Survey Methodology Bulletin* 33.

Gabler, S., Häder, S., Lehnhoff, I., and Mardian, E. (2012). Weighting for unequal inclusion probabilities and nonresponse in dual frame telephone surveys. In Häder, S., Häder M. and Kühne, M. (eds.), *Telephone Surveys in Europe*. Berlin: Springer, pp. 147–167.

Goodman, L.A. (1952). On the analysis of samples from k lists. *Annals of Mathematical Statistics* 23: 632.

Heckathorn, D.D. (1997). Respondent-driven sampling: a new approach to the study of hidden populations. *Social Problems* 44: 174–199

Heckathorn, D.D. (2002). Respondent-driven sampling II: deriving valid population estimates from chain-referral samples of hidden populations. *Social Problems* 49: 11–34.

Hedges, B. (1978). Sampling minority populations. In Wilson, M. (ed.), *Social and Educational Research in Action*. London: Longman.

Kalton, G., and Anderson, D. (1986). Sampling rare populations. *Journal of the Royal Statistical Society Series A* 149: 65–82.

Keeter, S., Kennedy, C., Clark, A., Tompson, T., and Mokrzycki, M. (2007). What's missing from national landline RDD surveys? The impact of the growing cell-only population. *Public Opinion Quarterly* 71: 772–792.

Kish, L. (1965). *Survey Sampling*. New York: Wiley.

Lynn, P. (1996). Weighting for survey non-response. In Banks, R. (eds.), *Survey and Statistical Computing 1996*. Chesham: Association for Survey Computing.

Lynn, P. (2005). *Weighting*. In Kempf-Leonard, K. (ed.), *Encyclopedia of Social Measurement*. London: Academic Press, pp. 967–973.

Lynn, P., and Lievesley, D. (1991). *Drawing General Population Samples in Great Britain*. London: National Centre for Social Research.

Moser, C.A., and Kalton, G. (1971). *Survey Methods in Social Investigation*, 2nd ed. Aldershot: Gower.

Nicolaas, G., Calderwood, L., Lynn, P., and Roberts, C. (2014). Web surveys for the general population: how, why and when? *National Centre for Research Methods Report*. http://eprints.ncrm.ac.uk/3309/

Nicolaas, G., and Lynn, P. (2002). Random digit dialling in the UK: viability revisited. *Journal of the Royal Statistical Society Series A* 165: 297–316.

Tyldum, G., and Johnston, L. (2014). *Applying Respondent Driven Sampling to Migrant Populations*. Basingstoke, UK: Palgrave Macmillan.

26

Interviewing

Mark Hughes

Introduction

Interviews play an important part in our lives, for example determining if you are accepted for a university course or if you are successful in gaining a new job. There are similarities across interviews, but in considering research interviews there are also differences. The aim of this chapter is to introduce you to the practices of research interviewing (subsequently referred to as *interviews*) and the informed choices which help to determine that they are conducted successfully. There is no one best way to conduct an interview, and consequently there is no universal definition of an interview. However, common goals of all interviews are to find out through interviewing what we cannot directly observe and to enter into the other person's perspective (Patton, 2014).

The chapter commences with a discussion about the appropriateness of interviewing as a research method for your research project. In writing up your research, you will need to offer a justification for the choices that you make in terms of both your choices about the research methods that you use and equally the research methods that you choose not to use. These discussions are couched in terms of the case for and against interviewing. If you decide that interviewing is the most appropriate choice, the next choice relates to the varieties of interview which can be conducted, ranging from the very structured to the very unstructured, and other potential opportunities include elite interviewing, group-based interviewing and internet interviewing. Instead of thinking of interviewing as a single event, it is informative to think of interviewing as a process of linked activities with such processes requiring you to make further choices. The dynamic human interactions of an interview often overshadow analysis, but analysis is equally important in effectively conducting interviews. These discussions will prepare you to make informed choices, and this chapter concludes with final reflections upon interviewing as a research method.

Research Methods for Postgraduates, Third Edition. Edited by Tony Greenfield with Sue Greener.
© 2016 John Wiley & Sons, Ltd. Published 2016 by John Wiley & Sons, Ltd.

Table 26.1 Strengths of interviewing (Marshall and Rossman, 2015)

Fosters face-to-face interactions with participants
Useful for uncovering participants' perspectives
Data are collected in a natural setting
Facilitates immediate follow-up for clarification
Useful for describing complex interactions
Facilitates discovery of nuances in culture
Provides for flexibility in formulating working hypotheses
Provides information on context
Facilitates analysis, validity checks and triangulation

The Case for and against Interviewing

There are good reasons to choose interviewing as your research method, yet equally there are good reasons for not choosing interviewing. Whilst interviewing is a very popular research method, you need to evaluate its appropriateness for your research and be prepared to justify your choice. In Table 26.1, the case is made for interviewing in terms of strengths of this research method.

The strengths identified in Table 26.1 highlight why interviewing has proved to be such a popular research method which may reflect human preferences for interacting with others. However, the interpersonal nature of interviewing also raises challenges for anyone considering using this method of data gathering, and Table 26.2 highlights such challenges.

The challenges depicted in Table 26.2 are very real; in particular, interviewing can potentially be far less structured and far more ambiguous than, say, a questionnaire survey. It is important to make a self-assessment about your own ability to manage and facilitate what at times can be a dynamic and ambiguous process. The challenges in Table 26.2 encourage early consideration about ethical issues interviewing raises which include:

- Is that relationship non-manipulative?
- Is there the potential for reciprocity?
- Is there the potential for pain and anguish when the person shares painful experiences? (Marshall and Rossman, 2015)

Table 26.2 The challenges of interviewing (Marshall and Rossman, 2015)

Possible misinterpretation due to cultural differences
Depends on cooperation of key informants
Readily open to ethical dilemmas
Difficult to replicate
Data more affected by researcher presence
Too dependent upon participant openness and honesty
Too artistic an interpretation undermines the research
Depends on the researcher's interpersonal skills

Increasingly, institutions have required researchers to formalise their interactions with interviewees through codes of practice, protocols and contracts. These moves towards transparent informed consent are overseen by ethics committees which safeguard the interests of both interviewees and interviewers. You need to think carefully about these issues and talk this through with relevant stakeholders and, where appropriate, follow your institution's protocols.

Varieties of Interview

An ideal type of interview which suits all situations does not exist. Instead, you need to choose the most appropriate variety of interview for your research project, and many variations have been used. One way to think of the different varieties is in terms of a continuum of formality capturing options ranging from very informal to very formal interviews. Patton (2014) identified three different varieties of interview – the informal conversational interview, the general interview guide approach and the standard open-ended interview – and each variety has its own merits.

When using the informal conversational interview, the questions which emerge from the immediate context are asked in the natural course of things: there is no predetermination of question topics and wording. A major advantage of this naturalistic style is that the salience and relevance of questions are enhanced. A major disadvantage is that different information may be collected from different people as a consequence of different conversations.

The general interview guide approach involves topics and issues being covered and specified in advance, in outline form. The interviewer then decides upon the sequence and wording of the questions in the course of the interview. For example, if the researcher was interested in the drivers of organisational change, they might list potential drivers such as reliability, flexibility, profitability, cost cutting and efficiency. This list would then form the starting point for the interview. A major advantage is that the outline increases the comprehensiveness of the data and makes data collection more systematic for each interviewee. A major disadvantage is that important and salient topics may be inadvertently omitted.

The standardized open-ended interview requires the exact wording and sequence of questions to be determined in advance. All interviewees are asked the same basic questions in the same order. This is what many people understand as an interview. A major advantage is that respondents answer the same questions, thus increasing comparability of responses, and consequently data are complete for each person on the topics addressed in the interview. A major disadvantage is that there is little flexibility in relating the interview to particular individuals and their circumstances.

These different interview approaches offer a range of options for the researcher. At the outset of research, informal conversational interviews can be effective for refining and tightening the focus of your research. However, as your research progresses more standardized approaches can be used. Also, it is possible, within an interview, to use more than one type of interviewing behaviour. For instance, a personal preference is not to use too structured an approach at the beginning or the end of an interview and to use the interview guide approach in the middle of the interview.

There are a number of other varieties which might be particularly suitable for your research; telephone interviewing, group interviewing and elite interviewing are introduced and evaluated here in terms of their strengths and weaknesses. Telephone interviewing may be used when interviewees are geographically dispersed, also telephone interviewing can be used as a means of following up a questionnaire survey or earlier interviews. Bryman and Bell (2011) identify the following advantages and limitations of telephone interviewing. They can be far quicker and cheaper to administer and easier to supervise. The geographical remoteness removes a potential source of interviewer bias. In terms of limitations, they exclude people who do not have access to a telephone and they are unlikely to be sustained beyond 20–25 minutes. Potentially, response rates can be lower. Asking questions about sensitive issues can be more problematic. Developments in telephone communications such as call screening may act as an impediment. It is not possible to observe behaviour during an interview. The quality of data may be inferior to that of other forms of interview.

Group interviewing may be appropriate where interviewees belong to a particular group or when collaboration of interviewees is an objective of the research, which may reflect contemporary interest in group and team-based working in organisations. The advantage is that group discussion may encourage new insights which might not have been forthcoming in a one-to-one interview. The disadvantage is that noisy or powerful individuals may dominate the group.

Elite interviewing is a specialised treatment of interviewing that focusses on a particular influential type of interviewee. For Marshall and Rossman (2015), elite individuals are the influential, prominent and/or well-informed in the organization or community you are researching. They highlight the following advantages and disadvantages of elite interviewing. Valuable information as well as an overview of the organisation can potentially be obtained. Elite individuals may also have privileged access to histories, policies and plans. Disadvantages relate to gaining access to elites, requirements to alter the interview and maintaining control of the interview process.

Internet Interviewing

Information technology in general and the internet in particular are enabling new means of gathering and analysing data. Hine (2008), acknowledging communicating as a key use of the internet, suggests that it is natural for researchers to use the internet to communicate with research subjects through quantitative surveys and qualitative research online. As Mann and Stewart (2002) highlighted, researchers can interview disembodied people from across the earth, as well as the people frequenting the imagined environments of cyberspace.

> The excitement of working with an interviewing medium that is not constrained by boundaries of time and space, and offers digital data literally at one's fingertips, is matched by the growing realization that the virtual venue makes practical, legal, ethical, and interpersonal demands that move beyond the knowledge and expertise that researchers may have acquired in conducting off-line interview studies. (Mann and Stewart, 2002: p. 622)

Earlier in this chapter the case for and against interviewing was outlined, and in a similar manner there are costs and benefits of internet interviewing. Issues meriting consideration which are very dependent upon the research project are as follows:

- Who will be sampled, and how many people?
- What are the implications of this choice in terms of expense and time?
- What are your capabilities in terms of working with the digital data?

This choice is finely balanced and will be influenced by your research topic, your abilities and in this instance your research funding. For example, on an international comparative study with a limited budget, internet interviewing might be the best way forward. If you elect to undertake internet interviews, the following guidance from Hine (2008) is useful.

- Choose a comfortable medium for your interviewees, avoiding imposing your own preferences.
- Put people at ease by focussing upon building relationships.
- Avoid long lists of questions and asking too many questions.
- Offer acknowledgement and encouragement when you receive answers.
- Do not expect online interviewing to be quick.

Interview Questions

Whilst Bell and Waters (2014) believe that question wording may not be as important in interviews as questionnaires, they suggest the following best practices of question wording for questionnaires (see also Chapter 21 in this book). The content of interviews is very dependent upon the nature of your particular research and the variety of interview adopted. However, with these caveats in mind, the following generic guidance is useful when beginning to consider your interview questions. Patton (2014) highlighted different kinds of questions that can be asked of people:

1. **Experience/behaviour questions** – What a person does or has done, for example 'In a typical day, what would I see you doing?'
2. **Opinion/value questions** – These questions aim to understand the cognitive and interpretive processes of people, for example 'What do you think about…?'
3. **Feeling questions** – These questions aim at eliciting emotions, for example 'How did you feel about that?'
4. **Knowledge questions** – These questions inquire about factual information, for example 'What services are available?'
5. **Sensory questions** – These questions are about what is seen, heard, touched, tasted and smelled, for example 'What did he actually say?'
6. **Background/demographic questions** – These questions identify characteristics of the person being interviewed, for example 'How long have you worked in your current role?'

Thinking in terms of this kind of question offers a good initial framework for considering questions. Each kind of question can be asked in the past, present or future tense. However, do not place too much emphasis upon questions about the future as this requires an element of crystal-ball gazing. There is also merit in considering the sequencing of questions. Patton (2014) expressed a preference for asking questions about noncontroversial present behaviours, activities and experiences at the beginning of an interview, and background and demographic questions which can be uncomfortable for the interviewee at the end of the interview.

The interactive nature of an interview means that it is likely that at some point you will need to use probing questions. Legard *et al.* (2003) highlighted four potentially useful probes: amplificatory, exploratory, explanatory and clarificatory probes. Amplificatory probes encourage the interviewee to elaborate further, for example 'What was it that you liked about her attitude?' Exploratory probes explore the views and feelings of interviewees, for example 'What did you feel after …?' Explanatory probes are concerned with asking why, for example 'Why was the department closed down?' Clarificatory probes are concerned with clarity and precision, for example 'What do you mean by "cognitive noise"?'

Interviewing as a Process

It is best to think of an interview as a process, rather than an event. Whilst the interview depends upon the variety of interview, the setting, the focus of your research and your own qualities, successful interviews will involve a series of linked activities, rather than a single event. The following discussion draws upon the author's own experiences of interviewing (see Hughes and Berry, 2000, for further discussion) as well as the stages of an interview identified by Legard *et al.* (2003).

Pre-interview Preparations

Before you arrive, there will be interactions with your research subject potentially by email, letter or phone. These initial interactions typically focus upon negotiating research access, securing agreement to undertake the interview and agreeing a mutually convenient date, time and venue for the interview (see Buchanan *et al.*, 1988, for discussion of their own experiences). Interviewees will be forming impressions about your professionalism as the interviewer and your research from these earliest contacts. In this sense, the interview process begins before any questions have been asked.

In preparing for your interview, you will need to gather background information about your host organisations, check your interview approach and any equipment being used (particularly batteries), ensure that you have access to maps and plan to arrive in good time. It is good practice to write to your interviewee before the interview confirming details and telling them more about your research project. At this stage, you will sometimes be asked to provide questions in advance.

> Two principles inform research interviews. First, questioning should be as open ended as possible, in order to gain spontaneous information rather than rehearsed positions. Second, questioning techniques should encourage respondents to communicate underlying attitudes, beliefs and values, rather than glib or easy answers. The objective is that the discussion should be as frank as possible. (Fielding and Thomas, 2008: p. 249)

This is good advice, although it will be contingent upon the structured or unstructured nature of your interviews. If feasible, do try to avoid providing your full questions in advance as spontaneous responses often have more authenticity than rehearsed responses; however, this will be very dependent upon your research subject matter.

Upon Arrival

As with any social encounter, in the first few minutes impressions form which will influence the outcome of the subsequent interview. You may encounter interviewees who are either anxious about being interviewed fearing an inquisition or are hostile resenting the time the interview is going to take. In all circumstances, your professionalism will help set the tone of the subsequent interview. It is often the case that even a hostile interviewee will warm to the research once they learn more about your research and how you will conduct the interview. You may be offered a drink at this stage which is to be viewed as a hospitality ritual. The first few minutes really are about having faith in yourself and the merits of your research.

Introductions

At the outset of the interview, there is a need to establish the purpose of the interview. While the research will have been introduced at the time of setting up the interview, this non-threatening material can make for a neutral starting point, allowing both parties further time to orientate. The interview may be a new experience for the interviewee, and part of your responsibility as the interviewer is to guide them through this process. Whenever possible, it is worth agreeing the format of the interview with the interviewee; they will usually agree to your format, but you are reliant upon their goodwill.

Beginning the Interview

If you are going to record the interview, it is best to wait a few minutes before discussing this aspect of the interview (please see later discussion in this chapter). If your research topic involves asking sensitive questions, these sensitive questions should be avoided at the beginning of the interview. Ideally at the beginning of the interview, you are seeking to model how the interview will be conducted and guide the interviewee.

During the Interview

The conventional social rules of 'You speak, then I speak' are suspended in interviews. Invariably the interviewer's role is to listen and to actively listen. Wengraf (2001) advocates 'double attention', referring to both listening to the interviewee yet simultaneously managing the interview process. Interviewees occasionally give answers to questions which you might challenge in the course of a conventional conversation. However, part of this 'uneven conversation' requires you to listen to the interviewee's worldview without judging what they are saying. At times, this non-judgemental stance can be quite challenging for the interviewer. Patton (2014) captured the need for rapport with regards to the person being interviewed and neutrality with regards to the content of what that person is saying. In maintaining such a view of the interview process, the necessary rapport should develop without the interviewer introducing their own attitudes, values and beliefs, even if they disagree with the interviewee's attitudes, values and beliefs.

Ending the Interview

Ending an interview initially appears to be straightforward. However, if sufficient rapport has developed during the interview, an abrupt exit would be impossible and undesirable. Appropriate questions can send out subtle signals and a useful question at the end of the interview is to ask: are there any questions you would like to ask? As well as signalling that the interview is drawing to a conclusion, such questions sometimes lead to further information being provided. It is very dependent upon the research subject matter, but if you have raised sensitive issues and the interviewee wishes to talk them through with you, even if the time allocated has elapsed you should be willing to listen. Reassurances about confidentiality and how the data will be used are often given at the access stage or the beginning of the interview, but it is still worth repeating these reassurances at the conclusion of the interview.

Post Interview

During the hour after an interview, your mind should be very focussed upon your research topic. Notes made during the interview and directly after the interview can be very fruitful. Contact summary sheets and data accounting sheets have been recommended as a vehicle for summarising the main points of interviews and checking that research questions have been addressed (Miles and Huberman, 1994). Interviews can be intense experiences, requiring complete attention and frequent thinking on your feet. What you have learnt can be very informative, yet simultaneously the process can be exhausting. There can be a feeling of uneasiness after interviews, which it is difficult to rationalise, but probably stems from interviewees sharing their inner worlds with you. A letter of thanks can demonstrate appreciation for the time that the interviewee provided for questions and in the tradition of good research helps to keep the door open for future research.

Recording Interviews

The discussions in the 'Interviewing as a Process' section encourage you to think about how you are going to record the interview. Three potential options are to make notes, to audio record (digital or cassette) the interview or a combination of note taking and audio recording. Whilst you will have your own personal preferences, your interviewee will make the final choice about agreeing to be recorded or not recorded. Box 26.1 draws upon my own experience of interviewing.

Box 26.1   The recorded interview.

I had obtained access to seven senior managers at the head office of a major building society for a project about technological and organisational change. At the outset of one of the interviews, I asked if I could tape record the senior manager. The manager agreed and we conducted the interview with the tape recorder running. We concluded the interview and he offered to take me for lunch. Over lunch, the manager confided that he would have answered my questions very differently if the tape recorder had not been present. Instead of portraying a successful organisation, he would have described the low morale and problems within the organisation, but he had been unwilling to have these comments recorded. At the time of carrying out the interview, I had no idea of the trade-off that I had made.

The vignette in Box 26.1 highlights the compromises that sometimes arise when interviews are recorded. There are no hard and fast rules, and experienced researchers often have their own personal preferences. The following questions should help you to determine the appropriate form of recording for your research.

1. How sensitive are the issues you wish to address?
2. If the roles were reversed, would you feel comfortable being recorded?
3. Do you have the ability to record the interview in note form?
4. Does the type of interview that you are adopting require you to make notes to act as probes about certain topics as they arise?
5. Do you have the resources to have the recordings transcribed?
6. How are you going to analyse your interviews?
7. Are transcripts a prerequisite of your analysis?

Legard *et al.* (2003) favour audio-recording interviews because it allows the interviewer to give her or his full attention to listening to the interviewee and to be able to probe in depth, and because it is less intrusive and more neutral than note taking. If you decide to use a recording device, the following supplementary notes may be useful. Invariably, interviewees are willing to be audio recorded as long as their permission is requested tactfully and assurances particularly around confidentiality are offered. Institutional ethics protocols usually require you to store the recordings securely and determine how long the recordings will be kept for. While interviewees may agree to be recorded, there are occasions where they may request that the recorder is switched off and that they speak 'off the record'. While there may be a strong desire to switch the recorder on at the start of the interview, it is often desirable to avoid using the recorder for the first few minutes whilst rapport is developing. In terms of transcription, Patton (2014) offers helpful guidance on how to keep transcribers sane. Alternatively, Buchanan *et al.* (1988) advocated typing up your own transcripts as a means of keeping close to your data.

Analysing Interviews

There may be a desire to partition analysis from the conduct of the interviews in a manner similar to the methods of the quantitative survey researcher. However, Miles and Huberman (1994) were great advocates of interweaving analysis and data collection. They suggested five main stages of qualitative data analysis with the researcher having a range of options about how to conduct each stage.

1. Collect the data.
2. Data reduction.
3. Display data.
4. Draw conclusions.
5. Verify findings.

By interweaving data collection and data analysis, it is possible to test the effectiveness of your interviewing and make amendments where necessary. The implications for interviewers are as follows. You should be thinking about analysis when interviewing. If your interview approach is not providing insights into your research subject, be willing to make changes to how you conduct the interview. Interviews typically generate a large volume of data; as a rule of thumb a one-hour interview can generate a 30-page transcript. It will be necessary to reduce your data; there are many options, but a favourite method is to only analyse the data that speak directly to your research questions. The next stage is thinking creatively about how you display your data. Miles and Huberman advocated matrices as a creative way of displaying data. The data display should allow you to begin to draw conclusions from your data, and the final stage is to verify your findings. The ideal is to verify your findings with your interviewees either using the transcripts and/or a summary of your preliminary findings. Please be mindful of what you ask of your interviewees; you are reliant upon their goodwill.

Historically, before advances in computer software, data reduction was undertaken by coding with emphasis placed upon reading and re-reading texts as a whole and noting ideas (Bazeley and Jackson, 2013). Manual analysis of interview data is still favoured by some researchers, with some researchers even preferring to listen to the recording of the interview, rather than read a transcript. However, the development of qualitative analysis software such as NVivo has provided valuable assistance in analysing interviews. NVivo now offers a fairly user-friendly means of analysing interview data, and it is particularly useful when you are conducting many interviews, at which point the amount of data generated can become overwhelming (please see the 'Internet Resources' section for further guidance and other software packages).

Final Reflections upon Interviewing

The unifying theme of this chapter has been that there is no one best way to conduct an interview; instead, the researcher needs to make informed choices when undertaking interviewing as a research method. In this sense, your preparations and reflections upon issues raised in this chapter are as important as the time spent asking your questions. Bell and Waters (2014) conclude their informative chapter on interviewing with a checklist of questions to consider when planning and conducting interviews.

Interviewing is potentially a very fruitful research method for both the interviewer and interviewee. The interviewer can gain fascinating insights into the topic that they are researching, and the interviewee has an opportunity to share their thoughts and opinions with an active listener. Whilst the potential is considerable, there are responsibilities attached which have been discussed in this chapter. People's accounts of aspects of their lives are fascinating, as each respondent has his or her own unique story to tell. However, you must be interested in people if you are to conduct effective interviews. Patton (2014) emphasises that to be a good interviewer you must like doing it, being interested in what people have to say, and believing that the thoughts and experiences of those people being interviewed are worth knowing.

I conducted over one hundred interviews with managers on different research projects, and those experiences have informed the writing of this chapter. I found both the content and process of those interviews fascinating and informative with regards to what I was researching. I found myself interviewing a leasing manager about financing advanced manufacturing technology on a hot Friday afternoon in the Home Counties when I had a realization that I was no longer interested in what the interviewee had to say. At that point I stopped conducting interviews, which is a good point to end this chapter.

Internet Resources

CAQDAS provides practical support, training and information in the use of a range of software programs designed to assist qualitative data analysis:

http://www.surrey.ac.uk/sociology/research/researchcentres/caqdas/

Online QDA offers a set of learning materials which address common issues of undertaking qualitative data analysis (QDA):

http://onlineqda.hud.ac.uk/index.php

Useful resources developed with specific reference to interviews in educational research at the University of Plymouth, but many of the points made have wider relevance:

http://www.edu.plymouth.ac.uk/resined/interviews/inthome.htm

References

Bazeley, P., and Jackson, K. (2013). *Qualitative Data Analysis with NVivo*, 2nd ed. Thousand Oaks, CA: Sage.

Bell, J., and Waters, S. (2014). *Doing Your Research Project: A Guide for First Time Researchers in Education, Health and Social Science*, 6th ed. Milton Keynes: Open University Press.

Bryman, A., and Bell, E. (2011). *Business Research Methods*, 3rd ed. Oxford: Oxford University Press.

Buchanan, D., Boddy, D., and McCalman, J. (1988). Getting in, getting on, getting out and getting back. In Bryman, A. (ed.), *Doing Research in Organisations*. London: Routledge.

Fielding, N., and Thomas, H. (2008). Qualitative interviewing. In Gilbert, N. (ed.), *Researching Social Life*. Thousand Oaks, CA: Sage.

Hine, C. (2008). The internet and research methods. In Gilbert, N. (ed.), *Researching Social Life*. Thousand Oaks, CA: Sage.

Hughes, M., and Berry, A. (2000). Learning by doing: a case study of qualitative accounting research. *Accounting Education* 9(2): 157–174.

Legard, R., Keegan, J., and Ward, K. (2003). In-depth interviews. In Ritchie, J., and Lewis, J. (eds.), *Qualitative Research Practice: A Guide for Social Science Students and Researchers*, 2nd ed. London: Sage.

Mann, C., and Stewart, F. (2002). Internet interviewing. In Gubrium, J.F., and Holstein, J.A. (eds.), *Handbook of Interview Research: Context and Method*. Thousand Oaks, CA: Sage.

Marshall, C., and Rossman, G.B. (2015). *Designing Qualitative Research*, 6th ed. Thousand Oaks, CA: Sage.

Miles, M.B., and Huberman, A.M. (1994). *Qualitative Data Analysis: An Expanded Sourcebook*. Thousand Oaks, CA: Sage.

Patton, M.Q. (2014). *Qualitative Evaluation and Research Methods*, 4th ed. Thousand Oaks, CA: Sage.

Wengraf, T. (2001). *Qualitative Research Interviewing*. London: Sage.

27

Measurement Error

Roland Caulcutt

Introduction

Perhaps you are concerned that measurement error might threaten the success of your research project. Then you should read this chapter to see how other researchers have assessed the quality of their measurements. However, even if you have no such concerns about your measurements, don't turn to Chapter 28 just yet. Finding the energy to study this chapter could well reduce the risk that your research will fail.

Scales of Measurement

If your research is in the physical sciences, you may encounter measurements of time or distance or weight or money. We are all familiar with such measurements and their associated units, with time being measured in seconds, distance in metres, weight in grams and money in pounds. But many researchers record data that do not have units of measurement. Let us, therefore, define *measurement* more broadly as 'the assignment of numbers to people or things'. With this broad definition you will find measurements in almost every research project, but you might wonder how useful some of these measurements are likely to be.

Next time you visit a shopping mall, you are likely to see a variety of numbers that qualify as measurements by our definition. For example, you might see '£2 per kilo' on the oranges, 'Level 5' on the elevator display, '63' on a bus, '8' on a football shirt or '184' on the door of a house. Numbers on football shirts are classed as *nominal-scale* measurements. The numbers are merely substitutes for names. Clearly you can't do much with nominal-scale measurements. It would be very foolish, for example, to assume that a number 8 shirt was twice as good as a number 4 shirt. House numbers can be much more useful than shirt numbers. They allow a postman to stack letters in the best order for delivery. However, they do not tell the postman how far he will need to walk from one house to the next. House numbers are classed as *ordinal-scale* measurements. They permit ordering, but they do not have a unit of measurement that permits subtraction to calculate walking distance, for example. The numbers on the elevator display would also be classed as ordinal-scale measurements if the levels, or stories, were not equal in height. Money measured in pounds and weight measured in kilograms do have units and would be classed as *ratio-scale* measurements. With such measurements we can carry out the common arithmetic operations of addition, subtraction and division. Thus we might conclude that the price of one coat was £40 more than the price of another, or we might decide that one bag of groceries weighed twice as much as another bag. For the sake of completeness I must mention a fourth scale of measurement known as the *interval scale*. Interval-scale measurements have a unit of measurement but do not have a true zero. Because the zero is false, it is not valid to calculate ratios by dividing one measurement by another. Temperatures in degrees Celsius and temperatures in degrees Fahrenheit are interval-scale measurements.

Four Researchers Who Need to Measure

Anne Roberts is a research student who hopes to submit her PhD thesis next year. She is studying the influence of learning styles on academic performance in GCSE exams. She has assessed the learning styles of a large sample of young people who have recently taken GCSE exams in various subjects. She has also measured the intelligence of each person.

George Hardy works for Hardcoat plc which manufactures paint. For three years he will act as a leader of process improvement teams, during which time his job title will be 'Blackbelt'. His first project requires him to investigate the factors that influence the abrasion resistance of gloss paint. He plans to carry out experiments to assess the effect of changes in the relative levels of several ingredients on the ability of manufactured paint to resist abrasion.

William Brown is a mechanical engineer employed by Travis Engineering which manufactures components for motor vehicles. Travis's customers fully realise that vehicles of high quality and reliability can only be produced from precisely engineered components. Thus, the monitoring of hundreds of component suppliers is a major concern. Suppliers are required to demonstrate their ability to meet strict dimensional and performance specifications, but first they must prove their ability to measure these criteria. William Brown has been asked to assess the performance of a new laser-measuring device which is capable of measuring several dimensions simultaneously and very quickly.

Jane Smollet is a trainee manager with Sustainable Foods which makes and distributes vegetarian-ready meals. She will spend her first two years working on projects in various departments, and then she will study part-time towards an MBA, in parallel with her first managerial role. Currently, she is working in Customer Services Department and has been asked to investigate customer dissatisfaction with late deliveries. She is delighted to find that lots of relevant data have been recorded over the past two years. For example, an existing database contains 'promised delivery date', 'actual delivery date', 'promised delivery quantity' and 'actual delivered quantity' for thousands of orders.

How Good Are Their Measurements?

Anne, George, William and Jane would be wise to consider the quality of their measurements. Perhaps they should start by asking, 'How were my measurements produced?'

I maintain that all measurements come from measurement processes. I believe that every measurement you have ever encountered was produced by a measurement process. Furthermore, I feel confident that you will understand this chapter more easily if you can get your head around the concept of a process.

A *process* is a combination of people, equipment and procedures that work together to produce products and/or services. Processes take in materials and information from suppliers, convert these into products/services, then pass these to customers. Suppliers and customers can be described as stakeholders. Other stakeholders of the process include workers, managers and shareholders.

A measurement process produces measurements, of course. The people, the equipment and the procedures within the process will influence the quality of the measurements produced. To make this clear let us consider the people, the equipment and the procedures within the measurement processes that our four researchers will be dependent on.

The measures of intelligence that **Anne Roberts** obtains will depend on the particular intelligence test that she uses, and upon the person who administers the test. This person should follow the procedure in the test handbook, of course. You may wonder what equipment is involved in measuring a person's intelligence. I suggest that the scores given to a person taking the test might be influenced by the room temperature, the chairs and tables, the lighting, the toilet facilities and so on.

The measures of abrasion resistance that **George Hardy** obtains will depend on the equipment used and upon the person who operates it. In practice a metal test piece is coated with the paint to be tested, then, after the paint has dried, the test piece is fitted into the abrasion machine that repeatedly rubs the surface. The laboratory has two such machines. The old one has been in use for several years and is thought to be unreliable. The new machine looks much better, but many people in the laboratory suspect that it is little better than the old machine.

William Brown has seen the new laser-measuring equipment in operation. It is very easy to use, and it is certainly very fast. But will the measurements be very accurate? William intends to assess the performance of the new equipment by using a standard approach known as *measurement systems analysis* (MSA). We will look at MSA in the 'How Can We Best Describe the Performance of a Process?' section.

Perhaps **Jane Smollet** is more fortunate. Her data have been gathered already, and they now sit in an Excel file in her laptop. She does not have to commit time and effort to gathering new data. However, she may find that her good fortune has a downside, as we shall see later in this chapter.

How Can We Best Describe the Performance of a Process?

Before we consider the performance of measurement processes, let us look at a simple process that throws darts at a dartboard. The objective is to get the darts as close as possible to the bull's eye, which is in the centre of the dartboard. It is reasonable to assess the performance of the throwing process by the distance of the darts from the centre of the board. In fact we will examine four such processes, labelled A, B, C and D in this diagram:

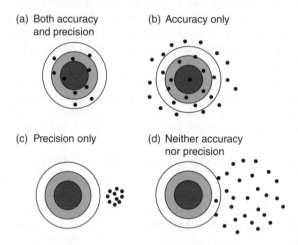

Each of the four throwing processes has performed differently to the other three, and we can describe the differences using two words, precision and accuracy. *Precision* tells us how close the darts were to each other. *Accuracy* tells us how close the darts were to the centre of the board. These two words are often used when describing the performance of measurement processes.

A good measurement process is one that produces measurements that are close to their true values. (In many situations we will not know the true value, but we may be able to get an estimate of the true value by using a better measurement process.) However, regardless of our knowledge of true values, we may be able to learn more about our measurement process by measuring something *more than once*. For example, **Anne Roberts** could measure the intelligence of a sample of people, and then re-measure their intelligence four weeks later. Unfortunately, **George Hardy** would have more difficulty re-measuring abrasion resistance as the first pass through the measuring process will destroy the test piece. But he could make double-length test pieces, cut them in two and assume that the two halves were identical. **Jane Smollet** already has her data. The data do not contain any repeat measurements. **William Brown** is very familiar with repeat measuring. The metal and plastic components his company produces can be measured over and over again. Because of the relative simplicity of these components and the possibility of re-measuring them, we will focus on William's data to illustrate how the performance of a measurement process can be assessed. But, first, we will look at a measurement process in psychology.

Assessing a Measurement Process in Psychology

Recall that **Anne Roberts** has measured the learning styles and intelligence of a large number of students. Before embarking on the measurement journey, she took advice from her supervisor on which tests to use. He recommended that she use the Peter Honey Learning

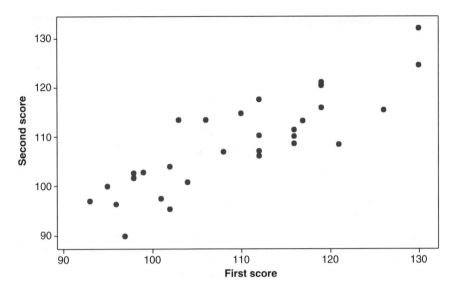

Figure 27.1 Thirty people take a test twice.

Styles Questionnaire and an intelligence test known as Raven's Advanced Progressive Matrices (Raven and Raven, 2008). Her supervisor claims that both tests have good reliability and validity. *Validity* focuses on the question 'Does the test measure what it is supposed to measure?', whereas *reliability* is concerned with how well the test measures whatever it does measure. Clearly, measurement error will reduce reliability.

The reliability of a test, or questionnaire, can be assessed in various ways. The most commonly used is known as the *test–retest method* in which a group of participants, or subjects, are tested and then retested at a later date. The handbook of Raven's matrices test suggests that the test–retest reliability is approximately 0.90. Anne does not doubt this, but she feels that she should take the opportunity to get more familiar with reliability, so she asks a group of her research subjects to re-sit the test four weeks after they were first tested. Her data are displayed in Figure 27.1.

Each point in Figure 27.1 represents one person. The position of a point on the graph is determined by the person's two scores. We see that the scores range from about 93 to 130, and that those people with high scores on the first occasion tend to get high scores on the retest. The best estimate of the test–retest reliability is the correlation coefficient of the data, which is 0.863.

If every student got the same score in the retest that he or she got in the first test, then all 30 points would lie on a straight line and the correlation coefficient would be 1.00, indicating extremely high reliability. But very few students got two identical scores, as we can see in Figure 27.2 where a 'line of equal scores' has been added. In fact, only one student was so consistent. Note that the four students who scored 112 on the first occasion recorded scores of 106, 107, 110 and 118 in the retest. Note also that the three students who scored 113 in the retest scored 103, 106 and 117 in the first test.

Anne Roberts feels that these graphs help her to understand the concept of test reliability. She now realises that the measurement processes she is using lack the reliability that would

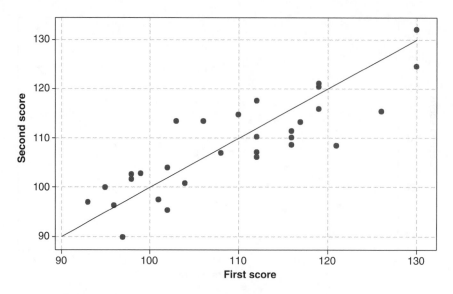

Figure 27.2   Only one person gets the same score on both occasions.

be needed to make precise predictions about individual students. But, she will have measurements on hundreds of students, so the measurement errors will 'average out' when she draws generalised conclusions.

Assessing a Measurement Process in Engineering Research

The data in Table 27.1 was produced by one of **William Brown**'s inspectors measuring the diameter of a metal shaft six times, using the new equipment. (The inspector was not aware that he was measuring the same shaft repeatedly. We want to assess the *normal* performance of the measurement process, not its *best* performance.)

We can see in Table 27.1 that the six measurements are not identical. (If the six measurements were identical, the mean would be equal to each measurement, and the standard deviation would be 0.000.) This variation from measurement to measurement is undesirable, and you might wonder, 'What is the cause of the variation?' You might suspect that the operator is to blame. William Brown has considerable experience with manufacturing processes and measurement processes, and he would be very reluctant to blame the operator. In fact, he has often said, 'There are no bad people, just bad processes'. Perhaps a different operator would have produced similar measurements. To identify any differences there might be, from operator to operator, he asks three more operators to make six measurements.

Table 27.1   Six measurements of the diameter of a shaft

Operator	Measurements (mm)	Mean	Standard deviation
Smith	20.2, 19.9, 20.4, 20.2, 20.1, 20.4	20.20	0.190

Table 27.2   Six measurements of shaft diameter by four operators

Operator	Measurements (mm)	Mean	Standard deviation
Smith	20.2, 19.9, 20.4, 20.2, 20.1, 20.4	20.20	0.190
Jones	20.2, 19.5, 20.6, 19.4, 20.4, 19.9	20.00	0.486
Brown	20.5, 20.8, 20.6, 21.0, 20.6, 20.7	20.70	0.179
Lee	20.0, 20.1, 21.1, 19.9, 19.9, 20.2	20.20	0.456

The four means in Table 27.2 are *not* identical. Clearly we have some evidence that differences between the operators do exist. Furthermore, the four standard deviations are not identical, suggesting additional differences between the operators. The means indicate differences in bias or accuracy. The standard deviations indicate differences in precision. Perhaps these differences are more easily seen in Figure 27.3.

Examine Figure 27.3 very carefully. I would suggest that we can reasonably draw three conclusions from Figure 27.3:

1. Lee has one measurement that is far removed from his other five. We will refer to this measurement as an *outlier*. It may be advisable to remove this outlier if we intend to use Lee's data to assess the performance of the measurement process.
2. Jones's measurements are less precise than those produced by Smith, Brown and Lee. Jones shows relatively poor precision. It may be advisable to reject *all* of Jones's measurements if we wish to estimate the precision of the measurement process.
3. Smith, Jones and Lee have similar means, but Brown's mean is significantly different. Brown appears to be biased with respect to the other three. (Note that some people do not like the word *bias*. They prefer to use *accuracy* or *trueness*.)

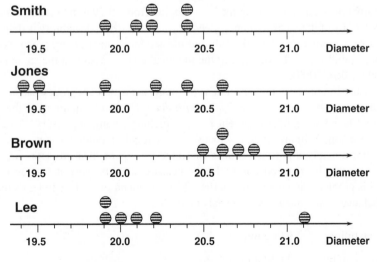

Figure 27.3   Measurements by four operators.

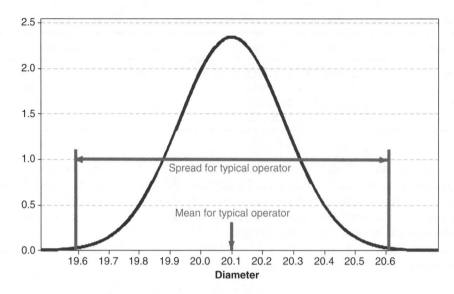

Figure 27.4   A normal distribution: mean, 20.1, standard deviation, 0.17.

Figure 27.3 and Table 27.1 tell us nothing about the true value of the shaft diameter. Perhaps the true value is about 20.1, in which case we could conclude that Brown is biased and the other three operators are not. However, if the true value were, say, 19.1, we might conclude that all four operators were biased or that the measurement process was inherently biased, regardless of which operator was involved.

To estimate the variation we might get from a 'typical' operator we could remove Lee's outlier, then combine the standard deviations from Smith, Brown and Lee to give us a 'typical' standard deviation of approximately 0.17. Multiplying this standard deviation by 6 gives us 1.02, which is the spread of measurements in Figure 27.4.

The spread of measurements in Figure 27.4, ranging from 19.59 mm to 20.61 mm, is known as the *repeatability* of the measurement process. More formally, repeatability is defined as 'the variation in measurements obtained, with one measurement instrument, when used several times by one appraiser, whilst measuring the identical characteristic on the same part' (from ISO/TS 16949; BSI, 2009).

William is also interested in the *reproducibility* of the measurement process, so he carries out a further analysis of the data in Table 27.2. *Reproducibility* can be defined as 'the additional variation introduced if the measurement process includes many appraisers'. To assess this additional variation, William attempts to use a statistical technique known as analysis of variance. Unfortunately this fails because the three groups of measurements are not equal in size. (Recall that he rejected one of Lee's measurements because it was an outlier.) To overcome this problem he replaces the outlier, 21.1, with a number close to Lee's mean, 20.1. With this balanced data, the analysis reveals that:

Repeatability	0.966 mm
Reproducibility	2.040 mm
R&R	2.270 mm

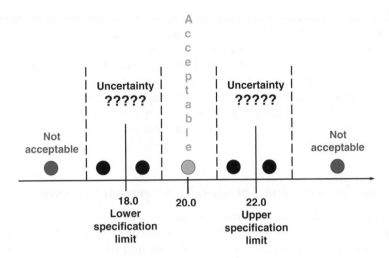

Figure 27.5   Measurement error creates uncertainty.

R&R is defined as the spread of measurements that would be obtained if many appraisers each measured one item. Clearly, the R&R of a measurement process depends on both the repeatability and the reproducibility. If both of these were zero, the R&R would be zero. Note that the reproducibility, 2.040, is larger than the repeatability, 0.996, suggesting that much of the total variation is due to differences between appraisers. If William wishes to reduce the R&R, he should focus on training the appraisers and/or improving the procedure.

'So what?' you might ask. 'Does an R&R of 2.270 mm mean that the process is useless?' To assess the usefulness of the measurement process in the real world, we should compare the R&R with the width of the customer's specification for the diameter of the shaft. The specification is 20.0 + 2.0 mm. So, the specification has a width of 4.0 mm and the R&R has a width of 2.270 mm, which is 57% of the specification. When lots of shafts are made and their diameters measured, William would hope that *all* of the measurements will lie within this specification (i.e. between 18.0 and 22.0 mm). But the measurement errors fill 57% of the specification when we measure just *one* shaft. So the real variation of the manufactured shafts would need to fit in the remaining 43% of the specification. This unsatisfactory situation is illustrated in Figure 27.5, which contains two bands of uncertainty. The uncertainty bands are the same width as the R&R, 2.270 mm.

Because of the uncertainty, William can be confident that a shaft is satisfactory only if its measured diameter lies in the acceptable band, which occupies only 47% of the specification. If William were to show Figure 27.3 to his customer, the reaction would not be favourable. The customer would be unwilling to buy shafts from a supplier who was not capable of measuring diameters with sufficient precision. On the evidence in Figure 27.5, William cannot regard the new measurement equipment as fit for purpose. He will need to continue using the existing process, even though it is much slower.

A Procedure for Assessing Engineering Measurement Processes

William Brown has assessed the performance of the new measurement process, but he has not used the procedure recommended in International Standard ISO16949:2009, Section 8

(BSI, 2009). This standard has superseded an earlier document, QS9000 (BSI, 2003), which was written by Ford, General Motors and Chrysler, to guide the multitude of component suppliers on whom the three vehicle manufacturers depend.

The procedure suggests that William should select *10* shafts from normal production, and instruct *two*, or *three*, operators to measure the diameter of each shaft. Then the 10 shafts should be presented, in a different random order, for the two or three operators to re-measure the diameters. The 40/60 measurements should then be analysed to assess the precision of the measurement process. This data analysis would normally be carried out using statistical software such as Minitab.

A Procedure for Assessing Chemical Measurement Processes

George Hardy wishes to assess the abrasion resistance of gloss paint. He is not an engineer. He is a chemist. If you showed him the above procedure he might say, 'No. This is wrong'. He and his colleagues adopt a different procedure when they wish to assess the performance of a measurement process. Their procedure is contained in BS ISO 5725-2: 1994 (BSI, 1994).

To understand why different procedures are used in the engineering industries, on the one hand, and in the chemical industries on the other hand, you need to appreciate how different their products are. The engineering industries supplying the motor vehicle manufacturers produce shafts, wheels, cables, metal panels and so on, whereas the chemical manufacturers produce bulk chemicals in liquid, powder or even gaseous form. So, in the chemical industry George Hardy will carry out an 'inter-laboratory trial' rather than the 'measurement system analysis' that William Brown favoured.

The procedure for an inter-laboratory trial, recommended in BS ISO 5725 (BSI, 1994), is as follows. Recruit between eight and 15 laboratories. For each level of concentration, send two identical samples to each laboratory. Analyse the resulting measurements, to estimate the repeatability and the reproducibility of the measurement process, at each level of concentration.

Clearly this procedure differs from that used in engineering. Additional confusion can arise because of differences in terminology used in the two industries. For example, chemists do not speak of R&R. Furthermore, repeatability and reproducibility are defined differently with emphasis, in the chemical industry, on the 'difference between two measurements' rather than the 'spread of many measurements' used in engineering. For example, the definition of *repeatability* is: 'The value, less than or equal to which, the absolute difference between two test results may be expected to be, with a probability of 95%. The two results should be obtained under repeatability conditions, of course (ie. With the same equipment in the same laboratory by the same operator using the same equipment within a short interval of time)'.

Errors in Data Taken from Computer Files

Back in the old days, when measurements were read from meters, simply tapping the glass would cause the measurement to change. Everyone who had ever tapped a glass fully realised that measurement errors existed. When meters were replaced by digital indicators, tapping no longer had any effect and some people concluded that measurement errors no longer existed. They were wrong, of course.

Now we have computers everywhere and some people believe that data stored in computers contain no errors. They are wrong, of course.

Recall that **Jane Smollet** is leading a project to investigate customer dissatisfaction with late deliveries. She is delighted to discover that the data she intended to collect already exist in a database on a computer. However, she should bear in mind that these data were recorded by some other person for some other purpose. It is unlikely that the data entry person was aware of Jane's requirements and may have failed to carry out checks that Jane would have seen as necessary. Before proceeding with her data analysis, Jane would be wise to interview the people who routinely enter the delivery dates and delivery quantities into the database from which she extracted the data. It is not unknown for data to be entered incorrectly; also, it is not unknown for data to be *deliberately* entered incorrectly, in order to deflect management attention from 'bad news' that could result in a shot messenger.

Measurement Uncertainty

Researchers have been concerned about measurement errors for many centuries. The procedures discussed in this chapter for assessing the repeatability and reproducibility of measurement processes have been in widespread use for many decades. More recently, attention has focussed on a rather simpler concept known as *measurement uncertainty*. This is described in EURACHEM (2000).

If you have measured something once, then your single measurement is your best estimate of the true measurement that you would have obtained with a perfect measurement process. But, you realise that your measurement process is not perfect. So your measurement is very unlikely to be exactly equal to the true measurement. However, you can calculate confidence limits for the true measurement.

Let us take a practical example from Table 27.1. Suppose Smith had stopped measuring after he got his first measurement, which was 20.2 mm. He could use this as the centre point for a confidence interval, which is 20.2 ± 0.34 mm. He could then report that the true diameter was between 19.86 and 20.54 mm. To calculate these confidence limits, I have used the repeatability standard deviation (0.17 mm) from Figure 27.4. The standard deviation gives us an estimate of the inherent variability of the measurement process. Without this standard deviation, Smith could only report '20.2 mm' without any indication of the uncertainty in the measurement.

The confidence interval concept is delightfully simple. However, if we think a little deeper, we might wonder if the repeatability standard deviation is appropriate. Perhaps we should use the reproducibility standard deviation which includes the variation between operators. The reproducibility standard deviation is larger and would give a wider confidence interval, in recognition of the possibility that a different operator might have made the measurement.

Summary

It is difficult to think of anything good to say about measurement error. It is also difficult to think of anything good to say about researchers who pretend that measurement error doesn't exist. But it is much easier to admire the researcher who attempts to assess how large his or her measurement errors are likely to be.

Different procedures are in use in different industries and different disciplines. These approaches are described in detail in the References. Perhaps they are not the most exciting documents you will ever read, but they do distil the experience accumulated by many researchers who were concerned about the measurements emerging from their measurement processes.

References

British Standards Institution (BSI). (1994). *BS ISO 5725-2: 1994. Accuracy (trueness and precision) of measurement methods and results. Basic methods for the determination of repeatability and reproducibility of a standard measurement method.* London: BSI.

British Standards Institution (BSI). (2003). *QS9000. Measurement Systems Analysis.* London: BSI. (Note: This has been superseded by BSI, 2009)

British Standards Institution (BSI). (2006). *BS EN ISO 4259: 2006. Petroleum products. Determination and application of precision data in relation to methods of test.* London: BSI.

British Standards Institution (BSI). (2009). *PD ISO/TS 16949: 2009 Quality management systems. Particular requirements for the application of ISO 9001:2008 for automotive production and relevant service part organizations.* London: BSI, sect. 8.

EURACHEM. (2000). *Quantifying Uncertainty in Analytical Measurement*, 2nd ed.

Raven, J., and Raven, J. (eds.) (2008). *Uses and abuses of intelligence: studies advancing Spearman and Raven's quest for non-arbitrary metrics.* Unionville, NY: Royal Fireworks.

Part IV
Data Analysis

28

Elementary Statistics

David J. Hand

Introduction

Statistics is an unusual word because it has two meanings. It refers both to data describing some phenomenon and to the science of collecting and analysing those data. The first of these meanings is, of course, subject specific: what the numbers are, what they mean and the implications of them will depend on what they describing. In contrast, the other meaning, sometimes expanded to *statistical science*, describes analytic methods which can be applied in any domain. It is this second meaning which is the concern of this chapter and Chapter 29.

Statistics or statistical science is a vast subject. It arises – and indeed is used – in all walks of life. That assertion might come as a surprise to some readers, but consider the following: the development of pharmaceuticals products is undertaken using (legally mandated) statistical experiments called *clinical trials*; epidemiology, the study of disease outbreaks such as SARS or bird flu, is fundamentally statistical; Google, Amazon and other successful internet companies make very extensive use of statistical analyses to model how their users behave; credit scorecards, assessing the riskiness of applicants for financial products, are predictive statistical models; manufacturers all make heavy use of statistics to evaluate and improve the reliability of their products; astronomy has become a fundamentally statistical discipline, as electronic data capture technology has led to the creation of vast astronomical databases; fraud detection, in science, banking, elections and elsewhere, is based on statistical methods of anomaly detection; the discovery of the Higgs boson was a fundamentally statistical exercise; all civilised countries have a national statistical agency collecting data and compiling statistics describing the state of the economy and society; and so on. The list of uses of statistics is endless, and it is true to say that no aspect of life remains untouched by the discipline. Not unnaturally, given the breadth of statistics, a wide variety of ideas, methods and tools have been developed, perhaps originally with particular application domains in mind, but then often diffusing out to be applied in other domains.

Given the above, it will be obvious that the discipline of statistics is far too large for any single person to be able to be an expert in its entirety. In particular, it follows that all we can

Research Methods for Postgraduates, Third Edition. Edited by Tony Greenfield with Sue Greener.
© 2016 John Wiley & Sons, Ltd. Published 2016 by John Wiley & Sons, Ltd.

hope to do in these few pages is try to orient you: to give you an idea of the motivation behind statistical methods, to show you some very basic tools and to give you pointers to further reading which will provide the details of what you wish to know. Before we get down to it, however, a word or two about the nature of modern statistics is appropriate – if only to correct some widespread misunderstandings.

Statistics has long suffered from a bad press in two regards. One is the view that statistics can, like politicians, bend the facts to suit any purpose. And the second is that it is a tedious and boring subject. The first of these criticisms arises because it is easy to mislead people who are statistically naive. By misrepresenting the facts – presenting only some of the numbers rather than their entirety, focussing on only particular aspects of the results and generally distorting the data to be analysed – of course, one can mislead. But here it is the improper use of statistical methods which is the appropriate target of the criticisms rather than the statistical methods themselves. It is unfortunate that the mud flying around from such misuse has sometimes adhered to the tools being used rather than to those using them improperly.

The second criticism may once have had some truth. In the days when even a relatively simple analysis involved endless hours of mechanical numerical manipulations by hand, how could the subject be regarded as exciting? But things are not like that nowadays. Now we have computers to take over the tedium. Computers enable us to concentrate on the higher level tasks, the seeking for patterns and structures in data as well as the comparison of our theories with the data, without subjecting ourselves to the tedium of endless arithmetic – the computer, one might even say, has transformed statistics into a creative discipline.

A geologist once remarked to me that he envied me. 'Statisticians', he said, 'have the best part of scientific research. They don't have to put up with the boredom and repetition of collecting the data. They come in at the most exciting stage – when one is looking at the data one has collected'. He was partly right: it is certainly true that the statistician is, almost by definition, present when the data finally divulge their secrets. But he was forgetting the role that statisticians also have in deciding how to collect data (e.g. in experimental design and survey sample design, so that the most accurate results can be obtained in the shortest time for the least cost), and he was also failing to recognise that many, perhaps most, practising statisticians prefer to get some hands-on experience as it gives them a better idea of the aims and difficulties of the study. Nevertheless, what he said had some truth in it. And when you have reached the stage of analysing your data, of looking to see if the information it contains matches the ideas you have had, you will also find that statistics is there at the most exciting stage of your work.

The power of the computer means that we do not have to dwell on the algebra and arithmetic of statistical methods. For all but the most basic of operations on very small datasets, you will use a calculator or computer. Hence, in this chapter, we have attempted to focus on statistical concepts and the properties of statistical methods rather than on the mechanics, which electronics will do for you.

This is the first of three chapters on statistics. It introduces the basic ideas. Chapters 29 and 30 cover, respectively, more advanced statistical methods and some types of computer software.

Scales

Data come in various forms, and it is useful to distinguish between several of them. First, we can distinguish between *numerical* and *non-numerical measurements*. Examples of the former are a person's age or weight, and the size of a family. Examples of the latter are the position of a mark on a scale indicating one's extent of agreement with some statement, and scores of mild, moderate or severe on a pain scale. Numbers can, of course, be used to score the latter examples, but they do not have quite so strong an empirical force as the numbers used in the former cases, and sometimes one must be careful about how one analyses such data.

A second distinction we can make is between *continuous* and *categorical data*. Continuous data are measurements which can (at least, in principle) take *any* value within a certain range. Age, weight and the agreement score above are examples of this: weight not only takes discrete kilogram counts but also could include any fractional part of a kilogram. Categorical variables, in contrast, can take only one of (normally) a few values: the size of the family and the mild/moderate/severe pain scale illustrate this. Sometimes it is not clear in which class a variable lies (e.g. age might be recorded just to the nearest year), but this does not usually cause any problems.

It is often useful to divide categorical variables further, into *nominal* and *ordinal variables*. The former have no natural ordering – like different religions, for example – while the latter do (as in the pain scale example).

Issues of measurement are discussed in detail in Hand (2004; see also Hand, 2008).

Basic Measures

Study of a raw table of numbers may not be the most convenient way to see what the data are telling us. Consider Table 28.1, for example. This (from Frets, 1921) shows head lengths (in millimetres) of 25 men. Even in such a small table, without careful study it is not easy to see what sort of head size is typical, what size is unusually large or small, or how wide the range of head sizes is. It is not easy to see if large extremes are rarer than small extremes. It is not easy to see if there are striking exceptions to the general size of head. Since all of this is the case with only 25 values, the difficulty of coping with larger datasets will be obvious. And, moving beyond the mere difficulty of seeing what is going on in a raw data table, nowadays very large datasets, containing millions or billions of values, are increasingly common (e.g. in consumer banking, in astronomy, in particle physics, in bioinformatics etc.), to the extent that it may not even be possible to display a table showing the data. We need to find some way to summarise and perhaps display the data, so that they reveal rather than conceal their secrets.

Firstly, let us find a way to describe 'typical' head length. A number of statistics are in common use. (And this introduces us to a third use of the word *statistics*: as the plural of *statistic*. A statistic is some summary value calculated from or derived from a set of data.) We shall consider three such statistics which describe typical values here: the *mean*, the *median* and the *mode*. These are different kinds of *averages*.

Given a sample of n numbers adding to a total T, the mean of the sample is that number (often denoted) $\bar{x}$ such that n copies of it also add up to T. Obviously, x is smaller than the largest

Table 28.1   Male head lengths (in millimetres)

191	174	189	181	208
195	190	197	175	186
181	188	188	192	183
183	163	192	174	197
176	195	179	176	190

number in the sample, and it is larger than the smallest one. It is an average or *representative* value. Numerically, it can be found simply by calculating the total in the sample and dividing by the sample size. For Table 28.1, we find that the mean is: $(191 + 174 + ... + 190)/25 = 185.72$.

The *median* of a sample is the midpoint of the sample: half the sample values are smaller than it, and half are larger. In fact, we have to specify things a little more precisely than this. We order the sample in terms of size, and if there is an odd number of values in the sample, then the median is the middle one. In the above example, the reordered sample is 163, 174, 174, 175, ... 197, 208, and the middle one has value 188. The median is thus 188. If, however, there is an even number of values in the sample, then there isn't a 'middle one'. In this case, we define the median as being halfway between the two middle ones. For example, in the ordered sample of six numbers (10, 14, 15, 17, 18, 18) the median is halfway between 15 and 17. That is 16.

Finally, the *mode* is the value which occurs most often in the sample. In our example of 25 lengths above, there is in fact no single value which is the most common. Several values occur twice, but no value occurs more than twice. Such a dataset it said to be multimodal – it possesses several modes. In circumstances like this, the sample mode is of limited use as a summary of the data. (More sophisticated methods based on identifying the values of the several modes can be useful, but they are beyond the scope of this chapter.)

Sometimes averages such as those above are termed *measures of location*. This arises from the idea of viewing the sample values as points plotted on a number line. The mean, median and (if it exists) the mode are also points on the line, showing, in some sense, the 'average' position of the sample of points. Such a representation cannot be used with nominal data (since they have no natural order).

The median tells us that value which has half the sample values below it. Sometimes it is also useful to know what value has a quarter (i.e. 25 percent) of the sample values below it. This value is called the *lower quartile*. A similar definition, with 75 percent of the sample below it, gives the *upper quartile*.

So much for measures of location. These tell us the rough size of the values in the sample we are dealing with, but they do not tell us how widely distributed those values are. If we are using the mean to indicate the rough size of the values in our sample, perhaps it is the case that all the sample values are very similar to the mean value. Or, in contrast, perhaps they are widely dispersed with some very large values compensating for some very small values. We need to supplement the mean (or median or mode, or whatever 'average' we decide to use) with another value giving an idea of the spread or dispersion of the sample.

One simple measure of dispersion is the range. This is simply the difference between the largest and smallest values in the sample. For the data in Table 28.1, the range is $208 - 163 = 45$. As a measure of spread, the range is easy to interpret and understand. But it has the disadvantage that it is not very stable. Another sample of 25 values drawn from the same population is quite likely to have a substantially different range because, in calculating the range, only the two most extreme values are considered.

An alternative measure of dispersion which overcomes this problem and is less variable from sample to sample is the *interquartile range*, defined as the distance between the upper and lower quartiles.

Yet another measure of spread, which also eases this sample variability shortcoming, and which is the most popular measure of dispersion, is the *standard deviation*, often abbreviated to *SD*. This is most straightforwardly defined as the square root of the variance, so first, we have to define *variance*. The variance of the sample is the average (in the sense of mean) size of the squared differences between the sample values and their mean. In the data in Table 28.1, this is $((191 - 185.72)^2 + (174 - 185.72)^2 + \ldots + (190 - 185.72)^2) / 25$.

With a sample of size n, often the divisor $(n-1)$ is used in place of n when calculating the average of the squared differences to give the variance. We need not go into the theoretical reasons for this here – and, in any case, for even moderately large n, the difference will be negligible. In our example above, and using the $(n-1)$ divisor, the sample variance is 95.29. It follows that the sample standard deviation is the square root of this, namely 9.76. Taking square roots means that the sample SD is measured in the same units as the raw data. (The variance will be in square units – square kilograms, square inches and so on – because of the squaring operation in its definition.)

So far, we have discussed measures of location and measures of dispersion. One other simple summary statistic, important because it often indicates when common statistical methods may be unreliable (so one has to think more carefully), is *skewness*. The skewness of a distribution of values is a measure of asymmetry. For example, a skew distribution might have many small values and few very large values (positively or right skewed) or perhaps lots of large values bunched together and just a few small values (negatively or left skewed). Income and wealth distributions, with a very few people earning or owning a lot, tend to be positively skewed. Negatively skewed distributions are less common. We remarked in this chapter that the computer and calculator allowed us to side-step the arithmetic details. For this reason, we shall not give a formula for skewness. If you want such details, then you can find calculations described in the books in the References.

You will notice that we have slipped the word *distribution* in here. This is a technical term in statistics which fortunately has much the meaning one might expect it to have – the way in which the values are distributed: are they bunched together, are there some extreme outlying values and so on? One can speak of the distribution of a sample, and also of the distribution of a population – the set of all possible values which might have been chosen for the sample.

One of the attractive features of modern statistical technology is the ease with which plots and diagrams can be produced. Again, the computer has taken all the drudgery out of the exercise. An example is given in Figure 28.1. The data summarised here (Cox and Lewis, 1966) are time intervals between 800 successive pulses along a nerve fibre, measured in seconds. The diagram shows a histogram, produced by grouping the observations according to their size and plotting vertical bars whose heights indicate how many fall into each group.

Simple Distributions

We have already introduced the term *distribution*. Some distributions are particularly important in statistics, perhaps because data often follow such a distribution, or because they have attractive mathematical properties (quite what this means, we will see in a moment!).

Bernoulli Distribution

The *Bernoulli distribution* arises when there are just two possible outcomes, with respective probabilities p and $1-p$, for example. A classic example is the toss of a coin – where, if the coin is 'fair', both p and $1-p$ will equal $\frac{1}{2}$, since the probability of obtaining a head equals

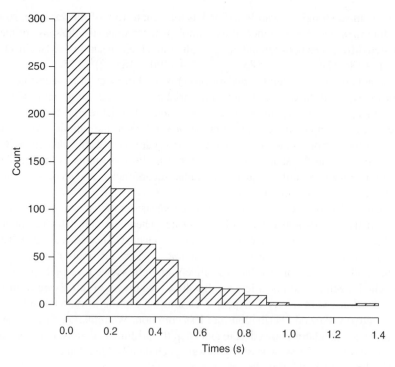

Figure 28.1   Histogram of 799 interpulse waiting times.

the probability of obtaining a tail. Other examples are an application to college, which can be successful or unsuccessful, and whether it will rain or not on your birthday, again with two values: rain or no rain.

Binomial Distribution

An extension of this is the binomial distribution. Suppose that, instead of one coin toss (which would have a Bernoulli distribution), we tossed the coin 20 times. Then the total number of heads would be a random variable: if we were to repeat the exercise, we would not be surprised if the number of heads was different, and indeed it could be any number between 0 and 20. The number of heads would thus itself have a distribution, telling us the probability of getting each number of heads in the 20 tosses. This is the *binomial distribution*. Other examples in which the binomial distribution arises might be how many of five applicants from a particular school will be successful in getting into college, or on how many of your next six birthdays it is likely to rain.

For situations like this to yield binomial distributions, certain conditions have to be satisfied. In particular, we require that the individual events (each coin toss in the example) have the same *probability of occurring* and are *independent*. If the probability that the coin came up heads differed from throw to throw (e.g. because someone tampered with it during the sequence of throws by attaching chewing gum to one side), then the binomial distribution would not be an adequate model for the physical situation. *Independence* is a rather more complicated – but very important – concept. Two events are said to be independent if the occurrence of one does not affect the probability that the other will occur. This is likely to be satisfied for tosses of a coin. But suppose we had asked about the number of the next 10 days on which it rained. Rain on consecutive days is not independent: if it was dry today, it is more likely to be dry tomorrow than if it was wet today. The binomial distribution is unlikely to be a good model for this situation.

Poisson Distribution

The binomial distribution is one possible model for *counts*. Another one is the *Poisson distribution*. Consider a time interval of a given length – one second or one minute, say – and events which occur at random (e.g. the radioactive decay of particles, or arrival of people at a queue). Then the Poisson distribution gives the probability that no events, one event, two events and so on will occur in the time interval (provided certain conditions are satisfied, such as, again, independence between events). This same abstract situation occurs in many situations – for example, it can describe the number of misprints on pages of a book (with, hopefully, pages with no misprints having by far the highest probability, and pages with larger numbers having smaller probabilities).

Normal Distribution

So far, we have only considered discrete distributions – the outcome could take only one of a discrete set of values (e.g. 0, 1, 2, …). In other situations, any value can occur (perhaps from a certain range) – and, as above, these are called *continuous distributions*. The most important example of these is the *normal* or *Gaussian distribution*. This is a particular shape of unimodal (one-peak) distribution, often referred to as *bell-shaped*, which tails off symmetrically to high values and to low values. This symmetry means that scores which are r units above the mean have exactly the same probability of occurring as scores which are r units below the mean. The most likely values are those around the mean, and very large or small values are very unlikely.

The normal distribution is often a very good approximation to empirical distributions which occur in real life, and this is one reason for its importance. Another is that, when large samples are involved, the normal distribution is often a good approximation to the distributions of statistics calculated from data. For example, suppose that a sample of 1000 people were asked their ages and the mean of the 1000 values calculated. If this was repeated – 100 times, say – we would obtain 100 slightly different mean values, and we could study the distribution of these 100 values. This distribution is likely to be very close to a normal distribution.

This sort of thing often occurs, especially when the means (or other statistics) are based on large samples (like the 1000 in our example). This striking property of the normal distribution makes it of fundamental importance in statistical theory: if exact distributions cannot be worked out, often a reasonable approximating normal distribution can be used.

Estimating Parameters

The statistics calculated from samples can often be regarded as estimates of 'parameters' of the populations from which the samples were drawn. For example, consider the population of the United Kingdom. This entire population will have an age distribution, and it will have an average value (a defining characteristic, or *parameter*, of the population). We could discover this average value by asking everyone their age and calculating the average (though we would have to move quickly – people are getting older even as we undertake the exercise, and the population is changing as people are born, die, immigrate and emigrate). Alternatively, we could, much more quickly and easily, take a sample of far fewer people from the population and simply find the average of the sample. This sample average is a statistic – and, as we have seen, its value would vary from sample to sample. However, it turns out that the variation in such a sample average is *inversely* related to the square root of the sample size. So, by taking a large enough sample, we can make the variation in our sample average as small as we like. Put another way, we can make the sample average as close as we like to the population average. Alternatively, we can specify an interval, called a *confidence interval*, such that the probability that the interval contains the true value can be as high as we like.

Testing Hypotheses

The basic principles of sampling can be used to test theories in the following way. Suppose we want to compare two treatments for some disease – let's call them 'treatment A' and 'treatment B'. We give A to one group of people and B to another group (e.g. the two groups being selected by *random allocation*, so as to ensure that we are not subconsciously assigning the sicker people to one treatment). And we compare the recovery rates in the two groups. If the two treatments were equally effective (the supposition that they are equally effective is the so-called *null hypothesis* in this example), we could work out the probability of obtaining any particular difference between the proportions in our sample who recover under A and under B.

In particular, we can work out this probability for differences as large as, or larger than, that which we *actually observed* from our experiment. Overall, this would tell us what would be the probability of obtaining an experimental value as large as, or larger than, that which we actually observed if the two treatments were equally effective. If that probability was very small, we would justifiably wonder if our null hypothesis (that the two treatments were equally effective) was true. If it was small enough, we might feel justified in rejecting that null hypothesis, favouring instead the alternative hypothesis that there was a difference in the effectiveness of the two treatments. What is 'small enough' here will depend on the experimental situation and the investigators carrying out the work. Common values used are 1% and 5%, but there is nothing magic about these choices.

In the above, we used sample statistics to make an inference about a population value – about the difference between the proportion who would recover if everyone received A and the proportion who would recover if everyone received B. Such a technique is called a *hypothesis test* because we are testing a hypothesised value of a population parameter (in that example, the hypothesised value of the difference was 0). This basic idea can be extended to a vast number of situations. A fairly straightforward extension of the above situation allows one to compare the means of the two groups, using what is called a *t*-test.

Now suppose we have two categorical variables – the three-valued pain scale introduced above, say, and the sex (two-valued, or binary) of the patients in the study. Data arising from such a situation can be arranged in a 3×2 cross-classification called a *contingency table*.

We might wish to know whether resistance to pain differs between the sexes – whether the distribution of female patients across the three pain categories is the same as the distribution for male patients. (This question can be expressed in several alternative but equivalent ways. For example, is the ratio of the number of males to females the same in each of the three pain categories? Or, are the two categorical variables, pain and sex, independent?) A hypothesis test of this question begins by assuming that the true population distributions across the sexes are the same (the null hypothesis) and, based on this assumption, derives the probability that one would obtain a sample difference between the observed distributions as great as, or greater than, that which was actually obtained. Again, if this probability is sufficiently small, one will feel that the initial assumption (the null hypothesis of independence) is untenable. This forms the basis of a *chi-square test* for independence. The idea can be extended to contingency tables of any size.

Sometimes, particular tests make fairly stringent assumptions about the distributions involved in the situation being investigated. For example, the *t*-test mentioned here assumes that the populations from which the samples are drawn follow normal distributions. Such tests are called *parametric tests*. Other tests – so-called *non-parametric* or *distribution-free tests* – relax these assumptions, and so are more generally applicable. Having said that, if the assumptions of the parametric test are justifiable, they are generally more powerful. That means that such tests are more sensitive to departures from the null hypothesis: more likely to detect such a departure when it is there to be detected.

Conclusion

Given the constraints on the length of this chapter, it is obvious that we could not go into great depth. This means that we have only been able to scratch the surface of the ideas we have outlined and present a very few central ideas. The books listed in the References and Further Reading go into more detail.

To conclude, however, there is a further general point which should be made. This is that modern statistics, as well as being a tremendously exciting discipline, is also a complex one. Research is very seldom a question of looking at the data, deciding what technique to apply, running the computer to produce a single numerical value and then writing things up. Typically, all sorts of complications arise as the complexities of the real world intrude. These might be things such as incomplete data (e.g. survey respondents who do not answer all the questions), distorted samples (e.g. people who are helped by the treatment deciding to drop out of the sample since they no longer feel unwell), incorrect measurements (e.g., instruments which truncate all values to a maximum) and so on. Because of this, you are advised to seek statistical advice *before* collecting your data, at the time you are designing your study and deciding what to measure and how to measure it, and when you are analysing it. Such a precaution could prevent a great waste of time and a good deal of mental anguish.

References

Cox, D.R., and Lewis, P.A.W. (1966). *The Statistical Analysis of Series of Events*. London: Chapman & Hall.

Frets, G.P. (1921). Heredity of head form in man. *Genetica* 3: 193–384.

Hand, D.J. (2004). *Measurement Theory and Practice: The World through Quantification*. London: Edward Arnold.
Hand, D.J. (2008). *Statistics: A Very Short Introduction*. Oxford: Oxford University Press.

Further Reading

Cox, D.R. (2006). *Principles of Statistical Inference*. Cambridge: Cambridge University Press.
Cox, D.R., and Donnelly, C.A. (2011). *Principles of Applied Statistics*. Cambridge: Cambridge University Press.
Wasserman, L. (2004). *All of Statistics: A Concise Course in Statistical Inference*. New York: Springer.

29

Further Statistical Methods

David J. Hand

Introduction

Chapter 28 outlined some basic statistical ideas and methods. Here, we move on to describe some more advanced techniques. Given their advanced nature, the best we can do is scratch the surface. However, we hope that this will be sufficient to indicate the sort of thing that can be accomplished using modern methods. Naturally, if you do have data which need such methods, you are advised to seek professional advice. Although statistical software is now very readily available, statistical understanding is not so easy to come by. A few minutes of discussion with an expert could save endless weeks of frustration.

We begin by distinguishing between techniques for prediction, where there is a criterion or response variable which has to be predicted from the other variable or variables, and techniques where the variables are all equivalent.

Regression Analysis

Regression analysis is a statistical model-building technique. It relates a single response variable to one or more predictor variables. The model, a mathematical equation, can be used as a summary of the relationship between the response and the predictors, and it can also be used to predict the value of the response given values of the predictors.

We shall illustrate using data on the output of wind-powered generators of electricity (Joglekar *et al.*, 1989). The predictor here is wind velocity (in miles per hour (mph)), and the response variable is *direct current output*. Then our regression equation will permit us to answer questions such as:

- On average, for the generator being studied, what extra current output results from an extra 1 mph in wind velocity?
- Given a wind velocity of 3 mph, what average current output should we expect?

Research Methods for Postgraduates, Third Edition. Edited by Tony Greenfield with Sue Greener.
© 2016 John Wiley & Sons, Ltd. Published 2016 by John Wiley & Sons, Ltd.

Table 29.1   Wind velocity and direct
current output

x	y	x	y
2.45	0.123	6.00	1.822
2.70	0.500	6.20	1.866
2.90	0.653	6.35	1.930
3.05	0.558	7.00	1.800
3.40	1.057	7.40	2.088
3.60	1.137	7.85	2.179
3.95	1.144	8.15	2.166
4.10	1.194	8.80	2.112
4.60	1.562	9.10	2.303
5.00	1.582	9.55	2.294
5.45	1.501	9.70	2.386
5.80	1.737	10.00	2.236
		10.20	2.310

Such a model is constructed as follows. We begin with a set of data. In this case, we will need a sample of measurements of wind velocities and the associated current outputs. Let us denote the values of the response variable, *direct current output* (in units unspecified in the paper, but possibly amps), by y, and the values of the predictor variable, *wind velocity* (in mph), by x. Then our sample provides us with a collection of pairs of values, one pair for each measurement occasion. The data are shown in Table 29.1. Now, let us conjecture that a suitable model relating output to velocity is that output is 0.25 times velocity. A glance at the data shows that this is certainly not a perfect model. The first pair has a velocity of 2.45, so that $0.25 \times 2.45 = 0.613$, which is not equal to the given output of 0.123. Nevertheless, examination of the other values shows that multiplying velocity by 0.25 gives current outputs in the right ballpark. But how can we find a better value?

What we need is some overall measure of goodness-of-fit between our model and the data. Each pair gives us a separate measure of goodness-of-fit: the difference between the output value predicted from the model (0.613 in the above example) and the observed output value (0.123 above). So, for this pair alone, its goodness-of-fit measure is $0.613 - 0.123 = 0.5$. But how can we combine the separate results for each pair into a single overall measure?

We could add them up, but this would have the problem that predictions which were too small (and, so, which had negative goodness-of-fit measures) would tend to cancel out predictions which were too large (and which therefore had positive goodness-of-fit values). This problem can be overcome by squaring the individual goodness-of-fit measures – so making them all positive – before adding them. This leads to an overall goodness-of-fit measure defined as the sum of squared differences between the output predicted from the model and the observed value.

We can calculate such a measure for a range of conjectured values for the constant relating output to velocity. For example, in addition to 0.25, we could try 0.2, 0.21, ..., 0.3, ..., 0.4. And then we could choose that which best fitted the data.

The sum of squared differences criterion is a common measure in statistics. One reason for its importance is that the optimum value of the constant, the one which minimises the sum of squared discrepancies, can be found analytically in the case of regression, avoiding the time-consuming search through possible values. In practice nowadays, of course, it would normally be found by a computer. The value of the constant which defines the relationship between output current and wind velocity is called the *regression coefficient* for the regression of output on velocity. Based on the sample of data we have, our computer will have given us an estimate of that coefficient.

We started the above example by saying that we needed a sample of pairs of measurements. This sample cannot simply be drawn at will, and, in order to yield a valid analysis, care must be taken when the sample is selected. We can distinguish two situations.

In the first situation, we suppose that the sample is a random sample of velocity–current pairs. In this case, hypothesis tests, as outlined in Chapter 28, can be carried out to see if the data conform to any conjectured value for the multiplying constant. So, for example, we could test the null hypothesis that there was no relationship between velocity and current (i.e. that the multiplying constant was zero). The hypothesis test would be asking the question 'How likely is it that we would obtain a sample giving an estimated multiplier as large as, or larger than, that which we actually obtained if there is no general (linear) relationship between velocity and current?' If the sample were not randomly selected, then the basis of the hypothesis test to answer this question may be invalid. For example, if we had deliberately chosen pairs of values which had low current associated with high velocity and high current associated with low velocity, then we might have found an apparent inverse relationship: more wind meaning less output, on average. But this would have been nonsensical in terms of statements about the overall relationship between wind velocity and current output.

The second way to choose the sample arises when you can exercise control over the predictor variable. For example, we might want to know whether increasing the concentration X of a chemical leads to an increase in product Y. To explore this, we could perform a series of experiments, measuring output Y for various *predetermined* levels of X. As it happens, the statistical inferences described in this section are also valid with this approach.

So far, we have described what is called *simple regression*. Simple regression involves a single predictor variable, which is wind velocity and chemical concentration in the above examples. Multiple regression extends this to model the simultaneous effect of several predictors on the response variable. So, to illustrate, in the above example we might have been concerned about the effects of wind velocity, air temperature and humidity on the current output. The mathematical model would then involve adding each of these predictor variables together after multiplying each of them by its own multiplying factor. In mathematical terms, the predicted value of the response would be expressed as a weighted sum of the values of the predictors, where the weights are the regression coefficients.

It might occur to you that an alternative would have been to fit three separate simple regression models, one for each of the three predictor variables. This, however, would mean something rather different, and it is important to understand the distinction between this and the multiple regression involving all three simultaneously. The regression coefficients from the separate models tell us how the response will change when there is unit change in one of the predictors. The regression coefficients from the multiple regression model tell us how the response will change when there is unit change in one of the predictors *if the other predictors were kept constant*. The multiple regression tells us the unique effect of each predictor over and above that of the others. It might be the case, for example, that humidity increases when air temperature increases, so that the simple regression of current output on air temperature would include an indirect effect via the impact on humidity. The temperature regression coefficient from the multiple regression eliminates this indirect effect. Furthermore, the approach involving three separate models does not allow one to explore whether there is a *synergistic* effect due to changes in two of the predictors. Multiple regression does allow this to be explored.

Multiple regression may seem straightforward, but it contains many hidden subtleties. On the other hand, this tool is a fundamental statistical tool, and a sound grasp of it is invaluable.

Logistic Regression

In the 'Regression Analysis' section, the model formulated to predict the value of a response variable had the form of a simple weighted sum of the predictor variables. This is probably the most ubiquitous statistical model that there is: it has a long history and has been used in just about every area of human endeavour. However, despite its power, it is not a universal answer. It has been extended in a number of ways, and this section and the next consider two of these extensions. Here, we look at what happens when the response can take values only in a certain range, and in the 'Discriminant Analysis' section, we look at what happens when the response is restricted to just two possible values.

Suppose we were exploring the relationship between the dose of a drug and the probability of being cured. One way we could set about this would be to make up several different doses of the drug and administer each of these doses to a group of patients. This is the same sort of method of drawing the sample as the second method described in the 'Regression Analysis' section.

People vary, so in each group we would expect some would recover and others would not. If the chance of cure increases with increasing dose, then we might expect a higher proportion of people to get better in those groups which were receiving the larger dose. This is beginning to look rather similar to the set-up described in the last section: we could, perhaps, model 'chance of recovery' as the response variable in a regression with 'dose' as the predictor.

This is a perfectly reasonable thing to do and is often done. However, it does have a drawback. This is that proportions – and probabilities – cannot be greater than 1 or less than 0. If we model the proportion recovering as some constant times the dose, then:

1. For sufficiently large doses, the predicted proportion recovering would be greater than 1.
2. For sufficiently small doses, the predicted proportion recovering might be less than 0.

To cater for this, so that our model will give only reasonable predictions, we modify it slightly. Instead of using the weighted sum (in our case, just the constant times the dose) to predict proportion recovering, we transform the model so that its predictions always lie between 0 and 1. The most common type of transformation used is the logistic transformation. If the raw, untransformed model predicts a value above 1, the transformed version predicts a value below 1 – but larger untransformed values correspond to larger transformed values. A similar thing applies to low predictions. We are, effectively, 'squashing' the predicted values, so that a plot of predicted values takes a flattened *S* shape over the range of dose values, never getting larger than 1 or smaller than 0. Because of the central role of the logistic transformation in such models, this technique is called *logistic regression*.

This basic model structure, a transformed linear combination of the predictor variables, is ubiquitous in statistics, with a variety of different transformations used for different contexts.

This generalisation of the linear form of regression is called, naturally enough, a *generalised linear model*.

Discriminant Analysis

Prognosis means determining the likely future outcome, and it is important for people who have suffered a head injury. It will depend on many factors, including age, response to stimulation and change in neurological function over the first 24 hours after the injury. We would like to predict the future outcome, for example recovery or not, on the basis of some of these variables. Again we shall build a model to do this, and again we will base our model on a sample of people. In particular, we will have a sample of people who have known predictor variable values (such as age) and whom we have followed up so we know their outcome.

One way to approach the problem is as follows. Firstly, consider the 'recovered' group. Such people will have a range of ages, a range of responses to stimulation and so on. In a word, they will have a distribution across the possible predictor variables. The same applies to the other, non-recovered, group. If these distributions differ, it means that for some combinations of age and response to stimulation, one of the two categories, recovery or non-recovery, is more likely than the other. By modelling the distributions, we can identify the combinations which are most likely to correspond to membership of each of the two groups.

There are many ways in which the distributions might be modelled, but the most common assumes a particular class of forms for the distributions. This method leads to what is known as *linear discriminant analysis*. It is called this because it leads to a predictive model which has the form of a weighted sum (a linear combination) of the predictor variables and which *discriminates* between the two groups: large values of this weighted sum are associated with combinations of the predictor variables which characterise one of the groups, and small values are associated with combinations which characterise the other group.

Analysis of Variance

In this chapter, we have described several predictive statistical techniques and pointed out that they all had the same underlying form – a (possibly transformed) weighted sum of the predictor variables. Another very common type of technique, which also has this underlying structure, though this is often concealed in elementary descriptions, is *analysis of variance*. This is aimed at describing the differences between groups of objects. (So it is closely related to the discriminant analysis of the previous section.) In fact, there are various different kinds of links between all of the techniques described in this chapter. That is one of the exciting things about statistics: its methods are not isolated tools; they form a complex and interlinked system of ideas and methods. (The notion that statistics is a collection of isolated tools, merely requiring pulling the right technique off the shelf, is so common a misunderstanding that it has been given a name: the *cookbook fallacy*.)

Suppose that each of the predictor variables is categorical – that is, they can each take one of only a few possible values. For example, age might have been partitioned into young, middle and old; sex will be male or female; and some other measure might be graded as bad, impaired or good. We shall suppose that the response variable is numerical. The cross-classification of the predictor variables forms a set of cells, for example, young males with impairment; old females with impairment. And now we can ask questions such as: do the males differ from the females? Does the response decrease as one moves from young to old? Does the effect of age differ between the two sexes?

To answer such questions, a linear model (a weighted sum of the predictors again) is constructed and hypothesis tests are carried out. However, since designs involving categorical variables in this way are so common, special ways of describing the results have been created, and such analyses are summarised by means of an analysis of variance table which shows the influence of (in this example) sex, age and the other predictors on the response, as well as how the effect of each of these predictors differs according to the levels of the others.

Other Methods

So far, all of the techniques we have discussed are predictive in the sense that they seek to determine the likely value of one variable given an object with known values of the other variables. Not all questions are like this, however. Another whole class of models is concerned with describing the relationships between variables and objects when no variable can be separated out as a response. *Principal components analysis* is one example of this. This technique allows us to determine which combinations of variables explain the most differences between the objects in the sample.

In a study of the patterns of consumption of psychoactive drugs (Huba *et al.*, 1981), data were collected on 1634 students showing the extent to which they had used cigarettes, beer, wine, spirits, cocaine, tranquillisers, drug store medications used to get high, heroin, marijuana, glue and other inhalants, hallucinogenics and amphetamines. Study of this dataset using principal components analysis showed that the greatest range of differences between the students could be explained in terms of the overall extent to which they used the substances. After this, most of the remaining differences between the students could be explained in terms of whether or not they used illegal substances. In effect, what principal components analysis has done is reduce the very complex array of data to a simple and comprehensible description of the main features which distinguish between the students.

Principal components analysis is often confused with *factor analysis*. Whereas principal components analysis seeks to find that combination of variables on which the objects are most different (in a formal sense), factor analysis aims to model the covariances between the variables – the extent to which high/low values of one are associated with high/low values of the other. Factor analysis is used to explore whether the relationships between variables can be explained in terms of unobservable variables – or *latent factors*. So, for example, one might conjecture that there is a single underlying factor (called *intelligence*) which accounts for the fact that there is a correlation between scores in different mental tests. Although principal components analysis and factor analysis often give similar results, they are really quite different classes of tool.

Time Series

All of the methods we have outlined in this chapter involve multiple measurements on each object. Typically, there will only be a few (or a few tens) of such measurements and there will be several or many objects. In a sense, such situations lie in the middle of a continuum, at one end of which lie univariate problems with single measurements on each object (methods for analysing data of that kind are described in Chapter 28). At the other end of this continuum lie *time series*.

Time series are characterised (at least, in their simplest form) by having just a single object but on which many measurements have been taken – the responses at each of a set of consecutive times. Time series are ubiquitous forms of data. Examples are stock closing prices at the end of each trading day, temperature at a particular location measured at midday each day, daily rainfall, an individual's body weight measured at 8:00 a.m. each day, and the country's Gross Domestic Product each year over the course of a century.

Time series are important for several reasons. *Forecasting* is an obvious one. It would be immensely useful if we could predict tomorrow's, next week's or next year's FTSE100 index! But, as well as forecasting, understanding is equally important: is GDP showing an underlying downward trend or are the short-term figures deceptive, and if a downward trend, what is causing it?

There are several approaches to modelling time series. Some focus on modelling them in terms of underlying components such as trend, seasonality and superimposed random terms. Others focus on the probability that a particular value will occur in the next period given that the current (and, perhaps, preceding) periods have the observed values. Yet others decompose the series into regular patterns, as in Fourier analysis. More complex models include the effects of other variables on the score at each time.

Other Techniques

Statistics is a vast domain, with methodological research going on all the time, driven primarily by two things over the past half century. One is the computer, which has revolutionised things in a number of ways. Computers have provided the ability to develop techniques which would have required (literally) years of arithmetic if done by hand (and think of the probability of making mistakes!), such as *neural networks, projection pursuit regression, radial basis function models* and *multivariate adaptive regression splines*. Taking this further, this raw calculation power has led to the development of statistical tools which no one had even conceived of before. An example is the *bootstrap*, which is based on repeated sampling and modelling from the available data, possibly involving building thousands of models, which are then combined into one. Another example is the use of simulation methods, such as *Markov chain Monte Carlo methods* to tackle problems which cannot be tackled analytically. In particular, these tools have enabled the practical application of so-called *Bayesian methods* of statistics. These are based on interpreting probability not as some objective attribute in the physical world (e.g. the property of a particular coin showing its probability of giving heads when tossed) but, epistemologically, as the strength of belief that a researcher has that an event will happen (e.g. a measure of the observer's belief that the coin is more likely to show heads than tails).

The second driver behind the development of statistics is that of new application areas. Recent examples are *genomics* and *bioinformatics* (leading e.g. to developments in multiple hypothesis testing and in regularisation) and the *internet* (which has been one of the main motivators for developing statistical methods for network data). More recently still, automatic collection of data via electronic measurement instruments, as well as data collected for 'administrative' purposes (e.g. supermarket transactions, medical records, social benefit payments etc.), has led to vast datasets – so-called *big data*. The opportunities there have only just begun to be tapped. What has already become apparent, however, is that solid statistical understanding is needed if valid and useful information is to be extracted from these large datasets. Massive datasets alone are insufficient – statistics has become even more essential, not less so.

References

Huba, G.J., Wingard, J.A., and Bentler, P.M. (1981). A comparison of two latent variable causal models for adolescent drug use. *Journal of Personality and Social Psychology* 40: 180–193.
Joglekar, G., Schuenemeyer, J.H., and LaRiccia, V. (1989). Lack-of-fit testing when replicates are not available. *American Statistician* 43: 135–143.

Further Reading

Hastie, T., Tibshirani, R., and Friedman, J. (2011). *The Elements of Statistical Learning Theory*. New York: Springer.
Krzanowski, W. (1988). *Principles of Multivariate Analysis*. Oxford: Clarendon.

30

Spreadsheets: A Few Tips

Tony Greenfield and Andrea Benn

Introduction

You have been warned in other chapters (15 and 16) about the temptation to try to do all your data analysis in a spreadsheet, notably Excel. The temptation is there because Excel is so widely used throughout universities, industry and commerce for assembling, editing and presenting data, for rapid calculations, business plans and domestic accounts. The warning is about analysis. Excel does provide some useful analysis and charting tools and relates well with Word for production of reports.

Nevertheless, Excel has so many advantages for any researcher that it deserves recommendation to you and some suggestions about how to use it to your advantage. If you plan to use a statistical package, you need not desert Excel. Consider using it, at least, for data preparation. Don't be afraid that, if you do this, you might have difficulties in using your statistical package. All standard packages can read Excel files. Also, their analysis results can be retrieved by Excel or Word so that you can create well-formatted reports. Unistat (Chapter 15) goes even better: it can be used from within Excel, as if its instructions are Excel instructions, and its results appear automatically in well-formatted reports on new worksheets.

In this chapter, therefore, I offer some guidance for a start and McFedries (2013) for further advice. Part One is to help you to enter and save your data in a clear structure, in a usable form and without errors. Achieve this and then you can read about advanced features. You will also be more confident about analysis using a statistical package. Many more tips could be offered, but this is my choice:

> Data source
>
> Planning
>
> Data entry
>
> Validation
>
> Data checking.

Research Methods for Postgraduates, Third Edition. Edited by Tony Greenfield with Sue Greener.

Part Two introduces some of the Data Analysis Tools that are supported by MS Excel for PC users and the equivalent tool supported by Apple for Mac users.

Analysis ToolPak (Windows)

StatPlus:mac LE (Apple)

Part One: Data Source

What is the source of the data? Have the data come from somebody else on sheets of paper? That's fine; you can plan your spreadsheets as if you had originated the data. But if the data come already entered into a spreadsheet on a disk, you may be in trouble. Look for that trouble before you start to do any analysis. Faults I have found include:

- String (alphabetic characters) mixed with numeric values in the same column. This will prevent the variables represented by such columns from being available in statistical analysis.
- Zeros used to represent missing values. This will make any analysis worthless. The worst case I have had of this was with some data from an industrial client. A stream of data had been logged automatically into a file; the instrument had recorded missing values as zeros. The company sent the data to me with the message: 'We have tried all standard methods of analysis and found nothing useful. Perhaps you can suggest a different approach'. All they needed were clean data.
- Columns that lack identifying labels.
- Decimal points missing or misplaced.
- Strings used where numeric codes would be better. The simplest are YES/NO and MALE/FEMALE, but I have met a study where a string of a dozen characters had been entered. These can be recoded if there are no errors, but there will be some.
- Blanks have been entered as first characters in text.

The section in this chapter on data checking contains advice on how to discover such faults.

If, however, you are the source of the data in the sense that you are designing your own study and will collect information from a survey, a laboratory, a field experiment or historical documents such as medical records or a 19th-century census, then you should design your data collection form to suit your data entry and storage structure. Do not do one in isolation from the other. Think about your data collection form at the same time as your data storage and analysis. Test the full process of data collection, storage and analysis in a pilot study before you embark on the study proper. Also, be sure that on every data collection form you record:

- Source of the data
- Time of data collection
- What the data represent
- Units of measurement.

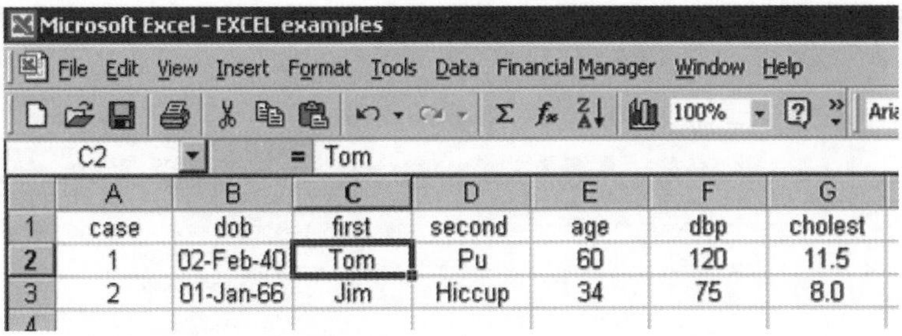

Figure 30.1   The first few rows of a spreadsheet with short labels for columns and unique case numbers for rows.

Planning

Rows and Columns

A spreadsheet has rows and columns. Stick to that arrangement when you organise your data. Designate columns to represent variables and rows to represent cases. For example, suppose we are studying a sample of people to discover an association between age, diastolic blood pressure and serum cholesterol, then these characteristics would be called *variables* and a column would be allocated to each. Provided there was no confidentiality problem, we might add two columns for first and second names. As a check on age, we might add a column to include the date of birth. Every row in the spreadsheet would represent an individual and would contain the values of those variables for that individual. Figure 30.1 shows the first few lines of this spreadsheet.

Unique Identity

Have one column, usually the first, which uniquely identifies each individual. In our example (Figure 30.1), I call each individual a *case* and start numbering the cases from one. You could equally use a staff number or a patient number.

Labels

Every variable should have a name that will be used to identify entries on the *data collection form* (DCF) and for reference in statistical analysis. Create that name when you design the DCF and use it to label the corresponding column on the spreadsheet. The name should start with an alphabetic character (A to Z), and it should have no more than eight characters in it. Few statistical packages will admit variable names of more than eight characters; also, a short name sits neatly at the head of the column. As examples (Figure 30.1), use: *dob* for *date of birth*, *dbp* for *diastolic blood pressure* and *cholest* for *serum cholesterol*.

Data Types

Variables values can be simply classed as strings of alphabetic characters, such as names of people, and numbers, such as age and blood pressure. The classes are refined to include dates,

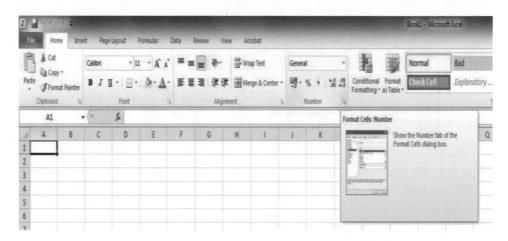

Figure 30.2   The general area for formatting cells: making the selection.

currencies, integers and decimal values. The classes are known as *data types*. When you first prepare your spreadsheet, before you enter any data, you can specify the data type of a variable or column. If you then try to enter a data value of the wrong type, Excel will warn you of an error. See also the 'Validation' section.

In our example, select column G which will contain the data values for the variable *cholest*. Do this by placing the mouse pointer over the G and clicking the mouse. Then, from the Home tab, choose the General section looking for Number at the bottom and then on the right-hand side an Option button to reveal the dialogue box Format Cells Number (Figure 30.2).

Click to confirm the instruction, and the dialogue box in Figure 30.3 appears.

In the left-hand pane, headed 'Category:', is a list of the data types from which you can choose one to allocate to your variable *cholest* which is noted at top right under 'Sample'. Choose to display data values with one place after the decimal point. Without this selection, a value such as 8.0 would appear on the spreadsheet as 8, so you might wonder, when checking your entries, if the correct value was present. With it, the zero will appear after the point. See Figure 30.1. Representation of negative numbers is unimportant since you will have no negative values.

Similarly, you can allocate appropriate cell formats for date of birth, first and second names (text) and case identifiers and age (numbers with no decimal places).

Data Entry

When you enter your data, copying from your DCF, you can either enter directly onto the spreadsheet, row by row, or use a *data form*.

Direct Entry

Keep Headings Visible
Copy directly from your DCF into a row of data, using the Tab button to skip from one column to the next. The more rows you enter, the further your current entry is from the first row that

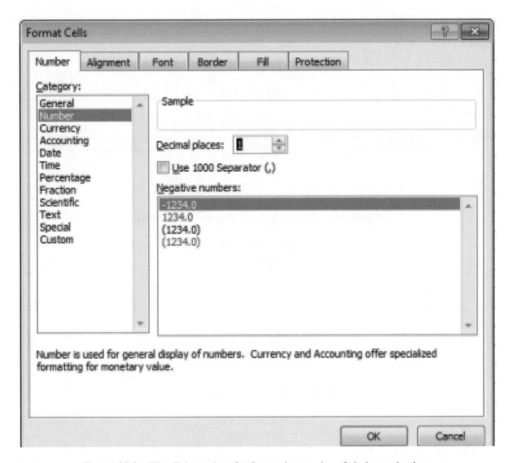

Figure 30.3   The dialogue box for formatting entries of cholesterol values.

contains the column labels. Eventually, the first row will disappear from the screen. You can keep the headings in view. Highlight the row below the headings (row 2), and select View from the main menu to reveal Freeze Panes (Figure 30.4). The headings in row 1 will always be displayed, no matter how many rows you have entered. You can remove this selection with View: Freeze Panes > Unfreeze Panes.

Pick from List
Excel offers a useful feature that saves time and avoids boredom and misspellings. As you enter text values for a string variable, Excel automatically generates a list behind the scenes. Figure 30.5 shows an example of a list of receptacle types. Now, if you type just the first letter of a receptacle, such as 'b', the full word, 'box', appears.

The list may grow long; there may be several clashes of first letters. It may be a list of counties, countries, foods or chemical elements. The solution is: select the next blank cell, right click the mouse and a drop-down menu appears (Figure 30.6a). Click on Pick a List, and the full list appears, in alphabetical order to make it easy (Figure 30.6b). Click on your choice.

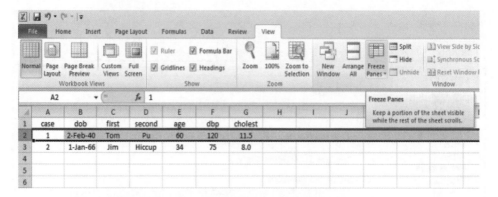

Figure 30.4   Selecting the headings to freeze panes.

Figure 30.5   A list is generated.

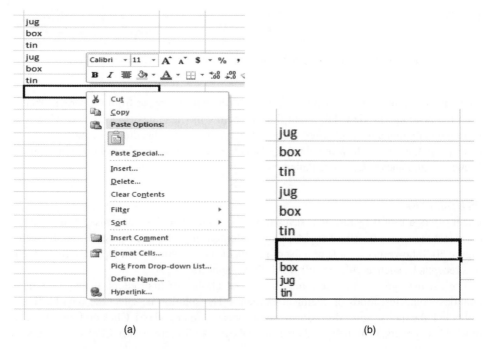

| | |
| (a) | (b) |

Figure 30.6   Pick from list: (a) the drop-down list; (b) alphabetical list.

Figure 30.7   Quick Access Toolbar.

Data Form

The data form (Figure 30.8) is an easy way to enter data into a spreadsheet. However, the Form button has not been included on the Ribbon, but you can still use it in Excel 2010 by adding the Form button 📧 to the Quick Access Toolbar, at the top left-hand side of the screen (Figure 30.7).

Add the Form button to the Quick Access Toolbar

1. Click the arrow next to the Quick Access Toolbar, and then click More Commands.
2. In the Choose Commands from the box, click All Commands.
3. In the list box, select the Form button, and then click Add.

You must put the first set of variable names into the first row of the sheet, after which you can then highlight any row in that sheet and click onto the Form button.

Benefits

It keeps the dataset together for each case and is another check to ensure that each entry is input to the correct cell.

The data form (Figure 30.8) is an easy means to view, change and delete records in the spreadsheet. You can also use it to find a specific record based on criteria that you enter. You can edit records in the text boxes just as you would edit them on the spreadsheet.

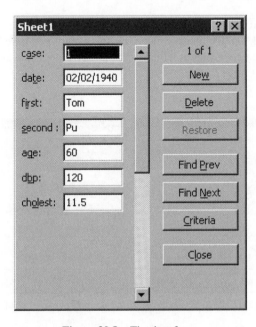

Figure 30.8   The data form.

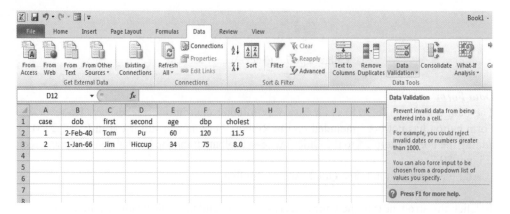

Figure 30.9   Data > Data Validation.

Validation

While you are entering data, either directly or via a data form, you need to know if you have made a mistake. Some mistakes can be flagged automatically by setting criteria. After you have specified the data type for each variable, click on Data > Data Validation (Figure 30.9) and the validation dialogue box will appear (Figure 30.10).

Repeat this procedure for each variable in turn. The result will be that if you ever enter a value of the wrong data type and outside the limits you have set, a warning message will appear telling you to enter a correct value.

Data Checking

Once you have entered your data, and especially if you have received data files from other people, you must check that the data are clean. Do this even if you have had validation checks in place. If anything can go wrong, it will. Several powerful tools are available to help you to check data.

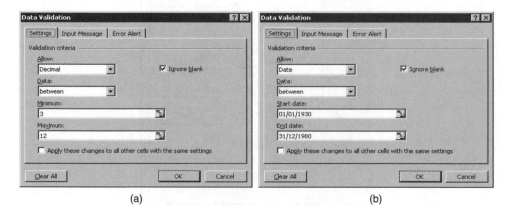

               (a)   (b)

Figure 30.10   The validation dialogue: (a) for a decimal number; (b) for a date.

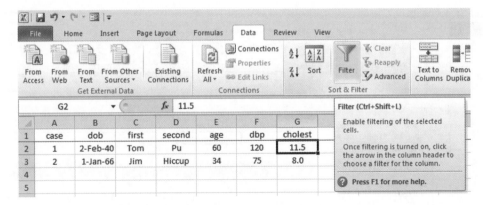

Figure 30.11 Data > Filter.

Autofilter

Highlight any occupied cell. Click on Data > Filter (Figure 30.11), and a downward arrow will appear at the top of every column. Click on the downward arrow in the column that you want to check. A drop-down list will appear.

Click on the downward arrow for the column that you want to check, click on Number Filter, then Custom Filter (Figure 30.12) and then enter the criteria for search.

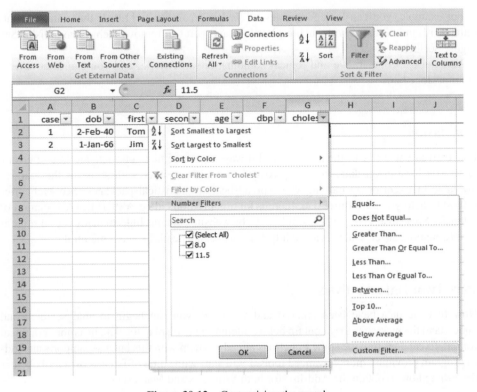

Figure 30.12 Customising the search.

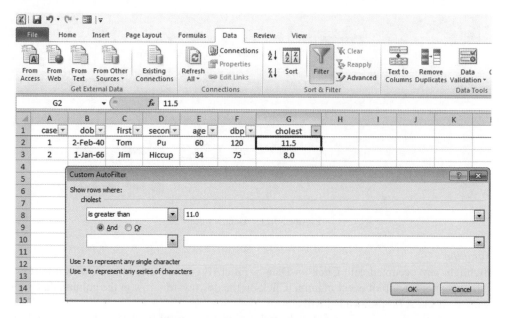

Figure 30.13   The autofilter procedure.

For example, you can look for values of *cholest* greater than 11.0. If there are, the autofilter will find all cases with such high values and you can remove them if you wish. (See Figure 30.13.)

You can use comparative and logical combinations of criteria, and you can filter several columns together.

Conclusion

I have given you a few of the more powerful ways to ensure that your data are free of errors. Once you have clean data, you can proceed to analysis, either with some of the features available in Excel or by moving your spreadsheets to statistical packages. See McFedries (2013) to help you with this. Also, there is an organisation called ASSUME (Association of Statistics Specialists Using Microsoft Excel). Its website contains links to many useful sources of information, articles and reviews: http://www.jiscmail.ac.uk/lists/assume.html.

Part Two: Data Analysis

Now that the data have been entered and validated, you can begin to analyse them and understand the messages they contain before attempting to display the data in a more graphical format. Excel does provide some Functions already to help with this but also supports an Add-In program providing more advanced tools. I will start with some of Excel's own tools before describing how to obtain the Add-ins, for both Windows and Mac users.

Figure 30.14   Using Excel's own functions.

From the Formula tab on the main ribbon, choose the *fx* Insert Function tab which will open a dialogue box (see Figure 30.14). From here, 'select a category' – Statistical – and a list of the available functions within Excel will be displayed. As you make a selection, Excel will show you the name of the function (STDEVA) and how to build the data for it to use (value1, value2): **STDEVA(value1, value2).** If you need more information about any function, clicking on the Help link at the bottom left of the dialogue box opens an Excel Help screen which provides a more detailed explanation and examples of what is required (see Figure 30.14). I have found these to be very thorough and easy to work with.

Pivot Tables

A pivot table summarises large amounts of data quickly. Excel has a wizard dialogue that will appear when you click on Insert > PivotTable and PivotChart (Figure 30.15). It is astonishingly

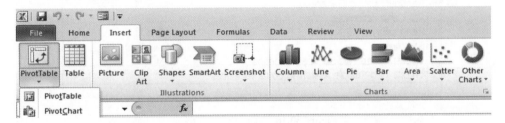

Figure 30.15   PivotTable > PivotTable or PivotChart, automatic chart procedures.

easy to use and produces excellent summaries, showing any functions of data that you choose: counts, means, standard deviations, minima and maxima. Try it, and within minutes you will be an expert. Apart from its value for checking data, it produces tables that will grace your reports.

Charts

A rule of data analysis is: first plot your data. It will give you a feel for the data that no amount of reading tables will give. Plotting is also a good check: you can easily see if there are any values that are out of range. Use Excel's automatic chart procedures, also available on the Insert tab (Figure 30.15).

Analysis ToolPak

The Analysis ToolPak is an Excel add-in program that provides data analysis tools for financial, statistical and engineering data analysis. It offers a variety of tools (Figure 30.16).

However, because it is an add-in, you may not find it installed when you first come to use it and will need to follow the instructions in Figure 30.17a,b before you can start.

In an open Excel workbook, look for the green File tab on the main ribbon bar (Figure 30.18). Click and look for Options, going a stage further to Add-Ins: File > Options > Add-Ins (Figure 30.17a). At this point, you need to ensure that the Manage box is showing Excel Add-Ins and then click Go … Check the Analysis ToolPak box, and choose OK to confirm (Figure 30.17b). The tools will then be found on the Data tab > Data Analysis (Figure 30.18).

Selecting Data Analysis from the Data tab will provide you with a list of available tools; scroll down until you find the one you need, and then click OK. This will open a dialogue

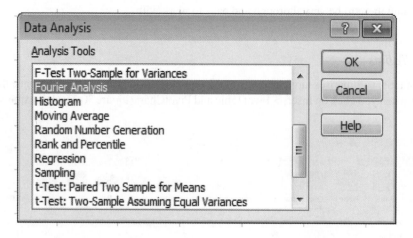

Figure 30.16   Data Analysis tools.

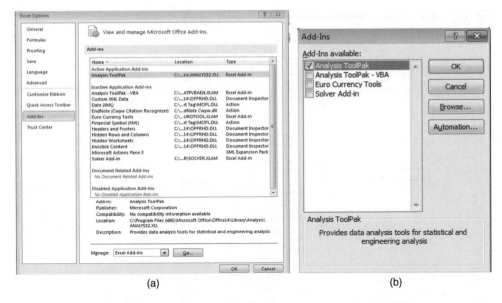

Figure 30.17 (a) Manage Excel Add-Ins; (b) select Analysis ToolPak.

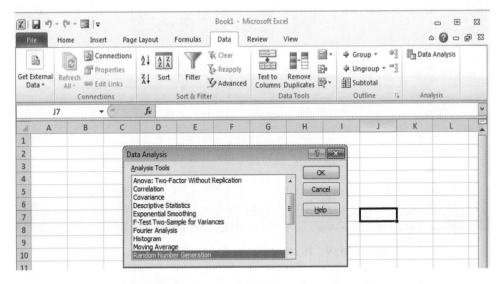

Figure 30.18 Data Analysis ToolPak now added to Excel.

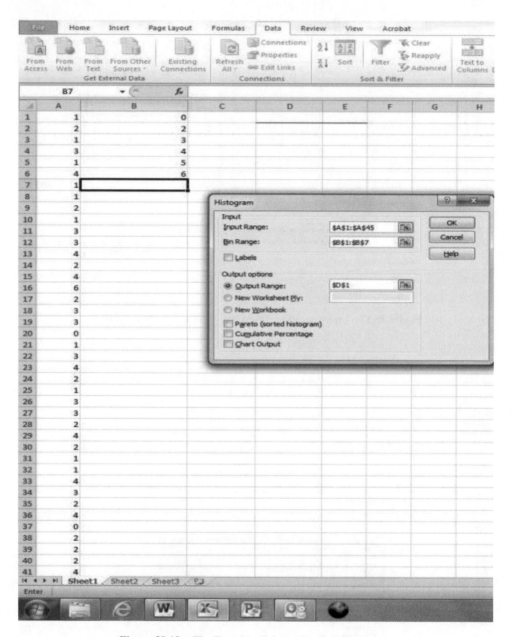

Figure 30.19   The Function dialogue box for Histogram.

box (Figure 30.19) which will ask you to input details of number ranges or cell references and of course will differ depending on the Function you are using. I have chosen to look at the Histogram option, simply because I cannot find this within the Excel provision.

The Help provision is also good and will provide an explanation of how the function works, what data it requires and how the results will be displayed (Figure 30.20).

The Analysis ToolPak, however, is not supported for Apple users, who are advised to download StatPlus:mac (Figure 30.21).

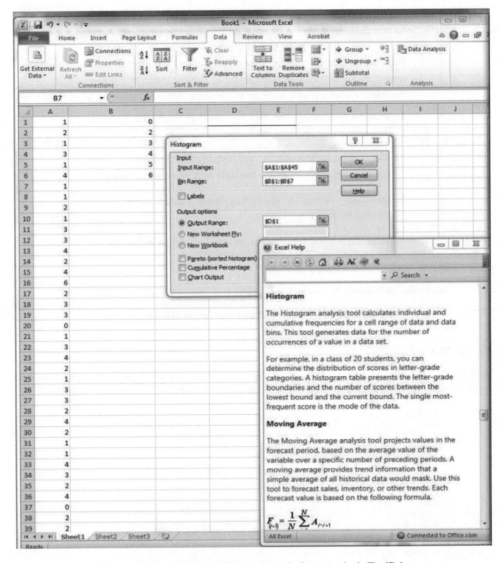

Figure 30.20   Using the Help option, via Data Analysis ToolPak.

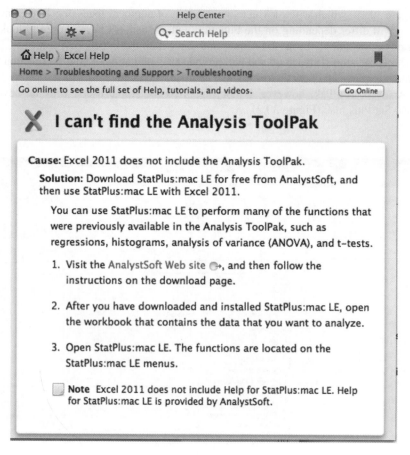

Figure 30.21   Statistical tools for Apple Mac Users.

Reference

McFedries, P. (2013). *Formulas and Functions for MS Excel.* Safari Books Online. http://my.safaribooksonline.com/
9780133260717?cid=shareLink

Part V

Special Tools

31

The Value of Mathematical Models

Andrew Metcalfe

Introduction

Maths is cool! During World Mathematical Year 2000, a sequence of posters designed at the Isaac Newton Institute for Mathematical Sciences was displayed month by month in the trains of the London Underground. The posters were designed to 'stimulate, fascinate – even infuriate!' But, the overriding aim was that they bring maths to life, illustrating the wide applications of modern mathematics in all branches of science – physical, biological, technological and financial. The April poster, Maths is Cool (Figure 31.1), explains that icebergs can become unstable as the ice below the surface melts and then topple so the process starts all over again. This phenomenon can be explained and modelled with catastrophe theory.

The American Mathematical Society has also produced a series of posters, Mathematical Moments, and related material to 'promote appreciation and understanding of the role mathematics plays in science, nature, technology and human culture'. Mathematical jokes can be found, for example, in *The Simpsons* (see http://en.wikipedia.org/wiki/Mathematical_joke, although I quite understand Bart's difficulty in getting it). The year 2000 play *Proof*, by David Auburn, won the 2001 Pulitzer Prize for Drama and a film adaptation, directed by John Madden, was released in 2005. Recent movies based on the lives of mathematicians include: *Agora* (2009), directed by Alejandro Amenabar and starring Rachel Weisz as Hypatia; *A Beautiful Mind* (2001), directed by Ron Howard with Russell Crowe playing John F. Nash; and *The Imitation Game* (2014), directed by Morten Tyldum, with Benedict Cumberbatch playing Alan Turing. Alain Goriely and Derek Moulton were asked to help with the mathematical aspects of the Warner Brothers movie *Sherlock Holmes: A Game of Shadows* in which Jared Harris plays Holmes's arch-foe, the mathematician Professor Moriarty.

Mathematical themes have inspired recent fictional books, including *Uncle Petros and Goldbach's Conjecture*, by Apostolos Doxiadis, first published in 1992; and *The Curious Incident of the Dog in the Night-Time*, by Mark Haddon, first published in 2000 and produced

Research Methods for Postgraduates, Third Edition. Edited by Tony Greenfield with Sue Greener.
© 2016 John Wiley & Sons, Ltd. Published 2016 by John Wiley & Sons, Ltd.

Figure 31.1   Maths is Cool. The other 12 posters can be seen at: http://www.newton.ac.uk/
wmy2kposters/. Reproduced with permission of Isaac Newton Institute: graphic design and text, A
Burbanks and HK Moffat; iceberg images, British Antarctic Survey.

as an award-winning play in 2013. Nicholas Fourikis's book *Hypatia's Feud*, published in 2011,
is a partly fictional account of the Greek mathematician Hypatia, who was born in Alexandria
in the fourth century, The MacTutor History of Mathematics archive at the University of St
Andrews is an extensive resource for mathematical history and includes a biography of Hypatia
(http://www-history.mcs.st-andrews.ac.uk/index.html).

In Australia, Burkard Polster and Marty Ross give an annual series of lectures at the
Melbourne Museum examining mathematical issues in an accessible and entertaining way,
including *Quasimodo's Cipher* and the mathematics of juggling. One of the museum lectures
can be found at http://www.mav.vic.edu/activities/public-lectures/364-rubik-cube.html, and
more material may be added. Ian Stewart, at the University of Warwick, and Keith Devlin,
at Stanford University, are well known for popularising mathematical ideas. Martin Gardner
is acclaimed for his recreational mathematics and physics books, such as the *Ambidextrous
Universe* (Gardner, 1991). Tom Lehrer, best known for his satirical songs during the 1950s, is a
mathematician, and mathematics features in some of his songs, such as 'New Math' which you
can hear on YouTube (http://www.youtube.com/watch?v=yQej8AOjEHc&feature=related).
Ian Stewart's Christmas Lectures at the Royal Institution can be found via its website (http://www.rigb.org/contentControl?action=displayContent&id=00000001776). Some
of Keith Devlin's articles can be found at http://www.maa.org/devlin/devlin_01_03.html. Steve
Mayer's website provides many other interesting mathematical links (http://dialspace.dial.
pipex.com/town/way/po28/maths/links.htm).

In addition, Persi Diaconis, a professor in mathematics at Stanford University and former
professional mathematician, has suggested that many mathematicians are magicians because
'Inventing a magic trick and inventing a theorem are very similar activities'. *The Manual
of Mathematical Magic* is co-authored by Peter McOwan and Matt Parker at Queen Mary,
University of London. It is a free download from www.mathematicalmagic.com. The project
is supported by the HEFCE More Maths Grads project and more recently through the National
HE STEM initiative. Alongside their mathematical careers, McOwan is a lifelong amateur
magician, and Parker is a fully trained stand-up comedian.

Mathematical Modelling

Mathematical modelling is the representation of aspects of the world around us with mathematical expressions and numbers. We all use simple mathematical models in our everyday lives. The most common example is arithmetic, which we need for calculating monetary transactions amongst so many other things, and which is part of our language. The numbers represent the physical currency, or electronic credits, but the example is atypical because no approximation need be involved. A more typical example is provided by the answer to the question: how much paint do you need to redecorate your living room? You would probably model the area to be painted by rectangles, work out their areas and sum the areas. Your rectangles alone will not provide precise information about the details of the room, such as the position of windows, but the simple model is quite suitable for calculating the volume of paint to buy. However, it would not be adequate for working out suitable sizes and positions for central heating radiators.

The mathematical models needed for research programmes are more complicated, but the sequence is common to all applications: stating the problem, formulating an appropriate mathematical model, obtaining the solution and interpreting the solution in the practical context. We should also monitor the accuracy of our solution, and then refine our model to improve subsequent predictions.

There are many reasons for developing mathematical models. Their use can be traced back to prehistory, when people cut notches in sticks to record counts. The invention of practical geometry is generally attributed to the Egyptians more than 4000 years ago. See, for example, the MacTutor History of Mathematics archive at the University of St Andrews.

The Egyptians knew how to calculate the areas of triangles, and so they calculated the areas of fields by dividing them into triangular shapes. The motivation for this was that the waters of the river Nile overflowed every year and swept away the land boundaries. As a consequence of geometry, farmers knew the extent of their land and the authorities could work out a corresponding taxation. Egyptians were also aware that a triangle with sides in the ratio 3:4:5 was right angled, and they made use of this fact when constructing pyramids, although there is no evidence that they ever considered proving that this result followed from a set of axioms. The general result, that the square on the hypotenuse equals the sum of the squares on the other two sides, was known to the Babylonians, about 1000 years before Pythagoras set up his school at Kroton (in southern Italy) in 532 BC, but there is no evidence that they proved it. The proof of the general case, referred to as Pythagoras's Theorem, is perhaps the best known of all mathematical theorems. The Pythagoreans were somewhat secretive, so it is not known for certain that Pythagoras was the first to prove the theorem that bears his name, but his school certainly made amazing contributions to mathematics. The most famous is a proof that the square root of two, which is the length of the hypotenuse in a right-angled triangle that has its other two sides of length 1, cannot be expressed exactly as a fraction (ratio of two integers). It is extraordinary that these mathematicians posed the question, and the method of proof by contradiction was another insight, yet, in modern notation, it can be encapsulated in a few lines of elementary algebra. A further surprise is that the difference between the square root of two and a very close approximating fraction can lead to quite different outcomes in the chaotic dynamic systems that have been proposed as models of weather.

Al-Khuwarizmi (AD 770–440) and his colleagues, the Bairn Musa, were scholars at the House of Wisdom in Baghdad. Their tasks included the translation of Greek scientific manuscripts, and they also studied, and wrote texts on, algebra, geometry and astronomy. The algebra treatise *Hisab Al-Jabr w'al-Muqabilah* is the most famous of Al-Khuwarizmi's works, and is the origin of the word *algebra*. The purpose of the book was to provide a means of solving practical problems of the time, which included those arising from trade, lawsuits and the construction of canals. His name was adopted in medieval Latin as 'algorismus' and later in English as 'algorithm' to mean 'a sequence of rules for solving a problem, usually, but not always, mathematical'.

Mathematical models have always been crucial for the physical sciences and engineering. The main objectives are explanation and prediction. Isaac Newton published his *Philosophiac Naturalis Principia Mathematica* in 1687 with the assistance of Edmund Halley. Newton showed how his principle of universal gravitation together with his three laws of motion explained the motions of the planets as well as the trajectory of a ball thrown on earth. In brief, he assumed that the force between two bodies is proportional to the product of their masses, and inversely proportional to the square of the distance between them, and that this force produces a change in their velocities. Newton developed the calculus to investigate the consequences of these assumptions. It is remarkable that such a succinct set of premises can explain nearly all observed motion, However, all explanations stop at some point, and the theory does not attempt to explain what might underlie the apparent inverse square law.

Newton was not the first person to offer a model for the orbits of satellites. By around 190 BC, Hipparchus had developed a theoretical model of the motion of the moon based on epicycles. He showed that his model did not agree totally with observations, and it seems to have been Ptolemy who was the first to refine the model to take these discrepancies into account. Hipparchus was also able to give an epicycle model for the motion of the sun, but he did not attempt to give an epicycle model for the motion of the planets. Ptolemy's epicycles give a geometric description of orbits but do not provide the depth of explanation offered by Newton's theory and are not so generally applicable. However, even Newton's model is not adequate for dealing with objects that move at speeds approaching that of light. Albert Einstein's Theory of Relativity overcomes this limitation, and when British eclipse expeditions in 1919 claimed to have confirmed his predictions, *The Times* ran the following headline (London, 7 November 1919):

Revolution in science — New theory of the Universe — Newtonian ideas overthrown

It would have been more accurate to say that Einstein's model explains phenomena that Newton's cannot, but the impact of the headline would have been lost. Einstein had written a popular exposition of relativity in 1916, and the 15th edition of the authorised translation was first printed in paperback in 1960 (Einstein, 1960).

Although much of the progress in mathematics has been inspired by a desire to solve practical problems, there are important exceptions. Extraordinary innovations, such as introducing i for the square root of -1 in the late 16th century for the solution of cubic and quartic polynomial equations, have turned out to have great practical value. The concept of i is of fundamental importance in mathematics and is extensively used in mathematical models of physical phenomena.

Until the end of the 20th century, the majority of mathematical applications have been predominantly in the physical sciences. In contrast, during the 21st century the emphasis may turn to biological, ecological and medical applications. The Wolfson Centre for Mathematical Biology established at the University of Oxford in 1983 was the first centre of its kind in Britain. The Centre for Mathematics Applied to the Life Sciences was set up in 2003 as a collaboration between the University of Glasgow and the University of Strathclyde, and there are similar centres at many universities throughout the world. The University of Michigan Mathematical Biology website lists 24 journals relevant to researchers in the area and invites more suggestions. There are many recent books covering advances in mathematical modelling in life sciences. Examples include the book by Britton (2003), the two volumes on mathematical biology by Murray (2002, 2003), the two volumes on mathematical physiology by Keener and Sneyd (2008, 2014) and edited volumes by Takeuchi *et al.* (2007a,b) and Formaggia *et al.* (2009) in Springer's Modeling, Simulation and Applications (MS&A) series.

Despite the global financial crisis, or possibly as a consequence of it, finance still provides scope for developments in mathematics. Felix Salmon, writing in *Wired* magazine (23 February 2009), said, 'One result of the collapse has been the end of financial economics as something to be celebrated'. As Li himself said of his own model, 'The most dangerous part is when people believe everything coming out of it'. The damage was foreseeable and, in fact, foreseen. In 1998, before Li had developed his copula function, Paul Wilmott wrote that 'the correlations between financial quantities are notoriously unstable' (http://www.wired.com/techbiz/it/magazine/17-03/wp_quant?currentPage=all).

Also, neither Li nor mathematics can be held responsible for the miss-selling of loans and failures of regulators that had much to do with the crisis (Keen, 2011). Other mathematical models have had a more successful record in finance. In 1973, Black and Scholes published a formula for pricing share options. Before its use became widespread, any dealers who happened to know of their formula were able to identify under-priced share options and make substantial profits. As a consequence, there was a certain mystique about the result. Fischer Black died in 1995, but Myron Scholes and Robert Merton went on to share the 1997 Nobel Prize in Economics for its derivation. Nevertheless, the Black–Scholes formula has also been associated with the financial crash (http://www.bbc.co.uk/news/magazine-17866646).

Ian Stewart says, 'It was abuse of their equation that caused trouble, and I don't think you can blame the inventors of an equation if somebody else comes along and uses it badly', a cautionary remark that applies to all mathematical modelling.

Mathematical models can make a valuable contribution to work in many disciplines. The Pythagoreans were fascinated by the relationship between mathematics and music. Links between mathematics and architecture go back to the ancient Egyptians and provide material for journals including *Archaeometry* and *Nexus Network Journal – Architecture and Mathematics*. Ron Knott has an interesting web page on the Fibonacci numbers which includes a discussion of the extent to which the golden section really occurs in architecture, art and music (http://www.maths.surrey.ac.uk/hosted-sites/R.Knott/Fibonacci/).

There are several fascinating examples of applications of mathematics in anthropology and history in Pearce and Pearce (2010). In their chapter on dating the last migration from Polynesia to New Zealand, they use stochastic models and statistical analysis to refute an established claim that the canoes of the *Heke* were not contemporaneous, and that the *Heke* was therefore not a historical event.

Sociological applications are more modern. Shannon and Weaver (1949) wrote a monograph entitled *The Mathematical Theory of Communication*, which has been relevant to electronic communications and linguistics. Poundstone (1992) provides an intriguing mathematical analysis of the 'Prisoner's Dilemma' and associated potential conflict situations. The availability and power of modern computers have led to mathematical models that are used for simulations rather than being manipulated to provide algebraic solutions. For example, a recent computer simulation that was demonstrated on a BBC science programme suggested that a crowd would disperse through an emergency exit more quickly if it had a central divider. People in the crowd were modelled by dots in a plane, and impasses were resolved by various sets of rules, which could include randomisation. Helbing *et al.* (2000) present a more sophisticated model of pedestrian behaviour, and suggest practical ways to prevent dangerous crowd pressures.

Logic and mathematics are inextricably linked, and an understanding of the principles of mathematical modelling would be an advantage for research studies in many areas of philosophy. For example, the best non-specialist account of Georg Cantor's work on infinite sets that I have read is given by Reid (1963). The history of mathematics is a well-researched subject, but it may be more surprising that mathematics can contribute to historical studies. Fairclough (2000) gives a nice example. Sometime around 1640, one Edward Somerset, the Second Marquis of Worcester, demonstrated a machine that gave the impression of perpetual motion to King Charles I and most of his court. The machine, which was constructed in the Tower of London, was a great wooden wheel, 14 feet in diameter, with 40 cannon balls strung at strategic intervals around the rim. It seems that the Marquis carefully avoided an explicit claim of having invented a perpetual motion machine, and merely invited his audience to 'Be pleased to judge the consequences'.

In 2000, the Clay Mathematics Institute announced seven Millennium Prize Problems with prizes of $1 million. Six remain to be solved, and three of these can be easily described in terms of mathematical modelling. The 'P versus NP problem' is to answer the question of whether every problem whose solution can be quickly verified by a computer can also be quickly solved by a computer. The 'Navier–Stokes existence and smoothness' is related to the modelling of turbulence, and the 'Yang–Mills existence and mass gap' concerns elementary particles in quantum physics. The Clay website is http://www.claymath.org/millennium/. Wikipedia provides discussion of the problems at http://en.wikipedia.org/wiki/Millennium_Prize_Problems, and you can find more detail in Devlin (2003).

Learning Mathematics

If you are working in any of the physical sciences, life sciences, social sciences or specialist fields in other disciplines including, for example, linguistics, archaeology and anthropology, there are at least three reasons why it will be worthwhile revising and broadening your mathematical knowledge:

1. An understanding of mathematics is necessary to understand much of the research literature.
2. Computer software provides a dazzling range of sophisticated techniques for analysing and displaying data. Comprehension of the underlying mathematical principles will enable you to choose appropriate methods for your research, to understand the computer output and to be aware of the limitations of the techniques you use.
3. Quite straightforward mathematical models can be used in new contexts, and such applications often lead to publications in reputable journals.

If you are an engineer, mathematician or physicist, you will already have had considerable experience of mathematical modelling, and have the necessary concepts to teach yourself new techniques relatively easily. Even so, finding a relevant course, at either the post-graduate or undergraduate level, should help you learn new methods more quickly. Kreyszig's book, which is now in its 10th edition (Kreyszig, 2011), is a comprehensive general reference work. Giordano *et al.* (2009) is an example of a text on 'mathematical modelling' rather than a specialist areas of mathematics. Your supervisor, and your research colleagues, will be able to recommend specific books for your research area. You can also browse the library shelves, and search the internet. Wikipedia articles generally provide a good introduction to a new topic with many references to detailed texts. You can often find useful course notes that instructors, and the institution that employs them, have generously made freely available. The Massachusetts Institute of Technology has set up **MIT OpenCourseWare** (OCW), a web-based publication of most of MIT's course content. MITOCW was announced in 2001, offered 50 courses in 2002 and 2000 courses, many relevant to mathematical modelling, in 2010.

If your background is in chemistry or the life sciences, such as biology and medicine, or the social sciences, such as business, economics, geography or sociology, it may be some time since you took any mathematics courses, although advanced mathematical options are often offered in these subjects. Studying mathematics up until your final secondary school year is a good basis for further reading, but you may have dropped maths at the first opportunity, and typical mathematics textbooks for the final secondary school years may be rather more detailed than is necessary for your purposes. A book that covers the essential material in the context of your subject area should be more relevant. An example is *Quantitative Methods for Business, Management and Finance* (Swift & Piff, 2010), and *Modeling the Dynamics of Life* (Adler, 2012). At a more advanced level, *An Introduction to the Mathematics of Biology — with Computer Algebra Models* (Yeargers *et al.*, 1996) is an interesting book that would be a good starting point for post-graduate research. The authors' objectives included making the subject relevant and accessible to students of either biology or mathematics. In the preface, they state that two major journals, *Mathematical Biosciences* and *Journal of Mathematical Biology*, had tripled in size since their inceptions 20 to 25 years earlier. For chemists, a recent title is *The Chemistry Maths Book* by Erich Steiner (2012), and for life scientists *Mathematics and Computing in Medicine and the Life Sciences* (Hoppensteadt and Peskin, 2001). There is tremendous scope for mathematical modelling in geography, geomatics and the earth sciences (see e.g. Gomez and Jones, 2010). The mathematical treatment of economics is referred to as *econometrics* and can be considered a specialist branch of probability and statistics (see e.g. Hendry, 2001).

It is often easiest to learn new mathematical material from lecture classes, and researchers are usually welcome to sit in on these. However, there may not be relevant courses at convenient times for you, you may need to learn the material more quickly and lecture classes are unlikely to cover your precise requirements. So, it is likely that you will have to teach yourself some mathematics. It may be a good idea to consult several books on a particular topic. If you find one author's exposition hard to follow, another's approach may be clearer to you or provide that extra detail that you need to understand the point. Also, books written for a general readership such as *Mathematics: The Science of Patterns* (Devlin, 1994), *Does God Play Dice?* (Stewart, 1997) and *Calculus Made Easy* (Thompson and Gardner, 1999) can be very helpful for understanding basic concepts. Whatever your speciality, it is worthwhile checking other fields. For example, marine technology and aeronautical engineering have much in common. The mathematical techniques associated with engineering control are also used in econometrics. Many research topics, in disciplines as different as civil engineering and medicine, involve the identification and analysis of non-linear systems, and there are many other such examples.

The advantages of taught courses over a textbook are: a good instructor will emphasise the essential aspects of a subject and may cover the gist an advanced textbook within a few lectures, you have the opportunity to ask the instructor questions and you can discuss matters with other participants on the course.

Mathematical Software

Modern computers have greatly reduced the need to learn the detail of mathematical methods, and it will often suffice to understand the general principles. For example, the details of algorithms for calculating the eigenvalues and eigenvectors of matrices can usually be ignored by non-specialists given the commercial and open source software that is readily available to perform the calculations in an efficient manner. However, software does not substitute for an understanding of the concept of eigenvalues and eigenvectors. It is usually far easier to understand these ideas for special cases, such as for 2×2 or 3×3 matrices when they have clear geometric interpretations, than for the general $n \times n$ case.

Mathematica, Matlab and Maplesoft are powerful software systems for numerical and symbolic computation and scientific graphics. Your choice of system is likely to be heavily influenced by what your research institution is willing to provide.

Alternatively, you can consider open source software. Octave and Scilab are similar in structure to Matlab. The terse array programming language J, developed by Kenneth E. Iverson and Roger Hui since the early 1990s, followed Iverson's work on APL. The R Project for Statistical Computing maintains R.

At a lower language level, Fortran is still commonly used for research work in engineering, despite the increasing use of C++. The latest version of Fortran is Fortran 2008. If you are writing code in languages such as C or Fortran, the Numerical Algorithms Group (NAG) library of subroutines is a valuable resource. Another source of algorithms is Numerical Recipes Software (www.nr.com).

Spreadsheet software may be appropriate if you are developing a product for general use. Excel is probably the most widely known (see e.g. Liengme, 2008), but there are others including open source software such as OpenOffice.

The web is the best source of up-to-date information on software and associated materials including books.

Literature Searches

When you have an idea for research, your first task should be to check the literature. Although it is disappointing if you find that others have already worked on the idea, it is much better to identify this work before you invest too much of your own time on the idea. Also, you may find that they have taken a different approach from what you intend following or they have not considered aspects of the problem that you have thought of.

Novel applications of mathematical modelling are likely to be published in journals relating to the application area. Web of Knowledge and Scopus are large databases that are very useful for all subjects. Then there are specialist databases, such as PubMed for medicine and the biological sciences, which you should also consult. Make sure you try searches over a wide range of keywords. For example, try searching on 'random process railways' and separately on 'stochastic process railways' in Web of Knowledge. You will find you pick up different papers.

If the mathematical modelling has included innovative mathematics, an account of this innovation may have been published in a mathematics journal that is included in the specialist database MathSciNet, which is provided by the American Mathematical Society, but doesn't make Web of Science or Scopus. For example, the *ANZIAM Journal Electronic Supplement* appears in MathSciNet but does not currently appear in Web of Knowledge. Also, always check Google Scholar – remember that you can make this search quite specific by using the drop-down Find Articles template.

I use both Google and Google Scholar, for conference papers, only some of which appear in databases; theses and abstracts of theses; and any manuscripts that are posted on the internet but have not been published, following peer review, in a journal. A renowned example is the 1970 paper by Hammersley and Clifford (1970). Such manuscripts are usually referred to as *preprints*, and they are typically a written version of a talk given at a conference. Preprints often, but not always, eventually appear in a journal after peer review. The electronic preprint archive arXiv was set up by Paul Ginsparg in 1991 for distributing theoretical high-energy physics preprints, but it now includes all areas of mathematics, computing, physics and quantitative biology.

At the end of your literature search, you will have to re-evaluate your idea. Ideally, it will be new ground. In other cases you will find that others have looked at the problem, but that you can refine your original idea and continue with the research and achieve a publication in a respectable journal – your supervisor will be able to advise you. Sometimes you will just have to think up another idea.

You should continue to monitor the research literature throughout your research. I encourage post-graduate students to publish results, in journals or the proceedings of well-established conferences, as they go along with the aim of establishing some priority for their ideas. However, your supervisor will advise you whether this is a good publishing strategy for your research. Moreover, many universities now allow PhD candidates to submit a thesis based on published papers together with a substantial introduction. This is becoming a popular option.

Reading Papers with Mathematical Content

Reading mathematics in papers is usually hard work unless you are very familiar with the techniques used. This is partly because the mathematics is often presented in a very formal manner, with many intermediate steps omitted. In addition, the final logical order may not correspond to the intuitive way in which it was developed. It is best to start by reading the article through fairly quickly to get an overview.

If then you decide you need to read it thoroughly, it may help to start by looking at special cases. For example, an exposition in terms of n-dimensions may become far clearer in one, two or three dimensions. Try not to be put off by the use of technical terms. These are often a generalisation of concepts that may be familiar to you. Also, Wikipedia usually provides an excellent introduction to mathematical topics, with references and links should you need more detail. Try it with 'tesseract' (http://en.wikipedia.org/wiki/Tesseract).

If you reach an impasse, it may be possible to leave the impenetrable section out and to proceed with the remainder of the paper. A difficult piece may seem quite understandable if you return to it on another day. Also, other people will be more inclined to help with a specific section of a paper than with the paper in its entirety, particularly if they can do so without reading all the paper. If you have a substantial query that you have not been able to resolve with colleagues', or a supervisor's, advice, you can always try contacting the author. A final tip is that some people find it worthwhile enlarging mathematical derivations when they photocopy a paper.

You may find that the relevant material from published papers has been incorporated into a recent book, in which case reading the book is likely to be easier, and more convenient, than reading the original papers. You can search for recent books, by author or by subject, on the internet.

Promoting Applications of Mathematics

There are many associations that promote applications of mathematics. The American Mathematical Society and the Isaac Newton Institute were mentioned in the Introduction. The Society for Industrial and Applied Mathematics (SIAM) was inaugurated in Philadelphia in 1952. Their website is at www.siam.org/nnindex.htm.

The goals of SIAM are to:

- advance the application of mathematics to science and industry;
- promote mathematical research that could lead to effective new methods and techniques for science and industry; and
- provide media for the exchange of information and ideas among mathematicians, engineers and scientists.

These goals haven't changed, and they remain as relevant today as they were in 1952.

The Institute of Mathematics and its Applications was founded in England in 1964, with similar objectives. Their website is at www.ima.org.uk. Its bulletin, *Mathematics Today*, contains general interest articles that describe novel applications. They are certainly not restricted to engineering and science. For example, Moiseiwitsch discusses the link between mathematics and art in his 1999 article (Moiseiwitsch, 1999).

Advanced Courses

In the United Kingdom, the London Mathematical Society (LMS; http://www.lms.ac.uk/content/forthcoming-short-courses), the Isaac Newton Institute for Mathematical Sciences at the University of Cambridge (www.newton.ac.uk) and the International Centre for Mathematical Sciences (ICMS) in Edinburgh (www.icms.org.uk) all offer short courses, and web seminars, from time to time.

For example, the LMS advertised in 2012:

- Stochastic Modelling in Biological Systems, 18–23 March 2012, Oxford; and
- Continuum Mechanics in Biology and Medicine, 17–22 June 2012, UCL.

In Australia, the Australian Mathematical Sciences Institute offers advanced courses (www.amsi.org.au). In particular, government funding enabled the establishment of an access grid room network to allow collaborative teaching of advanced mathematics at remote sites. Institutions in many other countries will offer similar opportunities.

Many of the courses are aimed at research students working in the area of mathematical modelling, rather than exclusively research mathematicians. If such courses are relevant to your research, apply early!

Types of Mathematical Models

I have found it convenient to give more details of applications of mathematics under three chapters: deterministic models (Chapter 32), stochastic models (Chapter 33) and optimisation (Chapter 34). The division between deterministic and stochastic is rather arbitrary: some aspects of signal processing are deterministic rather than stochastic, and some control theory is presented in a deterministic context whereas other formulations take account of random disturbances and measurement errors. Modification of deterministic models to take account of random variation may be relatively straightforward, for example by using a non-linear least-squares routine to fit the model to noisy data, or involve substantial mathematical innovation such as stochastic calculus. Also, some, if not all, of the optimisation methods can be thought of as models. For example, stochastic dynamic programming can be used to optimise the management of water resources, including reservoirs and hydroelectric power generation, and it relies on a mathematical model of the system. I have omitted explicit mention of statistical methods because they are discussed elsewhere in this book. However, fitting stochastic models to data is a statistical problem.

The following tables categorise some broad areas of mathematics, by deterministic, stochastic or optimisation models, with examples of applications in physical, life and social sciences. The maths areas overlap and the choice of areas and examples is quite subjective, but I aim to show the variety of applications and to encourage you to add novel applications of your own.

Deterministic Models

Maths area	Physical	Life	Social
Linear dynamics	Vibration of rotors (Chapter 32, this volume)	Growth rate of Loblolly pine and Norway spruce (Valentine *et al.*, 2012)	Social drivers of African elephant movement (Boettiger *et al.*, 2011)
Non-linear dynamics	Response of suspension bridge to earthquake (Kim and Thai, 2010)	RNA replication (Sardaynes *et al.*, 2012)	Human migration (Tabata *et al.*, 2011)
Control theory	Autopilots (Sharma *et al.*, 2012)	Heart rate biofeedback effect on performance of basketball players (Maman, 2012)	Robust management policies (Kasa, 2002)

Maths area	Physical	Life	Social
Catastrophe theory	Stable phases of boson stars (Kleihaus *et al.*, 2012)	Agricultural ecosystem health (Su *et al.*, 2012)	Dispute negotiation (Chow *et al.*, 2011)
Partial differential equations	Rain water runoff on building façades (Blocken and Carmeliet, 2012)	Pituitary cells (Osinga *et al.*, 2012)	Crowd dynamics (Dogbe, 2010)

Stochastic Models

Maths area	Physical	Life	Social
Markov chains	River flow in arid regions (Fisher *et al.*, 2010)	Speech recognition (Juang and Rabiner, 1991)	Forest fires (Boychuk *et al.*, 2005)
Point processes	Fibre-based materials (Gaiselmann *et al.*, 2012)	Population dynamics (Raghib *et al.*, 2011)	Epidemics (Wang *et al.*, 2012a)
Time series and signal processing	Control of a multi-cell power supply (Munoz *et al.*, 2012)	Blue whale calls recognition (Bahoura and Simard, 2012)	Addressing employment after the Great Recession (Neumark and Troske, 2012)
Spatial processes and image processing	Spatial rainfall (Cowpertwait, 2010)	Brain neuroimaging (Rasmussen *et al.*, 2012)	Automatic light beam controller to assist drivers (Alcantarilla *et al.*, 2011)
Stochastic calculus	Dust particles in a plasma (Asgari *et al.*, 2011)	Evolution of morphological forms in biology (Kloeden and Lorenz, 2011)	Stock market – European options pricing (Wang *et al.*, 2012b)
Simulation	Wave energy (Ozkop *et al.*, 2012)	Plant spread or extinction (Fennell *et al.*, 2012)	Social knowledge evolution (Kostas, 2012)

Optimisation

Maths area	Physical	Life	Social
Calculus	Shape optimisation of hydrogen fuel cell (Al-Smail and Novruzi, 2012)	Optimal airflow pattern for breathing (Hancao and Haddad, 2011)	Optimisation of multi-path routed network (MacLaren *et al.*, 2009)

Maths area	Physical	Life	Social
Descent algorithms	Airfoil geometry (Landman and Britcher, 2000)	Chemical reaction rates (Jackels *et al.*, 1995)	Debt dynamics (Semmler and Sieveking, 2000)
Linear programming	Gas transmission (De Wolf and Smeers, 2000)	Emission of greenhouse gases from agriculture (De Cara and Jayet, 2000)	Group decisions (Contreras, 2012)
Dynamic programming and stochastic dynamic programming	Reservoir management (Fisher *et al.*, 2014)	DNA sequence analysis (Wheeler and Hughey, 2000)	External financing of firms (Cerqueti, 2012)
Critical path analysis	Programming techniques for microprocessors (Broberg *et al.*, 2001)	Nuclear materials safeguards (Booth and Isenhour, 2000)	Planning training within a large bank (Hertz *et al.*, 2000)
Genetic algorithms and simulated annealing	Optimisation of pumping and disinfection of water supply (Gibbs *et al.*, 2010)	Human posture recognition (Hu *et al.*, 2000)	Cooperative trade (Sherratt and Roberts, 1999)
Artificial neural nets	Battery life indicator for electric vehicle (Dai, 2012)	Processing electroen- cephalography signals (Peters *et al.*, 2001)	Detecting cybercriminal networks (Lau *et al.*, 2014)
Particle filters	Image enhancement (Tsai *et al.*, 2012)	Model selection in computational biology (Lillacci and Khammash, 2012)	Social interaction and group behaviour (Langdon and Poli, 2006)
Heuristic methods	Energy saving for automatic trains (Sheu and Lin, 2012)	DNA motifs with nucleotide dependency (Leung and Chin, 2006)	Visiting all stations on a large subway system (Drozdowski *et al.*, 2012)
Multi-objective Pareto optimality	Design of composite laminate (Nik *et al.*, 2012)	Organic contaminants in European rivers (Pistocchi, 2012)	Selection from applicants for a vacancy in an organisation (Druart and De Corte, 2012)

Typing Mathematics in Your Thesis

If you are used to Microsoft Word (or some other word processor) and its equation editor, you may be happy to use it for your thesis. An alternative, which has benefits if you will have many complex equations, is the document mark-up language LaTeX.

Donald Knuth initially released TeX in 1978, and Leslie Lamport released his higher level language LaTeX six years later. Although LaTeX was originally developed by computer scientists for the mathematical community, it is also used by researchers who need to use non-Latin scripts (http://en.wikipedia.org/wiki/LaTeX). LaTeX is open source, and Tony Roberts provides a helpful introduction (http://www.maths.adelaide.edu.au/anthony.roberts/LaTeX/index .html).

If you are a research student in a mathematics department, you may be expected to use LaTeX. But, if most of your colleagues use Microsoft Word, it may be easier to follow their lead and ask for advice if you need it. Most publishers, Wiley for example, provide both LaTeX templates and Microsoft Word templates, and LaTeX is generally preferred for camera-ready copy. A recent Web discussion (http://mappingignorance.org/2015/04/06/word-or-latex-typesetting-which-one-is-more-productive-finally-scientifically-assessed/) discusses some of the advantages and drawbacks of LaTeX. The sooner you start with LaTeX, the more likely you are to appreciate its advantages. There are several books on LaTeX including those by Lamport (1994) and Mittelbach and Goossens (2004), and the free the Not So Short Introduction to LaTeX 2 by Oetiker *et al.* (https://tobi.oetiker.ch/lshort/lshort.pdf). There are many other useful resources and aide-memoires on the Web.

Summary

We all use mathematics to some extent.

Many of the developments in mathematics over the centuries have been motivated by a desire to solve practical problems. However, abstract mathematical inventions, such as introducing the square root of negative numbers, have also turned out to be of great practical importance.

Mathematics is the means for quantifying research findings.

Imaginative applications of mathematics are not restricted to trained mathematicians.

There are considerable resources for learning more mathematics, whatever your mathematical background at the start of your postgraduate studies.

Multidisciplinary research attracts a substantial proportion of research funding. The subject of your first degree is not critical, provided you have the enthusiasm to study whatever background material is needed for the project.

Research in mathematics itself continues to be supported, and successful applications of mathematics will encourage funding bodies to maintain, or even increase, this commitment.

References

Adler, F.R. (2012). *Modeling the Dynamics of Life: Calculus and Probability for Life*. Scientists Cengage Learning.

Alcantarilla, P.F., Bergasa, L.M., and Jimenez, P. (2011). Automatic LightBeam Controller for driver assistance. *Machine Vision and Applications* 22(5).

Al-Smail, J.H., and Novruzi, A. (2012). Shape optimization of the hydrogen fuel cell cathode air channels. *Journal of Mathematical Analysis and Applications* 389(1): 293–313.

Asgari, H., Muniandy, S.V., and Wong, C.S. (2011). Stochastic dynamics of charge fluctuations in dusty plasma: a non-Markovian approach. *Physics of Plasmas* 18(8): 083709.

Bahoura, M., and Simard, Y. (2012). Serial combination of multiple classifiers for automatic blue whale calls recognition. *Expert Systems with Applications* 39(11): 9986–9993.

Blocken, B., and Carmeliet, J. (2012). A simplified numerical model for rainwater runoff on building facades: possibilities and limitations. *Building and Environment* 53: 59–73.

Boettiger, A.N., Wittemyer, G., and Starfield, R. (2011). Inferring ecological and behavioral drivers of African elephant movement using a linear filtering approach. *Ecology* 92(8): 1648–1657.

Booth, D.E., and Isenhour, T.L. (2000). Using PERT methodology in nuclear materials safeguards and chemical process control. *Environmental Modelling and Assessment* 5: 139–143.

Boychuk, D., Braun, W.J., and Kulperger, R.J. (2005). A stochastic forest fire growth model. *Environmental and Ecological Statistics* 15(2): 133–151.

Britton, N.F. (2003). *Essential Mathematical Biology*. New York: Springer.

Broberg, M., Lundberg, L., and Grahn, H. (2001). Performance optimization using extended critical path analysis in multi-threaded programs on multiprocessors. *Journal Parallel and Distributed Computing* 61: 115–136.

Cerqueti, R. (2012). Financing policies via stochastic control: a dynamic programming approach. *Journal of Global Optimization* 53(3): 539–561.

Chow, P.T., Cheung, S.O., and Yiu, T.W. (2012). A cusp catastrophe model of withdrawal in construction project dispute negotiation. *Automation in Construction* 22(SI): 597–604.

Contreras, I. (2012). Ordered weighted disagreement functions. *Group Decision and Negotiation* 21(3): 345–361.

Cowpertwait, P.S.P. (2010). A spatial-temporal point process model with a continuous distribution of storm types. Water Resources Research 46: W12507.

De Cara, S., and Jayet, P.A. (2000). Emissions of greenhouse gases from agriculture. *European Review of Agricultural Economics* 27: 281–303.

De Wolf, D., and Smeers, Y. (2000). The gas transmission problem solved by an extension of the simplex algorithm. *Management Science* 46: 1454–1465.

Devlin, K. (1994). *Mathematics: The Science of Patterns*. New York: Scientific American Library.

Devlin, K. (2003). *The Millennium Problems*. London: Granta.

Dogbe, C. (2010). Modeling crowd dynamics by the mean-field limit approach. *Mathematical and Computer Modelling* 52(9–10): 1506–1520.

Drozdowski, M., Kowalski, D., Mizgajski, J., Mokwa, D., and Pawlak, G. (2012). Mind the gap: a study of Tube tour. *Computers & Operations Research* 39(11): 2705–2714.

Druart, C., and De Corte, W. (2012). Designing Pareto-optimal systems for complex selection decisions. *Organizational Research Methods* 15(3): 488–513.

Einstein, A. (1960). *Relativity*, 15th ed. New York: Penguin.

Fairclough, T.J. (2000). The great weighted wheel. *Mathematics Today* 36(4): 107–113.

Fennell, M., Murphy, J.E., and Armstrong, C. (2012). Plant Spread Simulator: a model for simulating large-scale directed dispersal processes across heterogeneous environments. *Ecological Modelling* 230: 1–10.

Fisher, A.J., Green, D.A., and Metcalfe, A.V. (2010). Managing river flow in arid regions with matrix analytic methods. *Journal of Hydrology* 382(1–4): 128–137.

Fisher, A.J., Green, D.A., Metcalfe, A.V., and Akande, K. (2014). First-passage time criteria for the operation of reservoirs. *Journal of Hydrology* 519: 1836–1847.

Formaggia, L., Quarteroni, A., and Veneziani, A. (2009). *Cardiovascular Mathematics*. Berlin: Springer.

Gaiselmann, G., Thiedmann, R., and Manke, I. (2012). Stochastic 3D modeling of fiber-based materials. *Computational Materials Science* 59: 75–86.

Gardner, M. (1991). *The Ambidextrous Universe*. New York: Penguin.

Gibbs, M.S., Dandy, G.C., and Maier, H.R. (2010). Calibration and optimization of the pumping and disinfection of a real water supply system. *Journal of Water Resources Planning and Management – ASCE* 136(4): 493–501.

Giordano, F.R., Fox, W.P., Horton, S.B., and Weir, M.D. (2009). *A First Course in Mathematical Modeling*. Belmont, CA: Brooks/Cole.

Gomez, B., and Jones, J.P. (2010). *Research Methods in Geography*. New York: Wiley.

Hammersley, J.M., and Clifford, P. (1970). Markov fields on finite graphs and lattices. http://www.statslab.cam. ac.uk/~grg/books/hammfest/hamm-cliff.pdf

Hancao, L., and Haddad, W.M. (2011). Optimal determination of respiratory airflow patterns using a nonlinear multi-compartment model for a lung-rib-cage system. In *2011 American Control Conference – ACC 2011*, pp. 3524–3529.

Helbing, D., Farkas, I., and Vicsek, T. (2000). Simulating dynamical features of escape panic. *Nature* 407: 487–490.

Hendry, D.F. (2001). *Econometrics: Alchemy or Science*, 2nd ed. Oxford: Oxford University Press.

Hertz, A., Robert, V., and Berthod, V. (2000). Planning time usage and training within a large firm. *Rairo-Recherche Operationelle-Operations Research* 34(1): 61–83.

Hoppensteadt, F.C., and Peskin, C.S. (2001). *Mathematics in Medicine and the Life Sciences*. Berlin: Springer.

Hu, C.B., Li, Y., and Ma, S.D. (2000). Human posture recognition using genetic algorithms and Kalman motion estimation. *Chinese Journal of Electronics* 9: 457–461.

Jackels, C.F., Gu, Z., and Truhlar, D.G. (1995). Reaction-path potential and vibrational frequencies in terms of curvilinear internal coordinates. *Journal of Chemical Physics* 102(8): 3188–3201.

Juang, B.H., and Rabiner, L.R. (1991). Hidden Markov models for speech recognition. *Technometrics* 33: 251–272.

Kasa, K. (2002). Model uncertainty, robust policies, and the value of commitment. *Macroeconomics Dynamics* 6(1): 145–166.

Keen, S. (2011). *Debunking Economics*, rev. ed. London: Zed Books.

Keener, J.P., and Sneyd, J. (2008). *Mathematical Physiology II. Systems Physiology*. Berlin: Springer.

Keener, J.P., and Sneyd, J. (2014). *Mathematical Physiology I: Cellular Physiology*, 2nd ed. Berlin: Springer.

Kim, S.E., and Thai, H.T. (2010). Nonlinear inelastic dynamic analysis of suspension bridges. *Engineering Structures* 32(12): 3845–3856.

Kleihaus, B., Kunz, J., and Schneider, S. (2012). Stable phases of boson stars. *Physical Review (D)* 85(2).

Kloeden, P.E., and Lorenz, T. (2011). Stochastic morphological evolution equations. *Journal of Differential Equations* 251(10): 2950–2979.

Kreyszig, E. (2011). *Advanced Engineering Mathematics*, 10th ed. New York: Wiley.

Lamport, L. (1994). *LaTeX: A Document Preparation System User's Guide and Reference Manual*, 2nd ed. Reading, MA: Addison-Wesley.

Landman, D., and Britcher, C.P. (2000). Experimental geometry optimization techniques for multi-element airfoils. *Journal of Aircraft* 37(4): 707–713.

Langdon, W.B., and Poli, R. (2006). Finding social landscapes for PSOs via kernels. In *2006 IEEE Congress on Evolutionary Computation*, pp. 1639–1646.

Lau, R.Y.K., Xia, Y., and Ye, Y. (2014). A probabilistic generative model for mining cybercriminal networks from online social media. *Computational Intelligence Magazine IEEE* 9(1): 31–43.

Leung, H.C.M., and Chin, F.Y.L. (2006). Discovering DNA motifs with nucleotide dependency. In *2006 IEEE Symposium on Bioinformatics and Bioengineering*.

Liengme, B.V. (2008). *A Guide to Microsoft Excel 2007 for Engineers and Scientists*, 4th ed. London: Academic Press.

Lillacci, G., and Khammash, M. (2012). A distribution-matching method for parameter estimation and model selection in computational biology. *International Journal of Robust and Nonlinear Control* 22(10): 1065–1081.

MacLaren, W.J., Weber, S., and Wa, M.C. (2009). Optimal rate-delay tradeoffs and delay mitigating codes for multipath routed and network coded networks. *IEEE Transactions on Information Theory* 55(12) 5491–5510.

Mittelbach, F., and Goossens, M. (2004). *The LaTeX Companion*, 2nd ed. Reading, MA: Addison-Wesley.

Moiseiwitsch, B. (1999). Mathematics and art. *Mathematics Today* 35(6): 175–178.

Munoz, J.A., Espinoza, J.R., and Baier, C.R. (2012). Design of a discrete-time linear control strategy for a multicell UPQC. *IEEE Transactions on Industrial Electronics* 59(10): 3797–3807.

Murray, J.D. (2002). *Mathematical Biology: I. An Introduction*, 3rd ed. New York: Springer.

Murray, J.D. (2003). *Mathematical Biology: II. Spatial Models and Biomedical Applications*, 3rd ed. New York: Springer.

Neumark, D., and Troske, K. (2012). Addressing the employment situation in the aftermath of the Great Recession. *Journal of Policy Analysis and Management* 31(1): 160–168.

Nik, M.A., Fayazbakhsh, K., and Pasini, D. (2012). Surrogate-based multi-objective optimization of a composite laminate with curvilinear fibers. *Composite Structures* 94(8): 2306–2313.

Osinga, H.M., Sherman, A., and Tsaneva-Atanasova, K. (2012). Cross-currents between biology and mathematics: the codimension of pseudo-plateau bursting. *Discrete and Continuous Dynamical Systems* 32(8): 2853–2877.

Ozkop, E., Atlas, I.H., and Sharaf, A.M. (2012). A novel switched power filter-green plug (SPF-GP) scheme for wave energy systems. *Renewable Energy* 44: 340–358.

Pearce, C.E.M., and Pearce, F.M. (2010). *Oceanic Migration*. New York: Springer.

Peters, B.O., Pfurtscheller, G., and Flyvbjerg, H. (2001). Automatic differentiation of multichannel EEC signals. *IEEE Transactions on Biomedical Engineering* 48: 111–116.

Poundstone, W. (1992). *The Prisoner's Dilemma*. New York: Doubleday.

Raghib, M., Hill, N.A., and Dieckmann, U. (2011). A multiscale maximum entropy moment closure for locally regulated space-time point process models of population dynamics. *Journal of Mathematical Biology* 62(5): 605–653.

Rasmussen, P.M., Hansen, L.K., and Madsen, K.H. (2012). Model sparsity and brain pattern interpretation of classification models in neuroimaging. *Pattern Recognition* 45(6): 2085–2100.

Reid, C.R. (1963). *A Long Way from Euclid*. New York: Crowell.

Sardaynes, J., Martinez, F., and Daros, J-A. (2012). Dynamics of alternative modes of RNA replication for positive-sense RNA viruses. *Journal of the Royal Society Interface* 9(69): 768–776.

Semmler, W., and Sieveking, M. (2000). Critical debt and debt dynamics. *Journal of Economic Dynamics and Control* 24: 1121–1144.

Shannon, C.E., and Weaver, W. (1949). *The Mathematical Theory of Communication*. Urbana: University of Illinois Press.

Sharma, S.K., Naeem, W., and Sutton, R. (2012). An autopilot based on a local control network design for an unmanned surface vehicle. *Journal of Navigation* 65(2): 281–301.

Sherratt, T.N., and Roberts, G. (1999). The emergence of quantitatively responsive cooperative trade. *Journal of Theoretical Biology* 200: 419–426.

Sheu, J-W., and Lin, W-S. (2012). Energy-saving automatic train regulation using dual heuristic programming. *IEEE Transactions on Vehicular Technology* 61(4): 1503–1514.

Steiner, E. (2012). *The Chemistry Maths Book*, 2nd ed. Oxford: Oxford University Press.

Stewart, I. (1997). *Does God Play Dice? The New Mathematics of Chaos*, 2nd ed. New York: Penguin.

Su, S., Zhang, Z., and Xiao, R. (2012). Geospatial assessment of agroecosystem health: development of an integrated index based on catastrophe theory. *Stochastic Environmental Research and Risk Assessment* 26(3): 321–334.

Swift, L., and Piff, S. (2010). *Quantitative Methods for Business, Management and Finance*, 3rd ed. London: Palgrave Macmillan.

Tabata, M., Eshima, N., and Takagi, I. (2011). A mathematical modeling approach to the formation of urban and rural areas: convergence of global solutions of the mixed problem for the master equation in sociodynamics. *Nonlinear Analysis – Real World Applications* 12(6): 3261–3293.

Takeuchi, Y., Iwasa, Y., and Saho, K. (2007a). *Mathematics for Ecology & Environmental Science*. Berlin: Springer.

Takeuchi, Y., Iwasa, Y., and Saho, K. (2007b). *Mathematics for Life Science & Medicine*. Berlin: Springer.

Thompson, S.P., and Gardner, M. (1999). *Calculus Made Easy*. New York: St Martin's Press.

Tsai, H-H., Chang, B-M., and Lin, X-P. (2012). Using decision tree, particle swarm optimization, and support vector regression to design a median-type filter with a 2-level impulse detector for image enhancement. *Information Sciences* 195: 103–123.

Valentine, H.T., Makela, A., and Green, E.J. (2012). Models relating stem growth to crown length dynamics: application to loblolly pine and Norway spruce. *Trees – Structure and Function* 26(2): 469–478.

Wang, J-Z., Qian, M., and Qian, H. (2012a). Circular stochastic fluctuations in SIS epidemics with heterogeneous contacts among sub-populations. *Theoretical Population Biology* 81(3): 223–231.

Wang, X-T., Wu, M., and Zhou, Z-M. (2012b). Pricing European option with transaction costs under the fractional long memory stochastic volatility model. *Physica A – Statistical Mechanics and Its Applications* 391(4): 1469–1480.

Wheeler, R., and Hughey, R. (2000). Optimizing reduced space sequence analysis. *Bioinformatics* 16: 1082–1090.

Yeargers, E.K., Shonkwiler, R.W., and Herod, J.V. (1996). *An Introduction to the Mathematics of Biology*. Berlin: Birkhauser.

32

Deterministic Models

Andrew Metcalfe

Introduction

A good mathematical model will be as simple as possible, while including the essential features for our application. If we want to calculate an escape velocity for a rocket carrying a spacecraft, we might start by modelling it as a point mass. That is, its physical extent is ignored in the simplest calculation. But, the planet earth has an atmosphere and we need to allow for air resistance. As a first attempt we could model the rocket as a cylinder, and we might then improve on this by using a paraboloid to model the nose cone. We expect the improved model to give more realistic predictions. If there are astronauts on board, recovering them requires more detailed modelling. For instance, we need to model the heat distribution in the surface layers of the command module during re-entry. The US National Aeronautics and Space Administration (NASA) gives hundreds of questions that can be answered with straightforward mathematical techniques (http://spacemath.gsfc.nasa.gov). However, there are many challenging problems, related to space exploration, that are not so easily answered. The three-body problem, for example the motion of the moon under the gravitational influence of the sun and the earth, and the more general n-body problem still present research opportunities (http://plus.maths.org/content/mathematical-mysteries-three-body-problem, http://demonstrations.wolfram.com/PlanarThreeBodyProblem/). Canada's contribution to the International Space Station is the Mobile Servicing System, a complex manipulator that includes robot arms; mathematical models, with computer simulation of their solutions, were crucial for its design (http://www.maplesoft.com/company/publications/articles/view.aspx?SID=366&L=F).

The success of a model is usually judged in terms of the accuracy of predictions made using it, but we also try to capture something, at least, of the way we imagine the world to be. I have chosen the following examples of mathematical models to give some indication of the variety of techniques and areas of application.

Research Methods for Postgraduates, Third Edition. Edited by Tony Greenfield with Sue Greener.
© 2016 John Wiley & Sons, Ltd. Published 2016 by John Wiley & Sons, Ltd.

Mathematical Software

Mathematical software is virtually essential for developing mathematical models. Well-known systems for general applications are Maplesoft Maple, MathWorks, Matlab and Wolfram Mathematica. All include algebraic manipulation as well as numerical algorithms and sophisticated graphics. Although you are likely to have access to at least one of these systems, it may be worth knowing that there are open source, or otherwise free, alternatives. For example, Octave, and also SciLab, are similar to Matlab; Sage System for Algebra and Geometry Experimentation is similar to Mathematica; J is a high-level, general-purpose, high-performance programming language; R is an open source programming and graphics language that is very popular with the statistical community; and Maxima is a 'computer algebra system'. All these systems have their adherents, and the situation is continually evolving. At the lower level of programming languages, Fortran (the most recent standard is known as Fortran 2008) and C (C11 being the most recent standard) are widely used. Open source resources include gfortran (GNU Fortan compiler) and GCC C. A Wikipedia article gives an extensive list (see http://en.wikipedia.org/wiki/List_of_compilers).

Discrete and Continuous Variables

Although we perceive time and space to be continuous, and variables that vary over space or time (often referred to as *field* or *state variables*) as discrete or continuous, we do not have to model variables in the same way. For example, the size of a population of animals is a natural number, but if the population is large its size can be treated as a continuous variable. In contrast, the volume of water in a reservoir is a continuous variable, but we may choose to model it in discrete increments to give a discrete state space. Time is generally considered to be continuous, but provided we sample sufficiently quickly we can model time as a sequence of discrete steps. Modern equipment can sample at rates of speeds of tens of giga-samples per second (GS/s). The theoretically smallest possible time measurement is Planck time (roughly seconds) (http://en.wikipedia.org/wiki/Planck_time).

Differential equations and, more generally, partial differential equations model the state variable as continuous and time as continuous. There are many renowned partial differential equations in mathematics and physics. The wave equation, diffusion equation and Laplace's equation are linear partial differential equations that are covered in engineering mathematics textbooks such as Kreyszig (2010). Maxwell's equations of electromagnetism, the Navier–Stokes equations for fluid mechanics (e.g. Acheson, 1990) and the Schrödinger equation in quantum physics even make their way onto T-shirts and coffee mugs. But, you may find that textbooks, such as Jeffrey (2003) and Logan (2008), are more helpful to your research.

Discrete time models are based on difference equations (see e.g. Elaydi, 2005), also known as recurrence relations, in which the ratio of the difference to the time step is an approximation to the derivative. As many differential equations can only be solved numerically, we expect similar solutions in discrete time steps from either formulation. While this is often the case, you need to take care if there are instabilities in the models or in the system being modelled.

Linear and Non-linear Systems

The distinction between linear and non-linear dynamic systems is fundamental.

A system is linear if the response to a sum of input signals is the sum of the responses that would result from the individual input signals. If the input signal is a sine wave, the steady-state system response will be a sine wave at the same frequency, but with a change in phase. If the amplitude of the input signal is doubled, the amplitude of the response doubles. In algebraic terms a differential equation, which models a system, is linear if the sum of any two solutions is also a solution. The theory of linear systems is thoroughly worked out, and has been applied successfully to a wide range of practical problems. So, an attractive approach to modelling non-linear systems is to linearise, locally. However, this may not give satisfactory prediction.

Some algebraically simple sets of differential equations exhibit quite different responses if input conditions are very slightly perturbed. These are known as chaotic systems, and their responses may appear to be random despite being deterministic. The double pendulum provides a practical demonstration, and one of many on the internet is Steven Troy's (http://www.youtube.com/watch?v=U39RMUzCjiU&feature=related).

Ian Stewart's book (Stewart, 1997) is a fascinating introduction to chaotic dynamics, which discusses the links with fractals. His examples include: the tumbling of Hyperion, a moon of Jupiter; the weather; turbulence; and dripping taps. The vagaries of the tidal current under the bridge at Halkida, which connects Evia to the Greek mainland, might be explained by chaotic dynamics. Aristotle was greatly puzzled by the way the tide changed direction so many times during the day, and was not able to provide an explanation. Systems can exhibit many different deviations from linearity, and there is no general theoretical framework corresponding to that for linear systems. There is, therefore, great scope for research projects that involve modelling non-linear systems. Your starting point could be one of the textbooks by Strogatz (1994) or Thompson and Stewart (2002), but Mark Nelson at the University of Wollongong continues to update a list of reviews (http://www.uow.edu.au/~mnelson/books.html).

Figure 32.1, taken from ongoing work by Sarthok Sircar and Tony Roberts in the School of Mathematical Sciences at the University of Adelaide, is an example of non-linear modelling. They are solving the non-linear Fokker–Planck equation (a well-known equation which has

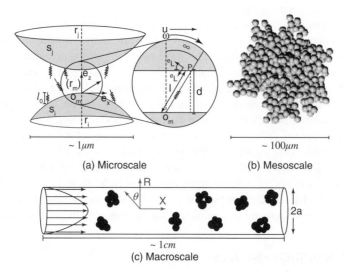

Figure 32.1   Modelling particles in fluid flows. Courtesy of Sarthok Sircar and Tony Roberts.

both the stochastic as well as the deterministic form) deterministically using a numerical scheme developed by Tony Roberts and colleagues. Figure 32.1 illustrates aspects of modelling aggregate ensembles of rigid particles with fluid flow inside a pipe. Applications include platelet aggregation inside a blood vessel and contaminant transport in water pipes.

Vibration Control

Although the simplest way to reduce unwanted vibration is to add dampers to the system, Victorian engineers designed clever mechanical devices, known as *tuned mass vibration absorbers*, which were tuned to counteract forces causing the disturbance. Some recent applications are described in Wikipedia (http://en.wikipedia.org/wiki/Tuned_mass_damper). The use of a tuned mass damper to render a wobbly footbridge more suitable for the general public is described by Jones *et al.* (1981), and a larger scale and more publicised use of such dampers is the Millennium Bridge, the footbridge linking the North Bank of the River Thames to the Tate Modern (http://en.wikipedia.org/wiki/London_Millennium_Bridge).

Jones *et al.* (1981) based their successful damper design on an assumption that the bridge can be modelled as a single-mode linear system (analogous to a mass on a spring). However, this assumption is certainly not realistic for large suspension bridges, because the cables supporting the deck oppose its downward movement but offer little resistance to its lifting. This is an example of non-linearity. In 1940, four months after the Tacoma Narrows Bridge was opened, a mild gale set up resonant vibrations along its half-mile centre span. It collapsed within a few hours (http://www.youtube.com/watch?v=3mclp9QmCGs). It was a slender bridge and, although some benign longitudinal oscillations had been allowed for in the design, the torsional vibrations that destroyed it were quite unexpected. Since the collapse, various mathematical models to account for the failure have been proposed, and relatively recently Billah and Scanlan (1991) and Doole and Hogan (2000) presented convincing non-linear models that account for the mechanism of the collapse.

Both dampers and vibration absorbers, which are made of small auxiliary masses and springs, are passive devices, insomuch as they do not require any auxiliary sensors, actuators or power supplies. This is still a significant advantage, despite microprocessors and advances in sensor and actuator designs. However, there are specialist applications where the advantages of active systems outweigh the drawbacks. The possibility of applying control forces with contactless electromagnets is crucial for some of these. Another advantage is their potential to supply energy when required as well as dissipate it, whereas a passive system can only dissipate and temporarily store energy. Active systems can also produce a force at a point in such a way that the force depends on signals received from sensors that may be far removed from that point. In contrast, passive systems produce a local force that is only related to local variables. A further advantage is that active systems can be adapted to different operating conditions, as these conditions occur, without any outside intervention. Applications include car suspensions, reduction of vibration of circular saws and other machine tools, high-speed centrifuges on magnetic bearings and the reduction of vibrations of rotating shafts. The modelling of rotating shafts illustrates many of the general issues which arise in mathematical modelling. The latest Range Rover Evoque offers a third-generation suspension system. This works by magnetising iron particles inside the suspension fluid to quickly adapt to road changes. Mercedes has developed active body control for its series E vehicles.

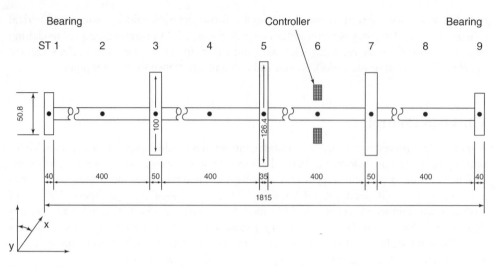

Figure 32.2   Model of rotor. Metcalfe and Burdess, 1992. Reproduced with permission from Springer.

Vibration Control of an Out-of-Balance Rotor

A diagram of the rotor, excluding the details of the bearings, is shown in Figure 32.2.

The rotor is a continuous body, but we were content to concentrate on the displacements of nine points on the rotor measured at nine stations (Metcalfe and Burdess, 1986, 1990, 1992). The first step is to imagine the rotor to be made up of nine pieces, with the mass of each piece concentrated at its centre of gravity, and with these masses connected by rods of negligible mass. The theory which describes this situation is well known (see e.g. Thomson, 2004), and the end result is a set of linear differential equations, with constant coefficients, which describe the free vibration of the undamped system. This model of the rotor has a system with 18 degrees of freedom (nine displacements in the x direction and nine in the y direction). We will return later to the question of whether 18 degrees of freedom will suffice. There will be nine characteristic frequencies (natural frequencies) and nine corresponding mode shapes in the plane containing the x direction and the axis of the shaft, and a similar set of nine frequencies and mode shapes for the y direction. The mode shapes corresponding to the first seven frequencies are shown in Figure 32.2.

If the rotor is flexed in some way and then released, the resulting motion will be modelled as some linear combination of the mode shapes vibrating at their natural frequencies. According to the model, this vibration will continue forever, because no damping has been allowed for. In practice, there will always be damping, caused by factors such as internal friction and air resistance, and the vibration will die down. However, such light damping has little effect on the natural frequencies or mode shapes.

The objective of our work was to reduce the vibration of a rotor caused by its not being perfectly balanced. This mass unbalance produces a disturbing force when it rotates, which will be harmonic at the frequency of rotation. If an undamped linear system is forced at a frequency that equals one of its natural frequencies, the theory predicts that the amplitude of vibration will increase without bound. This is known as *resonance*. Despite the light natural damping in structures, resonance is a real phenomenon that causes noise and wear, and it

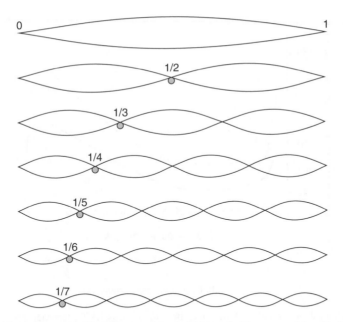

Figure 32.3 Mode shapes. Qef (public domain), via Wikimedia Commons. https://upload.wikimedia. org/wikipedia/commons/c/c5/Harmonic_partials_on_strings.svg

is potentially dangerous. Operating speeds of rotors must be well away from their natural (critical) frequencies. We were modelling a rotor in journal bearings, and the oil film provided some damping and increased the stiffness. Holmes (1960) presented an elegant model for a journal bearing, and showed that these effects depend on the speed of rotation. Our model was modified to include: the light damping; a sinusoidal disturbance at the shaft rotation frequency; and the controller with its transducer and actuator, which were positioned at station 6.

Although our model includes many differential equations, they are all linear and may be written as an equivalent system of 42 first-order linear differential equations, known as a *state–space description* (e.g. Kwakernaak and Sivan, 1972). There is a detailed theory of such systems, and of their control, most of which is implemented in Matlab and its associated toolboxes (see e.g. Nise, 2011).

In the engineering literature, a numerical solution to a mathematical equation, which purports to model a specific piece of equipment, is often referred to as a *(computer) simulation study*. This distinguishes it from a scale model, for example. Results of simulations at frequencies from 10 radian/s up to 240 radian/s in 10 radian/s intervals, with a low and high controller gain, are shown in Figure 32.4. The 'sum of squared displacements' is the sum of squares of the x and y displacements at the nine stations sampled over a 2 second interval, which is long enough for the transient response to become negligible. The vertical scale is 20 times logarithm base 10 of the ratio of the controlled to the uncontrolled response (dB). Thus, negative values reflect an attenuation, and any values above zero would correspond to a detrimental effect. At a speed of 80 radian/s, the vibrations are reduced by a factor of nearly 100. It may appear that further increasing the gain would improve the performance, but this would tend to make the controlled rotor system unstable.

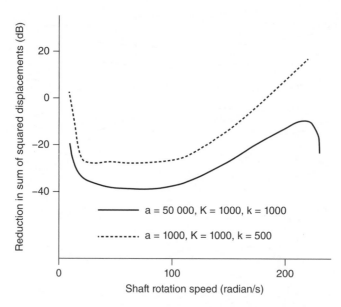

Figure 32.4 Control of a rotating shaft. Metcalfe and Burdess, 1992. Reproduced with permission from Springer.

It was not necessary, for this application, to model explicitly the rotation of the shaft. The harmonic forces generated by the shaft's rotation could be represented by the circular functions (sine and cosine). However, one drawback of a relatively simple linear model for the complex dynamics of an out-of-balance rotor is that it does not allow for the possibility of whirling (defined as the rotation of the plane containing the bent shaft and the line of centres of the bearings). Thompson (2004) explains how whirl can be modelled for a single disc. Whirl is a non-linear effect which comes under the classification of self-excited motion.

Another simplification was to think of the rotor as made up of nine point masses. This restricted the analysis to nine mode shapes and nine natural frequencies. We are assuming that the ignored modes are of no importance. Although high-order modes can often be neglected safely, they can have unexpected and potentially disastrous effects. Feedback control systems designed for a finite mode approximation to a physical system can be unstable in practice (Balas, 1982). This happens when the controller affects lightly damped modes which were ignored in the model, referred to as *control spillover*. The movements of these modes are detected by the transducers, *observation spillover*, and this may cause the controlled system to become unstable.

Unfortunately, it is easier to point out the hazards of ignoring high-order modes than it is to give general advice about how many to include. Individual skin panels in an aeroplane fuselage can resonate due to acoustic excitations from the jet engines. This phenomenon is known as drumming and can lead to fatigue failure, as in the original design of the de Havilland Comet which had square windows (http://news.bbc.co.uk/onthisday/hi/dates/stories/may/2/newsid_2480000/2480339.stm).

The mathematical model for the fuselage does not include high-frequency modes of individual panels. In most cases, the interaction between the global bending of the fuselage and natural frequencies of the panels is small enough for the latter to be analysed separately. But

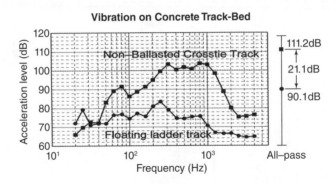

Figure 32.5   Vibration on a railway track. Courtesy of Roger Hosking.

it is unwise to rely on computer simulations, and laboratory experiments with test rigs, and scale models, are often the next step. The possibility of bugs in complex computer software, and hardware, should also be borne in mind.

The rotor example demonstrates the main concept of the finite element model, which is to divide a continuum into a finite number of bits that have simpler geometry than the original. Our bits were hypothetical point masses, joined by rods of negligible mass. In general, each bit (finite element) has a number of nodes that determine the behaviour of the element. Since displacements or stresses at any point in an element depend upon those at the nodes, we can model the structure by a finite number of differential equations describing the motions of the nodes.

Analysis of vibration is vital in many applications. Hosking and Milinazzo (2012) present a nice application of vibration analysis to ladder track for railways.

Finite Element Modelling

The finite element method is a computational method which is used routinely for the analysis of stress, vibration, heat conduction, fluid flow, electrostatics and acoustics problems. Practical applications rely on the power of digital computers, which have provided the impetus for the development of the method since its inception in the 1940s with the work of Alexandar Hrennikoff, Richard Courant and Olgierd Zienkiewicz.

Original work in elasticity, early in the 19thth century, led to differential equations that described the stress–strain relationship for materials, assumed to be linear. These equations can be solved for simple bar shapes. The essence of the finite element method is that quite complex and realistic structures can be made up from such simple elements. The use of continuous elements was the next step, and an example application is a cracked rotor (Imam *et al.*, 1989). John Appleby's website gives an ideal introduction to the method and his associated software FINEL, now in version 6 (http://www.staff.ncl.ac.uk/john.appleby/software.htm; Appleby, 1994–2012). This is easy to use, and it is capable of solving fairly large problems of static stress analysis, and field problems such as heat conduction and potential flow for two-dimensional (2D) and axisymmetric 3D situations. It can also be used to solve 2D and 3D frame structures.

In practice, we often use more than one mesh, of progressively increasing refinement, to help us assess the likely accuracy of the solution. If the increased resolution makes little

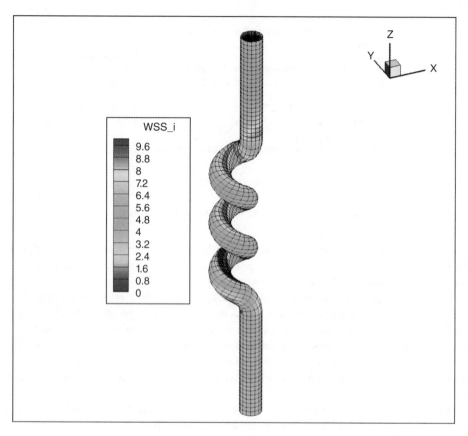

Figure 32.6   Wall shear stress in umbilical cord. Reproduced with permission of David Wilkie.

difference, it suggests that we may have reached convergence to a practical solution. It is important to realise that convergence may be very slow, so that we are much further from the limiting values than it appears; and convergence to a seriously misleading solution is also possible! We should then use a more accurate form of model, rather than increase the resolution of a simple one. We must have independent verification, if only from experience or an approximate estimate, before we can rely on any results. Wikipedia maintains a list of finite element software packages, including open source and proprietary systems (http://en.wikipedia.org/wiki/List_of_finite_element_software_packages).

Linear finite element analysis assumes that response is directly proportional to load, and that loads maintain their original directions as the structure deforms. This is adequate for many practical examples in which displacements and rotations are small, but not for analysing situations in which deflections are large and buckling may occur. As a consequence, nonlinear finite element analysis has been developed. Resources include Matthew Roberts's Non Linear Finite Element Toolbox for Matlab, which is available free under General Public Licence (http://nlfet.sourceforge.net); Peter McHugh's talk at the 2007 European Society of Biomechanics workshop (http://www.nuigalway.ie/micru/mchugh.htm); and the textbook by Crisfield *et al.* (2012).

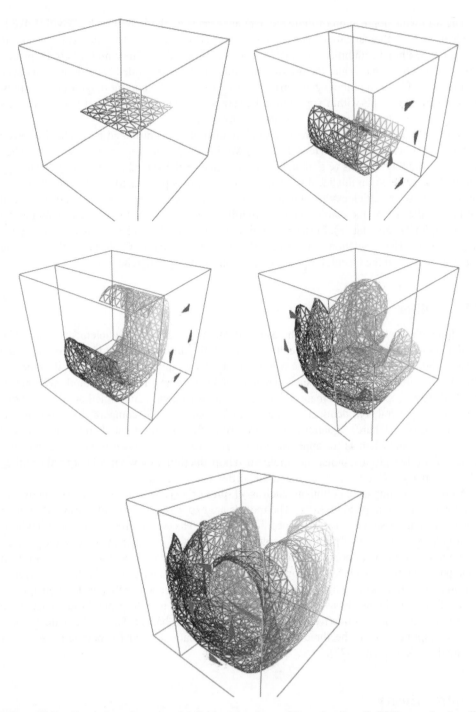

Figure 32.7   Simulation of non-inertial fluid mixing in a cubic cavity (Jewell, 2009; see also Jewell, 2012).

The following application of finite element analysis is work in progress by David Wilke in the School of Mathematical Sciences at the University of Adelaide. The context is examining blood flow within the umbilical vessels; WSS is wall shear stress. The umbilical cord typically takes on a helical shape, and the arteries and vein are likewise coiled. The project has been examining different umbilical geometries in order to assess the current diagnostic measures. He used finite elements within a program called oomph-lib which is developed at the University of Manchester. The flow for this particular run was steady.

Another example of finite element modelling of fluids is contributed by Nathaniel Jewell at the University of Adelaide (Jewell, 2009). Mixing processes tend to be non-inertial when the fluid is highly viscous (e.g. dough or honey) and/or the spatial scale is very small (e.g. biomedical devices). In this idealised simulation, stirring is performed in alternate directions in a clockwise or anticlockwise manner (red arrows). As stirring continues, the interfacial surface (indicated by the coloured mesh) rapidly expands in both surface area (quantitative) and complexity (qualitative). Surface evolution is continuously tracked using an adaptive triangular mesh. The simulation was performed using purpose-written C++ and Matlab software (computation) and the OpenGL open source graphics library (3D rendering).

Finite Difference Method

The finite element method applies the exact equations for the idealised elements to a model of the system, made up from these elements. The finite difference method approximates the differential equations describing the original system, by replacing derivatives with ratios of small, rather than infinitesimally small, changes in the variables. The two methods are conceptually different, and although many problems can be solved with either method, the solutions will not, in general, be identical. Gottardi and Venutelli (1993) compare the two methods for solving the Richards equation, which describes flow in unsaturated soil. The pros and cons of different schemes for approximating derivatives are discussed in books on the subject, Gerya (2009) for example, and briefly in Wikipedia (http://en.wikipedia.org/wiki/Finite_element_method#Comparison_to_the_finite_difference_method).

Modelling the impact of pollutants, such as oil spills or seepage from dumps, on groundwater is a topical area of applied research. The models are expressed in terms of partial differential equations, and these are solved for specific scenarios. The whole process can be thought of as simulating the effect of pollution incidents. The basis of the mathematical description of multiphase fluid flow in porous media is the conservation for mass and momentum for each fluid phase. Faust et al. (1989) present a two-phase flow model based on a 3D finite difference formulation. They use it to investigate the flow of immiscible, denser than water, nonaqueous fluids, from two chemical waste landfills near Niagara Falls, into the groundwater. One of the research issues was the effectiveness of clay as a geological barrier. The mathematical model was a simplification of the three-phase fluid flow equations used for petroleum reservoir simulation (Peaceman, 1977).

Control Theory

There are applications of control theory in almost all disciplines, and many textbooks such as Jacobs (1993), Barnett and Cameron (1985), Nise (2011) and Dorf and Bishop (2010). Dan Simon (2006) provides detailed coverage of the H-infinity criterion, minimising the worst case, and the method is supported by the Matlab Robust-Control toolbox. A particular feature of chemical processes is that there are often long time delays, due to thermal inertia, between applying control action and noticing its effect, and this potentially destabilising effect can occur in other contexts. Next, all this can be adapted for non-linear systems and Isidori (1995) is much cited. Professor Atherton's book is free (http://bookboon.com/en/textbooks/chemistry-chemical-engineering/an-introduction-to-nonlinearity-in-control-systems).

Robotics provides many challenging applications and is a very active research field. Just a few examples are the projects of the Robotics Research Group at the University of Adelaide (http://sites.mecheng.adelaide.edu.au/robotics/), the RiSE project's vertical wall-climbing robot (http://kodlab.seas.upenn.edu/~rise/newsite/) and presentations at the IEEE–RAS conference on humanoid robots in Osaka (http://www.humanoidrobots.org/humanoids2012/).

Catastrophe Theory (Singularity Theory)

René Thom's famous treatise on catastrophe theory, which is now considered part of singularity theory, *Stabilité Structurelle et Morphogénèse* (Thom, 1972), was the culmination of work, by him and others, over the preceding 10 years. He suggested using the topological theory of dynamical systems, originated by Henri Poincaré, to model discontinuous changes in natural phenomena, with special emphasis on biological systems. Poston and Stewart (1978) give an accessible account of the theory, some fascinating applications and citations of many more. Zeeman (1977) is a collection of influential papers. Wilson (1981) describes applications to urban and regional systems, and Burghes and Wood (1980) give a nice application to sales. The concept of structural stability, or insensitivity to small perturbations, plays an important part in catastrophe theory. Chaos is the name given to the pseudo-random behaviour of some deterministic systems, and their extreme sensitivity to small perturbations. Fractal geometry, particularly associated with Benoît Mandlebrot (e.g. 1982), is closely allied with chaos theory. Fractals exhibit a similar structure over a wide range of scales and provide realistic models for the outlines of coastlines, the shapes of trees and snowflakes, the surfaces of viruses and many other natural phenomena. The idea is that however much you enlarge portions of a coastline you will still see bays and headlands, even on individual rocks. You can see the Sierpinski Sponge constructed by the community at the Maths Learning Centre at the University of Adelaide (http://www.youtube.com/watch?v=W0uLbhRR-Hw). Fractals have been used to study the interaction of oil and water, in particular the phenomenon known as viscous fingering, in oil wells (http://www.youtube.com/watch?v=v9ARz3Cq0Ec). There are many popular books on the subject of fractals, and Feldman (2012), for example, gives an introductory mathematical account.

Wave Motion

Luke Bennetts and colleagues in the School of Mathematical Sciences at the University of Adelaide researched the interaction between ice floes and the sea, extending classic models of wave–body interactions. Figure 32.8 is from Skene *et al.* (2015) describing theoretical, and experimental, work on water wave overwash of a thin floating plate.

The sixth edition of Horace Lamb's renowned book on hydrodynamics (Lamb, 1932) is available as a Dover paperback. Drazin (1992, 2002) is a more recent book, Linton and McIver (2001) specialise on wave structure interactions, and several texts describe computational

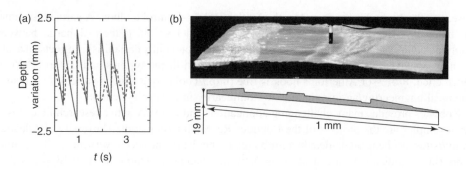

Figure 32.8   Comparison of experimental and modelled flows of water over ice. Skene *et al.* (2015).

fluid dynamics (CFD) with applications to ship motion (e.g. Anderson, 1995; Pozrikidis, 2011). Billingham and King (1998) is a useful text for wave motion. Also look up Mike Meylan's wiki on mathematical methods for water wave problems (www.wikiwaves.org). Jonathan Braithwaite has posted useful course notes, 'An Introduction to Hydrodynamics' (https://astro.uni-bonn.de/~jonathan/misc/hydro_notes.pdf).

Conclusion

I hope this chapter has given you a hint of some of the issues arising in mathematical modelling. There are very many deterministic mathematical models, and I have only mentioned a few, but there are abundant web resources which provide an introduction to most specialist mathematical areas. These include succinct introductions in Wikipedia, notes and slide shows that people have made available and entertaining video clips. It is prudent to be wary about material on the internet that has not been refereed, but in mathematics, at least, it is usually possible to verify that it is correct. Also, text in a refereed journal or technical book is not guaranteed to be correct, as Billah and Scanlan (1991) point out. Mistakes happen!

There are challenging research questions concerning the technical, mathematics and numerical methods for solution, that underpin deterministic mathematical models. New areas of application, especially in the biological (e.g. Murray, 2002) and social sciences (Bruggeman, 2013), offer research opportunities. Deterministic models are unlikely to fit data from social surveys and medical records as well as they do data from the physical sciences, and the fitting of such models is likely to present interesting issues. Deterministic models can always be embellished with the addition of random variation, and this is considered in Chapter 33.

References

Acheson, D.J. (1990). *Elementary Fluid Mechanics*. Oxford: Oxford University Press.
Anderson, J.D. (1995). *Computational Fluid Dynamics*. New York: McGraw-Hill.
Appleby, J.C. (1994–2012). *FINEL Software and Tutorial*. Newcastle upon Tyne: Department of Engineering Mathematics, University of Newcastle upon Tyne. http://www.staff.ncl.ac.uk/john.appleby/software.htm
Balas, M. (1982). Trends in large space structure control theory: fondest hopes, wildest dreams. *IEEE Transactions on Automatic Control* AC27: 522–535.

Barnett, S., and Cameron, R.G. (1985). *Introduction to Mathematical Control Theory*. Oxford: Oxford University Press.

Billah, K.Y., and Scanlan, R.H. (1991). Resonance, Tacoma Narrows bridge failure, and undergraduate physics textbooks. *American Journal of Physics* 59(2): 118–124.

Billingham, J., and King, A.C. (1998). *Wave Motion*. Cambridge: Cambridge University Press.

Bruggeman, J. (2013). *Social Networks*. London: Taylor and Francis.

Burghes, D.N., and Wood, A.D. (1980). *Mathematical Models in the Social, Management and Life Sciences*. Chichester: Ellis Horwood.

Crisfield, M., de Borst, R., Remmers, J., and Verhoosem, C. (2012). *Non-linear Finite Element Analysis of Solids and Structures*. Chichester: Wiley.

Doole, S.H., and Hogan, S.J. (2000). Non-linear dynamics of the extended Lazer-McKenna bridge oscillation model. *Dynamics and Stability of Systems* 15(1): 43–58.

Dorf, R.C., and Bishop, R.H. (2010). *Modern Control Systems*. Upper Saddle River, NJ: Pearson.

Drazin, P.G. (1992). *Nonlinear Systems*. Cambridge: Cambridge University Press.

Drazin, P.G. (2002). *Introduction to Hydrodynamic Stability* Cambridge: Cambridge University Press.

Elaydi, S. (2005). *An Introduction to Difference Equations*, 3rd ed. Berlin: Springer.

Faust, C.R., Guswa, J.H., and Mercer, J.W. (1989). Simulation of three-dimensional flow of immiscible fluids within and below the unsaturated zone. *Water Resources Research* 25(12): 2449–2464.

Feldman, D.P. (2012). *Chaos and Fractals: An Elementary Introduction*. Oxford: Oxford University Press.

Gerya, T. (2009). *Numerical Geodynamic Modelling*. Cambridge: Cambridge University Press.

Gottardi, G., and Venutelli, M. (1993). Richards: computer program for the numerical simulation of one-dimensional infiltration into unsaturated soil. *Computers & Geosciences* 19(19): 1239–1266.

Holmes, R. (1960). The vibration of a rigid shaft on short sleeve bearings. *Journal of Mechanical Engineering Science* 2: 337–341.

Hosking, R.J., and Milinazzo, F. (2012). Modelling the floating ladder track response to a moving load by an infinite Bernoulli-Euler beam on periodic flexible supports. *East Asian Journal on Applied Mathematics* 2(04): 285–308.

Imam, I., Azzaro, S.H., Bankert, R.J., and Scheibel, J. (1989). Development of an on-line rotor crack detection and monitoring system. *ASME Journal of Vibration, Acoustics, Stress and Reliability in Design* 111: 241–250.

Isidori, A. (1995). *Nonlinear Control Systems*. Berlin: Springer.

Jacobs, O.L.R. (1993). *Introduction to Control Theory*, 2nd ed. Oxford: Oxford University Press.

Jeffrey, A. (2003). *Applied Partial Differential Equations*. London: Academic Press.

Jewell, N.D. (2009). The development and stability of some non-planar boundary-layer flows. PhD thesis, University of Adelaide.

Jewell, N.D. (2012). A topological approach to three dimensional laminar mixing. *ANZIAM Journal* 51: 874–909.

Jones, R.T., Pretlove, A.J., and Eyre, R. (1981). Two case studies in the use of tuned vibration absorbers on footbridges. *The Structural Engineer* 59B: 27.

Lamb, H. (1932). *Hydrodynamics*. London: Dover.

Linton, C.M., and McIver, P. (2001). *Handbook of Mathematical Techniques for Wave/Structure Interactions*. London: Chapman & Hall.

Logan, J.D. (2008). *An Introduction to Nonlinear Partial Differential Equations*, 2nd ed. Chichester: Wiley.

Kreyszig, I. (2010). *Advanced Engineering Mathematics*, 10th ed. Chichester: Wiley.

Kwakernaak. H., and Sivan, R. (1972). *Linear Optimal Control Systems*. Chichester: Wiley

Mandlebrot, B.B. (1982). *The Fractal Geometry of Nature*. San Francisco: Freeman.

Metcalfe, A.V., and Burdess, J.S. (1986). Active vibration control of multi-mode rotor-bearing system using an adaptive algorithm. *Transactions of ASME Journal of Vibration, Acoustics, Stress & Reliability in Design* 108(2): 230–231.

Metcalfe, A.V., and Burdess, J.S. (1990). Experimental evaluation of wide band active vibration controllers. *Transactions of ASME Journal of Vibration & Acoustics* 112(4): 535–541.

Metcalfe, A.V., and Burdess, J.S. (1992). Wide band active vibration controller for an out-of-balance rotor. In Goodwin, M.J. (ed.), *Rotordynamics '92*. New York: Springer, pp. 327–334.

Murray, J.D. (2002). *Mathematical Biology: I. An Introduction*, 3rd ed. Berlin: Springer.

Nise, N.S. (2011). *Control Systems Engineering*, 6th ed. Chichester: Wiley.

Peaceman, D.W. (1977). *Fundamentals of Numerical Reservoir Simulation*. Amsterdam: Elsevier.

Pozrikidis, C. (2011). *Introduction to Theoretical and Computational Fluid Dynamics*, 2nd ed. Oxford: Oxford University Press.

Poston, T., and Stewart, I. (1978). *Catastrophe Theory and its Applications*. London: Pitman.

Schmidt, G., and Tondl, A. (1986). *Non-linear Vibrations*. Cambridge: Cambridge University Press.

Simon, D. (2006). *Optimal State Estimation: Kalman, H Infinity, and Nonlinear Approaches*. Chichester: Wiley.

Skene, D.M., Bennetts, L.G., Meylan, M.H., and Toffoli, A. (2015). Modelling water wave overwash of a thin floating plate. *Journal of Fluid Mechanics* 777.

Stewart, I. (1997). *Does God Play Dice? The New Mathematics of Chaos*, 2nd ed. London: Penguin.

Strogatz, S.H. (1994). *Nonlinear Dynamics and Chaos*. Boulder, CO: Westview.

Thom, R. (1972). *Stabilité Structurelle et Morphogénèse*. London: Addision Wesley Longman.

Thompson, J.M.T., and Stewart, H.B. (2002). *Nonlinear Dynamics and Chaos*, 2nd ed. Chichester: Wiley.

Thomson, W.T. (2004). *Theory of Vibration with Applications*, 4th ed. Boca Raton, FL: Taylor & Francis.

Wilson, A.G. (1981). *Catastrophe Theory and Bifurcation: Applications to Urban and Regional Systems*. Berkeley: University of California Press.

Zeeman, E.C. (1977). *Catastrophe Theory: Selected Papers, 1972–77*. Reading, MA: Addison-Wesley.

33

Stochastic Models and Simulation

David Green and Andrew Metcalfe

Introduction

The Agora Museum in Athens contains a piece of a Kleroterion, an allotment machine. Names of candidates for senate were placed in an array of slots (typically 50 rows by 11 columns). A funnel at the top contained bronze balls, which were coloured black or white, and a crank driven device released these, one at a time, down a tube. A white ball corresponded to selection. You can easily find out more from internet sites (e.g. http://www.alamut.com/subj/artiface/deadMedia/agoraMuseum.html).

So, the concept of random processes can be traced back at least as far as 2 BC. Although 'stochastic' is now synonymous with 'random', it is derived from the Greek *stokhazesthai* that is translated as 'aim at, guess'. This is quite appropriate because we use stochastic models to aim at values that might be taken by variables that are affected by chance events, so as to provide the best possible predictions despite the uncertainty. We can also quantify the uncertainty by using, for example, confidence intervals for unknown parameters, prediction intervals for individual outcomes, and in more complex cases a set of realisations from the fitted model. In contrast to short-term forecasts, stochastic models are also used to generate many possible long-term scenarios. Examples include evaluation of policies related to portfolio management (Ranne, 1999), the construction of flood defences (Environment Agency, UK, 2010) and climate change (Intergovernmental Panel on Climate Change (IPCC), 2007, 2012). Stochastic models are also used for computer simulations such as: modelling the arrivals of patients and the allocation of resources at a glaucoma clinic (Crane et al., 2011), modelling ground motion during earthquakes (Boore, 2003) and modelling the response of a wave energy system under extreme wave conditions (Yu and Li, 2011) (http://www.osti.gov/bridge). Economists use stochastic models to try to answer questions such as whether wage inflation causes price inflation or whether price inflation drives pay rises (Hess and Scweitzer, 2000) (http://www.clevelandfed.org/research/policydis/pd1.pdf). This raises the question of what is meant by *causality*. Philosophical discussion of the notion of causality, in the

Research Methods for Postgraduates, Third Edition. Edited by Tony Greenfield with Sue Greener.
© 2016 John Wiley & Sons, Ltd. Published 2016 by John Wiley & Sons, Ltd.

Western tradition, stretches back at least to Aristotle and continues today (https://en.wikipedia.org/wiki/Causality).

Economists use the concept of Granger causality (G-causality), which is a statistical concept based on prediction. A times series x is G-causal for y if past values for x provide information that improve forecasts of y beyond those based on past values of y alone. Physicists use stochastic models to describe how models at the molecular scale can describe large-scale phenomena, such as temperature and pressure, in a branch of physics known as statistical mechanics.

Random or Deterministic?

We all have a notion of what is meant by *random*, and how this differs from *deterministic*, but attempting to be precise about the distinction raises many interesting questions. Is the universe intrinsically random, or is it essentially deterministic and only apparently random because we cannot predict the future? If the universe is deterministic, can we have free will (e.g. Dennett, 2003; Kane, 2011)? These philosophical questions may seem far removed from mathematical modelling but similar questions do arise at a more prosaic level, for example in the contexts of random number generation, cellular automata and chaos theory (Stewart, 2002). How can we generate a sequence of random numbers to use in a computer simulation of a stochastic process?

Random Number Generation

There are many uses of random numbers in statistics. One important use is in the selection of a simple random sample from a finite population. Random numbers can also be used to generate data without performing the actual experiment. Historically three types of random number generators have been advocated for computational applications:

1. 1950s-style table look-up generators like, for example, the RAND corporations table of a million random digits, were the earliest methods for generating such random numbers and involved actual sampling of supposedly random events, such as dice throws, electronic pulses and so on;
2. generators that use such things as thermal *white noise* devices or *atmospheric noise*; and
3. algorithmic (software) generators.

Of these three types, only algorithmic generators have achieved widespread acceptance. The reason for this is that only algorithmic generators have the potential to satisfy all of the following generally well-accepted random number generation criteria. A random number generator should have the following desirable properties:

1. The sequence of numbers produced should pass all statistical tests that they are distributed uniformly on [0,1] and have no correlation with each other, denoted $U[0, 1)$. Consider the sequence $0.00, 0.01, 0.02, 0.03, \ldots, 0.99$, which is such that $x_n = (n - 1) \times 0.01$ for $n = 1, \ldots, 100$ (plotted in red). A random ordering of the exact same set of numbers is plotted in blue in Figure 33.1.

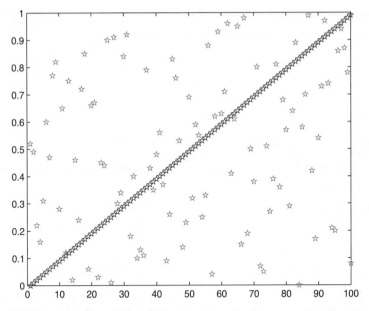

Figure 33.1 Two different ordered sequences of the same set of numbers.

2. The generator should be fast and avoid the need for a lot of storage. The storage issue is not as essential as it was in the past, but speed of generation is still a very desirable characteristic.
3. It is desirable that a given *stream* of numbers can be reproduced exactly. This is for two reasons:
   (a) It helps debugging and verification.
   (b) It makes for easier comparison of different systems.
4. There should be provision to generate different independent streams (sequences) of random numbers. This enables two or more sequences to be generated that are not correlated, as correlation *can be* very detrimental to simulation results.

For example, in a simple single-server queue (which is used for modelling many systems) (Figure 33.2), we have an arrival stream of customers who have inter-arrival times which are

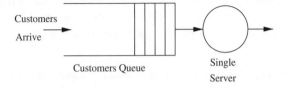

Figure 33.2 Schematic of single-server queue. A customer arriving to find the server free will immediately go into service; otherwise, they will join a waiting line and be served in order of arrival. Queueing theory says that if there is no correlation between these times, then the averaged occupancy of the system will be 0.5 customers. However, we can simulate this system in three ways to show the effects of serial correlation.

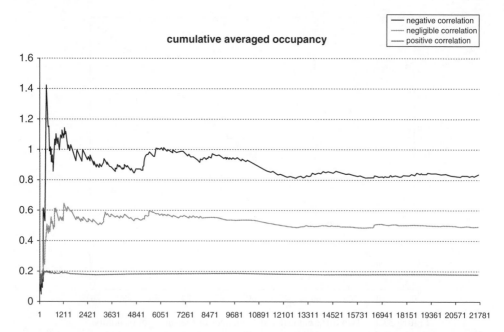

Figure 33.3   Effect of serial correlation of arrival times on average system occupancy. The vertical axis gives the averaged occupancy, whereas the horizontal axis gives the actual arrival times of the customers. To generate these times, we use random numbers U_i (uniformly distributed random numbers between 0 and 1, denoted $U[0, 1)$). In the blue trace, we have used the numbers U_i for the inter-arrival times and the numbers $1 - U_i$ for service times of customer i. In the green trace, we have used separate number streams U_i and U_j for customer i times m, and in the red trace we have used the numbers U_i for both the inter-arrival times and service times of customer i. This is an extreme case to show the effects. Sometimes when simulating systems it is actually beneficial to employ correlation, and hence the possible existence and effects of correlation are something that a practitioner must be aware of in designing simulation models.

random (exponentially distributed) with some mean value λ. The service times are similarly exponentially distributed with some mean value, say, $2 \times \lambda$.

In Figure 33.3, we record the cumulative average system occupancy for the situation where we have negative, negligible and positive serial correlation between the inter-arrival times for each customer and their associated service times.

Unfortunately, generating random numbers looks a lot easier than it really is. Indeed, it is fundamentally impossible to produce truly random numbers on any deterministic device. Von Neumann, a Hungarian and later American pure and applied mathematician, physicist and inventor, said it best:

> Anyone who considers arithmetical methods of producing random digits is, of course, in a state of sin.

The best we can hope for are pseudo-random numbers, a stream of numbers that appear as if they were generated randomly.

Let's now explore one of the first algorithmic techniques that was used to create a stream of pseudo $U[0, 1)$ random numbers, and that is still used today, called the *linear congruential generator* (LCG), introduced by Lehmer in 1951. In particular we will consider an infamous version known as RANDU, which has the following recurrence relation form given by:

$$Z_i = (65339 \times Z_i - 1) \bmod 2^{31}, \text{ with } Z_0 = (an\ odd\ integer).$$

This recurrence relation generates integer values and to get a pseudo $U[0, 1)$ random number, we divide each Z_i by 2^{31}. RANDU was distributed in the 1960s by IBM systems and has bad statistical properties:

> Its very name RANDU is enough to bring dismay into the eyes and stomachs of many computer scientists.
>
> Donald Knuth

The bad statistical properties can be demonstrated by showing that the numbers it generates do not even appear random. We do this by generating 10,000 sets of triples $(x, y, z) = (U_i, U_{i+1}, U_{i+2})$ and plotting the points (x, y, z) as a 3D scatterplot, rotating the view to see why Knuth and his colleagues were so dismayed. If the generator has produced numbers that at least appear random, they should appear as a cloud in 3D space and have no visible pattern, whereas for RANDU we see a very definite pattern of 15 hyperplanes (Figure 33.4).

Although hardware random number generators, based on thermal noise affecting voltage, are available, it is now usual to use pseudo random number generators (PRNGs). The output from a good PRNG should be virtually indistinguishable from a random sequence. Hardware has the disadvantage that it needs to be checked for possible biases, and PRNGs have the advantage that sequences are reproducible if we know the initial value. Hence, most often, we use a deterministic algorithm to model random behaviour, and there are many methods and indeed packages to check that the output from a PRNG is in fact not exhibiting bad behaviour like RANDU (http://csrc.nist.gov/groups/ST/toolkit/rng/batteries_stats_test.html).

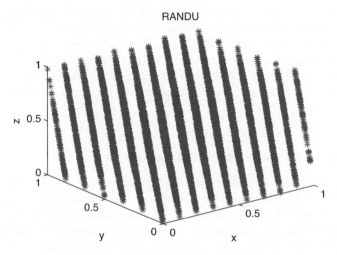

Figure 33.4 RANDU – a not so random PRNG.

Comparing Random and Deterministic Models for Arrivals

This example demonstrates the difference between queues with arrivals being random and independent, modelled by a Poisson process, and deterministic arrival times. We can calculate the relative proportion of time that a system has a particular number of customers present over a long period of time, which is called the equilibrium probability distribution. It also has the interpretation that it is the probability that, at an arbitrary time point (far away from any initial conditions), the system has a particular number of customers present. This means that a randomly arriving customer will see the equilibrium probability distribution, a result known as Poisson arrivals see time averages (PASTA).

Waiting times for customers depend on the number of customers in the system when the customer arrives. The distribution of the number of customers in the system when a customer arrives can differ from that reflected in the equilibrium probability distribution.

For example, consider a queue with mean inter-arrival time 1 and mean service time $\frac{1}{2}$, both of which are fixed values (there is no randomness). A plot of number in the system (vertical axis) against time (horizontal axis) looks like the following:

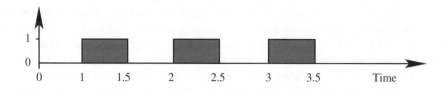

The system is occupied for half of the time, and so, if the system is observed at a random moment, the probability that it will have one customer is $\frac{1}{2}$. However, arriving customers do not observe the system at random instants, but come exactly one unit of time after the last customer. In fact, an arriving customer always sees an empty system. Hence, P (*an arriving customer sees no customers*) $= 1$, whereas P (*the system has no customers in it*) $= \frac{1}{2}$. The above simple example shows that arrival time distributions are not always the same as equilibrium distributions, in this case when we have a deterministic queue rather than a random queue.

The difference between a stochastic model and deterministic model is even more marked if the arrival rate equals the service rate. In the case of Poisson arrivals, the expected queue length increases with time and the system is unstable. In the deterministic case, the system could be set up so that an arrival coincides precisely with the end of the last service, and there would be no queue.

Synchronous or Asynchronous Simulation

A clock is central to any simulation over time. There are two strategies for discrete event simulation. One is to advance time by a small constant increment and check whether any events have occurred. This is known as fixed time increment or synchronous simulation, or continuous time simulation. The other is to simulate the occurrence of the next event and change the clock appropriately, and is known as variable time increment or asynchronous simulation.

Classifying Stochastic Models

A stochastic model is often based on a deterministic model, which can be a simple empirical relationship, and accounts for deviations between the model and data by postulating random errors. These errors encompass: inherent variation in the population being modelled; modelling error; and measurement error. A typical research project will involve thinking of reasonable models for the situation; fitting these models to existing data and choosing one or two that seem the best, from an empirical point of view; simulating future scenarios; and monitoring the success of predictions. The random errors are modelled by a probability distribution. Generation of random numbers from a given probability distribution is an essential part of any (stochastic) simulation. Ross (2006) gives a thorough account of simulation methods, and Kroese *et al.* (2011) is an excellent detailed reference guide.

We have chosen the following examples of stochastic models to give some indication of the range of both techniques and areas of application. Although we perceive time and space to be continuous, and variables that vary over space or time (field or state variables) as discrete or continuous, we do not have to model variables in the same way. For example, the Markov chain model for dam storage (Moran, 1954) treats both volume of water and time as discrete. That is, volume of water is measured in multiples of some fraction of the total capacity of the dam; and time jumps from the end of one dry season to the end of the next one. The stochastic calculus deals with continuous variables defined over continuous time, and financial applications remain a popular research topic. Martin Haugh, at the Center for Financial Engineering at Columbia University, has made his helpful introductory notes available to all (http://www.columbia.edu/~mh2078/). A popular textbook on the subject is Joshi (2008).

You may find the following summary of some of the main techniques of stochastic modelling by category helpful, although the categorisation is somewhat subjective. For example, the theory of signal processing techniques, such as spectral and wavelet analysis, may seem nicer when expressed in terms of continuous signals, but modern data analysis is almost always performed on the digitised signal. Provided the sampling rate is high enough, and gigabits per second (Gbps) rates can now be achieved, nothing is lost.

		Space–time	
		Discrete	Continuous
Field Variable	Discrete	• Markov chain • Cellular automata	• Point processes
	Continuous	• Time series models • Signal processing of digitised signals • Image processing • Kriging	• Field models • Spectral and wavelet theory for analogue signals • Ito calculus, Stratanovich calculus

There are three facets to stochastic modelling: the theoretical properties of stochastic models, which involve some esoteric mathematics; computer realisations of stochastic models, which require proficiency in some programming language; and the fitting of stochastic models to data.

There are many relevant textbooks, and internet resources increase all the time. Some book titles that we have found useful include: Ross (2003) and Norris (1997) for applied probability models, Lawler (2006) for an introduction to the mathematical theory and Chatfield (2004) for an introduction to time series.

Applications of Markov Chains

The Markov property is that the future depends on the present but not the past, given the present state. In a Markov chain, the process can be in any one of a discrete set of states after each time step. In brand loyalty models, the state is the brand and the time steps are times between purchases. Brown *et al.* (1997) ranked pension funds by their financial performance and defined the state as the quarter (within top 25%, …, within lowest 25%) in which a pension fund appeared at the end of each year.

More specialist topics include hidden state Markov chains (e.g. Latouche and Ramaswami, 1999; MacDonald and Zucchini, 2009; He, 2014), in which the state is not observed, which have been used for such diverse applications as the modelling of genome structure, speech recognition and river flow in arid regions (Fisher *et al.*, 2010). Markov chain Monte Carlo (MCMC) methods, such as the Gibbs sampler, are now routinely used for stochastic modelling – for example, a space–time model for Venezuelan rainfall data (Sanso and Guenni, 1999) and treatment monitoring of HIV-infected patients (Prague *et al.*, 2012). Casella and George (1992) is a good introduction to the Gibbs sampler, the Hastings–Metropolis algorithm (see e.g. Chib and Greenberg, 1995) is more general and Kroese *et al.* (2011) is an authoritative reference.

Applications of Point Processes and Simulation

In a point process, events, which result in a change of state of the process, occur at instants of time. There are many applications, including machine breakdowns and repairs and queuing situations. In a typical queuing model, the state is the number of persons in the system and an event is the arrival of another person or the completion of a service. It is possible to build up more complex models by superimposing point process models and, for example, rectangular pulses, as Rodriguez-Iturbe *et al.* (1987) do for rainfall at a point. Cowpertwait (1995) has extended these models so that the points, which represent the centres of storms, are distributed over both space and time (Burton *et al.*, 2008; Leonard *et al.*, 2008). If point processes are simulated by generating times between events, it is known as discrete event simulation (Ross, 2006). The alternative, which is commonly referred to as continuous time simulation, is to proceed by small time increments and simulate whether or not an event occurs in each time increment.

Telephone networks have provided the motivation for much of the work on queuing theory. When the telephone was first invented, you needed one line to connect you directly to each person you wished to communicate with. This led to a proliferation of phone wires (Figures 33.5 and 33.7). Then the telephone exchange was invented.

The Danish engineer and mathematician Agner Erlang (1878–1929) was one of the first to develop a mathematical theory of the telephone exchange (Figure 33.6).

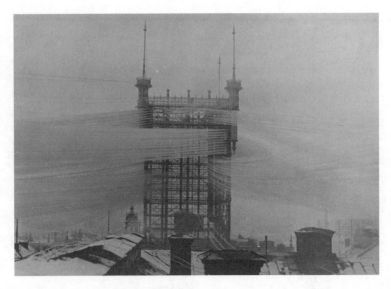

Figure 33.5 A telephone tower in Stockholm, Sweden, with 5000 connected lines. It was used between 1887 and 1913, but the tower stood there until 1953, when it fell down after a fire. www.flickr.com/photos/tekniskamuseet/6838150900/in/set-72157629589461917

Figure 33.6 'Telephone exchange Montreal QE3 33' by various photographers for Cassell & Co. – *The Queen's Empire*. Volume 3. Cassell & Co. London. (Licensed under Public Domain via Commons, https://commons.wikimedia.org/wiki/File:Telephone_exchange_Montreal_QE3_33.jpg - /media/File:Telephone_exchange_Montreal_QE3_33.jpg)

Figure 33.7   Artist's depiction (1890) of Broadway and John Streets in Manhattan before and after (circa 1910) the undergrounding of telephone wires around the time of the introduction of telephone exchanges. (Casson 1910).

Applications of Time Series, Signal Processing and Simulation

Signal processing covers a vast range of applications. A medical example (Miller *et al.*, 2001) involves recording the electromyograph (EMG) response to a pseudo-random sequence of taps applied to the biceps muscle. In this context, the biceps is adequately modelled as a single degree of freedom linear system and it follows, from the theory of spectral analysis, that the autocovariance function of the EMG response is an estimate of the impulse response of the biceps. The importance of the method is that it provides clinicians with a quick painless test to ascertain whether babies are suffering from cerebral palsy. Early detection increases the efficacy of treatment for the condition.

A much larger scale example (Metcalfe *et al.*, 2007) involved the estimation of the frequency response functions for the three translational and three rotational motions of a small ship, from data collected during sea trials. Hence, the risks of capsize in extreme sea states, and the accelerations experienced by crew at various stations on board, can be estimated.

The wavelet transform is an important development in signal analysis. Spectral analysis (see e.g. Hearn and Metcalfe, 1995) gives an average frequency description of random signals, but it is of limited use if statistical features of the signal change over time. Wavelets can be thought of as local Fourier analyses, which track the changes, and a musical score may be a helpful analogy. An early application was the analysis of earthquake records (Goupillaud *et al.*, 1984), and the book edited by Silverman and Vassilicos (2000) includes applications to astronomy, turbulence, the physiology of vision, acoustics and economics. Nason (2008) is a useful introduction which is based on the open source software R (R Development Core Team,

2011). Bakar *et al.* (2012) compare spectral and wavelet estimators of the transfer functions for linear systems.

A Random Field Model for Rainfall

The distribution of rainfall, over time and space, is essential information for designers of water resource projects ranging from flood protection to irrigation schemes. Ideally, and provided there were no long-term climate changes, statistics could be calculated from long records over an extensive network of rain gauges. In practice, rain gauge networks are often sparse or non-existent and, even in countries with good coverage, records for periods exceeding 50 years are relatively uncommon. Furthermore, records usually consist of daily rainfall totals, and for some purposes, such as assessment of the hydraulic performance and pollution impact of sewers, finer resolution, down to five-minute rainfall totals, is needed. For some purposes, it may be possible to progress with rainfall at a single site. Other applications need rainfall at several sites, and projects that are more ambitious require a rainfall field model. The development of rainfall field models, and their calibration from radar data, is an active research topic. The rainfall model described by Mellor *et al.* (2000) has been coupled with a deterministic rainfall-runoff model of the River Brue catchment in the South West of England and could be used for flood warning and the design of flood protection schemes (Mellor *et al.*, 2000).

The structure of the model is shown in Figure 33.8 (Mellor, 1996). An initial random field (lowest plane, i) is produced by randomly generated intersecting prisms, which move along their axes with constant velocities. This field is modulated by a function composed of sine waves (ii), which has an independent velocity, and introduces the effect of rain bands. This modulated field (iii) represents a time-varying spatial rate of an inhomogeneous Poisson process, which controls the births of rain cells. That is, where the field takes on a high value, there is a large probability of rain cells occurring, and where the field is low there is a smaller chance of a rain cell occurring. The rain cells themselves are described as parabolas of revolution (iv), whose heights represent rainfall intensity at the corresponding point on the surface. The peak of a rain cell grows and then decays over time (top plane, v), and all the rain cells are assumed to have the same velocity. In flood-warning applications, rain cells and rain bands are identified from radar data and an ensemble of scenarios is stochastically generated up to six hours ahead (Mellor *et al.*, 2000). Thus, the system gives limits that have any chosen probability of being exceeded rather than a single estimate.

Other Random Field Models

Another particularly active research area is the analysis of digital images from space probes, electron microscopes, brain scanners and so on. Noise and blur have to be removed, and there is a wide variety of statistical techniques that can be used. The images are usually considered in terms of discrete greyscale values, defined over a grid of picture elements, typically 1024×1024. The greyscale is sometimes treated as continuous. Some good intro-ductory papers can be found in the following internet directory: http://www.stats.bris.ac.uk/pub/reports/MCMC.

Kriging is a method for spatial interpolating between a few point values. It was developed in the mining industry, and this is reflected in some of the terms used, such as the *nugget effect*.

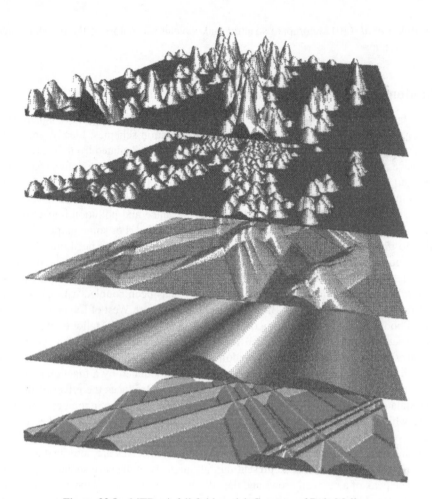

Figure 33.8   MTB rainfall field model. Courtesy of Dale Mellor.

Details can be found in texts on geostatistics, such as Isaaks and Srivastava (1989) or Davis (2002). The open source language R has many resources for spatial statistics.

How Can Maths Fight Influenza?

Occasionally, new strains of influenza are resistant to vaccines and can develop into pandemics. Recent such outbreaks are Spanish flu (1918–1920), Asian flu (1957–1958), Hong Kong flu (1968–1969) and swine flu (2009–2010). In order to identify and control such pandemics, health officials need to predict how fast new strains are spreading. A simple but effective mathematical model is a cellular automaton (e.g. Beauchemin *et al.*, 2005). The automaton consists of a grid of cells, that each assume one of a number of states, together with rules for how states might change as a consequence of neighbouring states. For example, the probability of infecting a neighbour and of recovering during a time step can be modelled. Then a computer simulation is set up from some initial distribution of infected individuals. For more information, see Wikipedia (http://en.wikipedia.org/wiki/Mathematical_modelling_of_infectious_disease, http://en.wikipedia.org/wiki/Cellular_automaton); try CounterPlague, part of the Millennium Mathematics Project (http://motivate.maths.org/content/DiseaseDynamics/Activities/e-CounterPlague); or see John Conway's Game of Life (http://www.bitstorm.org/gameoflife/).

House (2011) reviews some other mathematical models for epidemics that provide more biological insight.

Bayesian Modelling

A Bayesian approach to statistical analysis is particularly appropriate for stochastic modelling, because it allows for subjective assessment of probabilities. Lee (2012) is a nice introduction, and Gelman *et al.* (2013) is a more detailed reference.

Conclusion

There are many applications for stochastic models in all subject areas. In particular, we think there is scope for adding stochastic elements to physical-based deterministic mathematical models. You can aim to publish in journals of applied probability, journals of general mathematical applications and journals in specialist subject areas.

References

Bakar, M.A.A., Green, D.A., and Metcalfe, A.V. (2012). Comparison of spectral and wavelet estimators of transfer function for linear systems. *East Asian Journal on Applied Mathematics* 2(3): 214–237.

Beauchemin, C., Samuel, J., and Tuszynski, J. (2005). A simple cellular automaton model for influenza A viral infections. *Journal of Theoretical Biology* 232: 223–234.

Boore, D.M. (2003). Simulation of ground motion using the stochastic method. *Pure and Applied Geophysics* 160: 635–676.

Brown, G., Draper, P., and McKenzie, E. (1997). Consistency of UK pension fund investment performance. *Journal of Business, Finance and Accounting* 24(2): 155–178.

Burton, A., Kilsby, C.G., Fowler, H.J., Cowpertwait, P.S.P., and O'Connell, P.E. (2008). RainSim: a spatial-temporal stochastic rainfall modelling system. *Environmental Modelling & Software* 23(12): 1356–1369.

Casella, G., and George, E.I. (1992). Explaining the Gibbs sampler. *The American Statistician* 6(3): 167–174.

Casson, H.N. (1910). *The History of the Telephone*. Chicago: A. C. McClurg & Co.

Chatfield, C. (2004). *The Analysis of Time Series: An Introduction*, 4th ed. London: Chapman & Hall.

Chib, S., and Greenberg, E. (1995). Understanding the Metropolis-Hastings algorithm. *The American Statistician* 49(4): 327–335.

Cowpertwait, P.S.P. (1995). A generalise spatial-temporal model of rainfall based on a clustered point process. *Proceedings of the Royal Society A* 450(1938): 163–175.

Crane, G.J., Karnon, J., Kymes, S., Casson, R., Metcalfe, A., and Hiller, J.E. (2011). A discrete event simulation to optimise the allocation of constrained hospital resources for glaucoma *Value in Health* 14(3): A55.

Davis, J.C. (2002). *Statistics and Data Analysis in Geology*, 3rd ed. Chichester: Wiley.

Dennett, D. (2003). *Freedom Evolves*. New York: Viking.

Environment Agency, United Kingdom (EAUK). (2010). *Future Flooding in Wales: Flood Defences, Possible Long Term Investment Scenarios*. London: EAUK.

Fisher, A.J., Green, D.A., and Metcalfe, A.V. (2010). Managing river flows in arid regions with matrix analytic methods. *Journal of Hydrology* 382: 128–137.

Fisher, A.J., Green, D.A., and Metcalfe, A.V. (2011). Modelling of hydrological persistence for hidden state Markov decision processes. *Annals of Operations Research* 199(1): 215–224.

Gelman, A., Carlin, J.B., Stern, H.S., Dunson, D.B., Vehtari, A., and Rubin, D.B. (2013). *Bayesian Data Analysis*, 3rd ed. London: Chapman & Hall.

Goupillaud, P., Grossmann, A., and Morlet, J. (1984). Cycle-octave and related transforms in seismic signal analysis. *Geoexploration* 23: 85–102.

He, Q-M. (2014). *Fundamentals of Matrix-Analytic Methods*. Berlin: Springer.

Hearn, G.E., and Metcalfe, A.V. (1995). *Spectral Analysis in Engineering: Concepts and Cases*. London: Arnold.

Hess, G.D., and Schweitzer, M.E. (2000). *Does Wage Inflation Cause Price Inflation?* Cleveland: Federal Reserve Bank of Cleveland.

House, T. (2011). Modelling epidemics on networks. *Contemporary Physics* 53(3): 213–225.

Intergovernmental Panel on Climate Change (IPCC). (2007). *Fourth Assessment Report: Climate Change 2007 (AR4)*. Geneva: IPCC. http://www.ipcc.ch/publications_and_data/ar4/syr/en/contents.html

Intergovernmental Panel on Climate Change (IPCC). (2012). *Managing the Risks of Extreme Events and Disasters to Advance Climate Change Adaption*. Cambridge: Cambridge University Press.

Isaaks, E.H., and Srivastava, R.M. (1989). *Applied Geostatistics*. Oxford: Oxford University Press.

Joshi, M.S. (2008). *The Concepts and Practice of Mathematical Finance*, 2nd ed. Cambridge: Cambridge University Press.

Kane, R. (2011). *The Oxford Handbook of Free Will*. Oxford: Oxford University Press.

Kroese, D.P., Taimre, T., and Botev, Z.I. (2011). *Handbook of Monte Carlo Methods*. Chichester: Wiley.

Latouche, G., and Ramaswami, V. (1999). *Introduction to Matrix Analytic Methods in Stochastic Modelling*. Philadelphia: ASA-SIAM.

Lawler, G.F. (2006). *Introduction to Stochastic Processes*, 2nd ed. London: Chapman & Hall.

Lee, P.M. (2012). *Bayesian Statistics*, 4th ed. Chichester: Wiley.

Leonard, M., Lambert, M.F., Metcalfe, A.V., and Cowpertwait, P.S.P. (2008). A space-time Neyman-Scott rainfall model with defined storm extent. *Water Resources Research* 44(9).

Mac Donald, I.L., and Zucchini, W. (2009). *Hidden Markov Models for Time Series – an Introduction Using R*. London: Chapman & Hall.

Mellor, D. (1996). The Modified Turning Bands (MTB) model for space-time rainfall: I model definition and properties. *Journal of Hydrology* 175: 113–127.

Mellor, D., Sheffield, J., O'Connell, P.E., and Metcalfe, A.V. (2000a). A stochastic space-time rainfall forecasting system for real time flow forecasting I: development of MTB conditional rainfall scenario generator. *Hydrology and Earth System Sciences* 4: 603–615.

Mellor, D., Sheffield, J., O'Connell, P.E., and Metcalfe, A.V. (2000b). A stochastic space-time rainfall forecasting system for real time flow forecasting II: application of SHETRAN and ARNO rainfall runoff models to the Brue catchment. *Hydrology and Earth System Sciences* 4: 617–626.

Metcalfe, A.V., Maurits, L., Svenson, T., Thach, R., and Hearn, G.E. (2007). Modal analysis of a small ship sea keeping trial. *ANZIAM(E) Journal* 47.

Mikosch, T. (2000). *Elementary Stochastic Calculus with Finance in View*. Thomas Mikosch.

Miller, S., Clark, J., Eyre, J.A., Kelly, S., Lim, E., McClelland, V.M., *et al.* (2001). Comparison of spinal myotatic reflexes in human adults investigated with cross-correlation and signal averaging methods. *Brain Research* 899: 47–65.

Moran, P.A.P. (1954). Theory of dams and storage systems. *Australian Journal of Applied Science*, 5.

Nason, G.P. (2008). *Wavelet Methods in Statistics with R*. Berlin: Springer.

Newland, D.E. (1993). *Random Vibrations, Spectral & Wavelet Analysis*, 3rd ed. London: Longman.

Norris, J.R. (1997). *Markov Chains*. Cambridge: Cambridge University Press.

Prague, M., Commenges, D., Drylewicz, J., and Thiebaut, R. (2012). Treatment monitoring of HIV-infected patients based on mechanistic models. *Biometrics* 68(3): 902–911.

R Development Core Team. (2011). *R: A Language and Environment for Statistical Computing*. Vienna: R Foundation for Statistical Computing.

Ranne, A. (1999). The investment models of a Finnish pension company. *Vector* 15(4): 63–70.

Rodriguez-Iturbe, I., Cox, D.R., and Isham, V. (1987). Some models for rainfall based on stochastic point processes. *Proceedings of the Royal Society of London A* 410: 269–288.

Ross, S.M. (2006). *Simulation*, 4th ed. London: Academic Press.

Ross, S.M. (2003). *Introduction to Probability Models*, 8th ed. London: Academic Press.

Sanso, B., and Guenni, L. (1999). Venezuelan rainfall data analysed using a Bayesian space-time model. *Applied Statistics* 48: 345–362.

Silverman. B.W., and Vassilicos, J.C. (2000). *Wavelets*. Oxford: Oxford Science.

Stewart, I. (2002). *Does God Play Dice? The New Mathematics of Chaos*, 2nd ed. Malden, MA: Blackwell.

Yu, Y., and Li, Y. (2011). Preliminary results of a RANS simulation for a floating point absorber wave energy system under extreme wave conditions. Paper presented at the 10th International Conference on Ocean, Offshore and Arctic Engineering, Rotterdam, the Netherlands, 19–24 June, 2011.

34

Optimisation

Aiden Fisher and Andrew Metcalfe

Introduction

We are all familiar with the idea of optimisation. Suppose that you live in London, and that you wish to travel to San Francisco to attend a friend's wedding. Your objective is to arrive in time for the wedding at a minimum cost, with a constraint that you have only one week of leave from your work in London. You then either check websites yourself, or ask a travel agent to find the best possible (optimal) travel arrangements for you. You may then realise that you need to add some other constraints if you are not prepared to get up before 4 a.m. or want to fly direct. This is a typical optimisation problem, and much of our daily activity can probably be described in terms of optimisation problems, even if our solutions aren't optimal.

Optimisation is an area of mathematics designed to solve problems in which we aim to meet an objective in the most efficient manner given the constraints on our resources. For example, a train driver is given the objective of arriving at the destination with minimum fuel use, subject to constraints of speed restrictions and arriving on time. The driver's control actions are to: accelerate, hold speed, coast or brake. TTG Transportation Technology has developed a real-time driver advisory system, Energymiser®, that advises the driver which action to apply over segments of the journey to minimise fuel consumption subject to the constraints. Fuel savings from 8.9% for iron ore trains in Africa; to 10% or more for freight trains in Australia, the United Kingdom and India; to 14% for coal trains and over 20% for high-speed passenger trains in the United Kingdom, have been recorded (http://www.ttgtransportationtechnology.com/energymiser/). You can attempt getting a train to a station whilst minimising fuel yourself at http://scg.ml.unisa.edu.au/PCB/.

Optimisation relies on a mathematical model for the situation. It is common to discuss finding the least value of a function, but this is not a restriction on the problems we can optimise. Finding the greatest value of a function is equivalent to finding the least value of its negative, and aiming for some preferred value is equivalent to minimising the discrepancy between the function value and that preferred value. Optimisation has often been the motivation for

Research Methods for Postgraduates, Third Edition. Edited by Tony Greenfield with Sue Greener.
© 2016 John Wiley & Sons, Ltd. Published 2016 by John Wiley & Sons, Ltd.

developing mathematical models. A famous early application was the Brachistochrone problem (http://mathworld.wolfram.com/BrachistochroneProblem.html), posed by Johann Bernoulli in 1696, which is to find the curve between two points such that a ball rolling along the curve, under gravity and without friction, moves between the two points in the minimum time. Johann Bernoulli provided a first solution, and his work led to the invention of the calculus of variations.

Optimisation is the basis for fitting mathematical models to data. In 1805, Adrien-Marie Legendre first published his least-squares method for fitting an ellipse to observations of the path of planetary objects. The objective function in this problem is the sum of squared errors, defined as the difference between observations, which are subject to error, and predicted values from the ellipse. The variables in this context are the position of the foci and the length of the axes of the ellipse. The principle of least squares remains the most common criterion for optimising the fit of a model to data. Although optimisation plays an important part in many subject areas, including, for example, control theory and cost–benefit analysis, it has its own discipline of operations research.

Often the optimal solution may initially seem counterintuitive. During World War II, the RAF Bomber Command decided to increase the armour on their aircraft in order to increase their chances of returning safely. After conducting a survey of returning aircraft, the proposal was to place armour over the heaviest damaged areas. Patrick Blackett, a renowned physicist, instead argued that armour should be placed on the areas where none of the planes had been damaged. His reasoning was that the Bomber Command were looking at a biased sample of planes that had sustained damage but were still able to fly because they were not hit in a vital area. The work of Blackett and others during the war led to the development of the subject of operational (or operations) research (OR) which is now mainly concerned with the application of optimisation methods in industry and commerce. Several countries have professional societies focussed on operations research, in particular the OR Society in the United Kingdom (www.theorsociety.com) and the Institute for Operations Research and Management Science (www.informs.org) in the United States. These organisations, and others, publish many journals devoted to the discipline.

A common optimisation problem that we encounter every day is the shortest path problem. Here the objective function is to find the shortest path, in terms of either distance or time, between where we are and where we want to go. GPS systems, or Google Maps, offer solutions obtained by treating the map as what is known mathematically as a graph: a series of nodes and links between the nodes representing the possible paths. Algorithms are then used to find the optimal path between two nodes.

Optimisation problems become more complex with the introduction of constraints. 'How to cross Dublin without passing a pub' was a challenge set by the protagonist, Leopold Bloom, in James Joyce's *Ulysses*. Here we must go from a node on one side of Dublin to a node on the other side of Dublin without passing a link that has a pub adjacent to it. Rory McCann claims to have found a solution, shown below, that does not get closer than 35 m to a pub. But, it does pass restaurants that serve alcoholic beverages as well as the Guinness Brewery. Are further constraints possible? You can read more at McCann's blog: http://www.kindle-maps. com/blog/how-to-walk-across-dublin-without-passing-a-pub-full-publess-route-here.html.

When solving optimisation problems, there is often a trade-off between mathematically elegant solutions and computer-intensive solutions. However, brute force is rarely the best

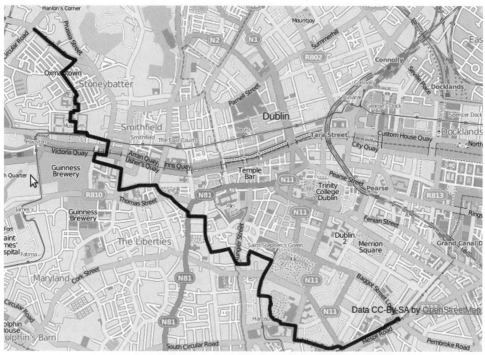

openstreetmap.org © OpenStreetMap contributors, CC BY-SA 2.0.

approach. George B. Dantzig, the Western[1] developer of linear programming, recounts the problem of finding the problem of minimal-cost adequate diet for the air force. Specifically, he aimed to find the cheapest way to meet nine dietary requirements with 77 food sources to choose from. With a computer that could test 1000 combinations per second, it would take five years to calculate by testing all possible combinations. Dantzig had an assistant solve the problem by hand (without even the use of a desk calculator) in a few days with an ingenious algorithm he developed.

Many optimisation problems are highly structured, and there are very efficient methods for their solutions. If you hope to solve optimisation problems with a large number of variables, it is essential to use the most efficient algorithm available, and, since the size of the problem usually increases factorially with the number of variables, this situation will not change with the introduction of more powerful computers.[2]

There are many techniques and ingenious algorithms for solving optimisation problems, or at least for coming close to doing so. These techniques range from calculus, that can provide

---

[1] Linear programming was first developed by Soviet mathematician Leonid Kantorovich in 1939 but was kept as a state secret.

[2] For example, the calculation of a determinant of an $n \times n$ matrix, from the permutation definition, requires the sum of $n!$ products of n terms. If n equals 100, $n!$ is about 9.33×10^{157}. A computer that can make a calculation in Planck time, 10^{-44} second, and that has been working since the universe began, NASA estimate 14×10^9, would only have achieved 4×10^{61} calculations. In contrast, when an algorithm for reducing determinants is used, the calculation takes just seconds.

algebraic solutions, to stochastic trial-and-error methods such as genetic algorithms. Although most of these techniques were originally applied to deterministic problems, they have been adapted to deal with optimisation problems that include stochastic terms in their definitions. The strategy behind trial-and-error methods is to find a solution that is close to optimal without having to consider all possibilities.

With many optimisation problems, a potential solution found by some search algorithm may only be a local minimum value rather than a global minimum; this is particularly true when using an iterative solution method. We may think we have an optimum solution when we have overlooked something better through the choice of our starting value and the shape of the objective function. One way to mitigate the situation is to start the algorithm from several different sets of values of the variables. Another strategy is to use a simulated annealing algorithm which allows a probability of jumping out of a local minimum. However, no algorithm can guarantee to find the global minimum of an arbitrary function (e.g. Griewank, 1981).

When constraints are imposed, the least value also does not necessarily correspond to a stationary point, because it may occur on the boundary of the set of values that the variables are allowed to take. For example, we might be asked to find the most economic cruising speed for a cargo ship given the cost of fuel, the efficiency of the engine as a function of the ship speed, the cost of storing fuel on board the ship, the cost of time at sea and the cost of delays in delivering the cargo. It could turn out that the most economic speed corresponds to the highest speed of which the ship is capable, in which case a higher speed would likely be more economical, but infeasible without a faster ship.

Another feature of any iterative method is cycling. Here the algorithm will continue to select between two or more variables of equal functional value, stepping over the locally minimal value with each iteration, and will not converge. Rules need to be introduced to avoid continual repetition of the same solution. In the remainder of this chapter, we review, very briefly, some of the ideas behind optimization.

Calculus

If the objective function is a continuous and differentiable function of the variables, stationary points can be found, in principle, by setting partial derivatives equal to zero. However, the resulting equations may have to be solved numerically. Constraints can be handled by using Lagrange multipliers, or more generally Karush–Kuhn–Tucker conditions which allow for inequalities. It is essential to check boundaries for least values that are not stationary points.

The calculus of variations deals with maximising functionals, which assign real numbers to functions. An important application is the general solution of the linear-quadratic optimal control problem, and its stochastic variant, the linear-quadratic-Gaussian optimal control problem, in which the system is subject to disturbance noise and the observations are subject to measurement noise (see e.g. Barnett and Cameron, 1990; Jacobs, 1993). An associated and mathematically equivalent problem is constructing an optimal observer which is usually referred to as a Kalman filter (Kalman and Bucy, 1961).

Descent Algorithms

For the most part, even if the function is continuous, other algorithmic techniques will need to be used, especially when dealing with large numbers of variables. Descent algorithms find

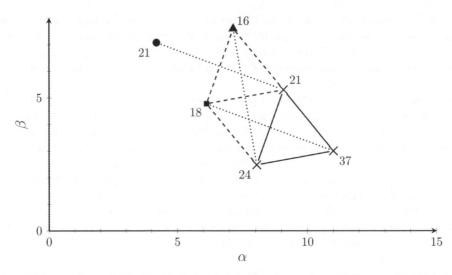

Figure 34.1 An example of a simplex algorithm for minimising a function of two variables α and β. The initial function evaluations are made at the points shown by crosses. As 37 is the greatest value, this point is reflected in the line joining the other two points to arrive at the square. The function evaluation at the square is 18, so the point with function evaluation 24 is reflected to the point shown by the circle, and the function evaluation at the triangle is 16. The cross with function evaluation 21 is reflected to the dot and so on.

a minimum value of a continuous function of several variables by calculating approximations to local derivatives and then proceeding in the direction of steepest descent until a minimum is found along the curve in this direction. One advantage of this method is that the function need only be locally differentiable. When working with constrained optimisation problems, Lagrange multipliers or Karush–Kuhn–Tucker multipliers will be needed as before; however, a caveat needs to be considered: the optimal solution may occur at saddle points, rather than local minima. As descent algorithms are designed to find the local minima, alteration will be needed.

The simplex method and modified simplex method of Nelder and Mead (1965) are somewhat simpler and less efficient, but generally effective and very popular, means of finding a minimum. They do not require the calculation of derivatives or of approximations to derivatives. If the objective function depends on two variables only, the simplex method is easily described. Begin by calculating the values of the objective function at the vertices of a triangle in the plane of the two variables. Then reflect the point corresponding to the largest value of the objective function in the edge joining the other two points. Now calculate the value of the objective function at this new point. Continue in this manner until a point reflects back to itself, and then reduce the size of the simplex until it is sufficiently small to stop.

Linear Programming

The linear programming (LP) problem is to minimise a set of linear functions of the variables subject to sets of linear inequalities. Typical examples include blending problems, transportation problems and resource allocation problems. The inequalities define a feasible region that is bounded by hyper-planes. The optimum corresponds to one of the vertices of this region, and the search can therefore be restricted to vertices. The simplex method (not the same as the Nelder and Mead simple method) for solution of the LP problem moves between vertices so that the value of the objective function improves or, at worst, stays the same.

When it comes to computing LPs, there are many free and proprietary software packages available. For a long time, CPLEX has been the standard and is free for academic use; however, other options are available. The package linprog in R is free and applies the simplex algorithm (see CRAN: https://cran.r-project.org).

Standard LINDO will handle up to 200 variables, and the extended version will go up to 100,000 variables. The Matlab's Optimisation Toolbox also includes algorithms for the efficient solution of linear programming problems. If commercial packages are unavailable, there are many open source programs and libraries for scripting languages available. Python has become the language of choice, and there are several LP modules available, usually with some code written in C or Fortran. GNU Linear Programming Kit (GLPK) is a C library supported by a number of open source programs.

A simple online program for small problems is available at: http://www.phpsimplex.com/en/index.htm.

Bean and Jewell (2013) describe an application of linear programming methods to management of the Murray River.

Mathematical Programming

Williams (1999) includes linear programming (LP), non-linear programming (NLP) and integer programming (IP) under the general heading of mathematical programming. The IP methods are important if the variables can take only a few integer values, and the case of all variables being either 0 or 1 (binary integer programming) is quite common (Chen, 2011). One approach to the NLP problem is to use piecewise-linear approximations to the non-linear functions.

Duality

Most optimization problems have a dual. In general, the dual problem consists of finding the maximum of a specially constructed function that has a single maximum (the convex conjugate). If the original objective function is itself convex, then the optimal solution of the dual is equal to the optimal solution of the original problem, otherwise, the dual problem gives a lower value for the minimum value of the original problem (and this may well be all that is needed). The dual problem is considerably easier to solve, as by construction it is always convex.

Linear programming problem duals often have a nice interpretation; for example, a typical resource allocation problem has a dual problem which is a resource valuation problem.

Dynamic Programming and Stochastic Dynamic Programming

The ideas behind dynamic programming were first formulated by Bellman (1957). The method can be used for a variety of problems, including scheduling of work in factories and optimal control. A simple example is that of a traveller who intends to travel from a city A to a city Z in several stages. At each step there are many possible intermediate destinations, and the costs of travel for all possible routes between stages are known. The most efficient way to minimise the total cost is to work *backwards* from Z. In general, if there are S states, the best decision from state i with t steps to go is to choose to move to state j to maximize:

Value of going from state i to j + Value of being in state j with t−1 steps to go.

Critical path analysis is a related problem, and programme evaluation and review technique (PERT) is a stochastic version of this.

Fair allocation of water from a network of reservoirs to households, to farms for irrigation and to industry is a vital issue. A typical operating policy will specify the amounts of water to be released to the various recipients each month. The decisions that make up the policy depend on the amount of water in the reservoir, the time of the year and the expected future inflows into the reservoir. These inflows are unknown and are described by a probability distribution. The determination of a policy that will optimise the benefits to the community is an example of a stochastic dynamic programming (SDP) problem (see e.g. Fisher *et al.*, 2012). Howard was a pioneer in the methods of SDP, and his book (Howard, 1960) is an elegant exposition of the principals involved. Simple SDP problems can be solved by using decision trees. An example of open source software is the R package MDPtoolbox (Puterman, 1994; Chadès *et al.*, 2014).

Simulated Annealing and Genetic Algorithms

These are not a problem type as such but rather stochastic solution tools that can be used for solving particularly difficult problems. When a problem has numerous local minima/maxima solutions, descent or hill-climbing algorithms will become stuck and will rarely yield the global optimal solution. Stochastic algorithms aim to get around this by moving the optimisation at random to other locations in the state space. However, there is no unique best way of doing this, and there is no guarantee that they will find a global optimum. They are customisable to any problem and are a versatile tool.

Simulated annealing is a descent algorithm with small probabilities of moving to another value entirely instead of the next calculated value. These probabilities become smaller as the value of the objective function decreases, although this rule is flexible. For example, you may wish to increase the probability of moving if the randomly selected point is significantly lower than the next point the descent algorithm moves to. When implementing simulated annealing, it is a good idea to keep a record of the best value obtained so far along with the current value.

Genetic algorithms (GAs) draw some inspiration from evolutionary biology, by coding the values of the variables as binary numbers and linking those to form a string of binary digits referred to as chromosomes. Strings are selected from an initial set with probabilities proportional to their fitness, which will increase as the objective function becomes lower in a minimisation problem, and are combined by crossing over sections of the strings. To prevent getting stuck in a locally optimal solution, strings are selected with a separate probability, called the mutation probability, and have parts of their genome randomly mutated. This is where GAs differ the most from evolutionary biology as the offspring in GAs are essentially costless, whereas biological evolution will tend to stick to locally optimal solutions with a very low mutation rate due to the large amount of resources needed to create offspring. Genetic algorithms are extremely flexible as crossover and cloning rules, along with probabilities and fitness measures, can be chosen to suit the problem. The heuristics of the GAs make sense when the objective function depends on many dichotomous variables, and they can be adapted for use with continuous variables.

Conditional Value at Risk

In the case of stochastic optimization, there will typically be a probability distribution of possible outcomes that fans out with the lead time of the projection (the ensemble). The expected monetary value (EMV) of the investment at a given time in the future will equal the mean of the ensemble. Expected monetary value has the limitation that it does not penalize risk. Conditional value at risk (CVaR), also known by other names such as expected shortfall, has been used in the finance industry to rectify this situation (Rockafellar and Uryasev, 2002). The CVaR is the expected outcome given that the loss is below some specified quantile of the distribution of outcomes, typically the lower 0.10 quantile. Despite its use in finance, applications of CVaR in other areas are infrequent. Webby *et al.* (2007) and Piantadosi *et al.* (2008) are two applications to water resources.

CVaR is a move away from expected monetary value towards minimising the impact of worst-case scenarios. The concept of H-infinity control, also known as robust control, is similar inasmuch as it aims for a minimum response over all possible disturbance frequencies. George Zames is renowned for his work in this area, and recent textbooks include those by Skogestad and Postlethwaite (2005) and Simon (2005).

Multi-objective Optimisation

Multi-objective optimisation, or Pareto optimisation, occurs when there is more than one objective function that needs optimisation using some or all of the same variables. One way

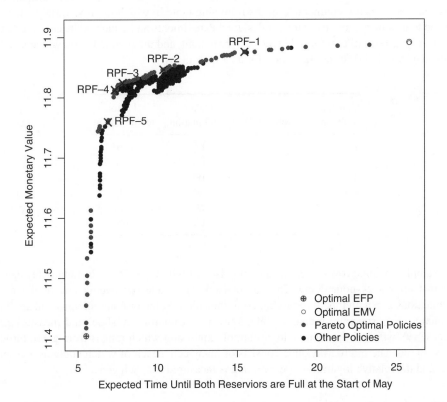

to do this would be create an objective function that is a weighted mixture of the objective functions to create a single objective function (called the aggregate objective function), but there is often no way to objectively decide upon the weightings. An alternative strategy is that the decision maker selects from a set of possible optimal solutions, known as the Pareto set. A solution is in the Pareto set so long as it is not Pareto dominated by another solution, that is, the solution does not perform worse than another solution for all of the objective functions. An advantage of dealing with the Pareto set is that trade-offs between the objective functions can be clearly displayed. Fisher *et al.* (2014) consider operating policies for two reservoirs in south-east England. The objectives are to maximise the expected monetary value of water supplied (EMV), and to reduce the expected time until both reservoirs are full at the start of May. The second objective is for environmental, aesthetic and recreational reasons, and quantification in monetary terms would be somewhat arbitrary. The plot below shows that policy RPF-1 corresponds to a substantial reduction in the expected time until both reservoirs are full for only a slight reduction in EMV (note the scales do not start at 0), and that policies up to RPF-4 provide further decreases in the time with associated decreases in EMV of less than 1%.

Case Study (Sanchez *et al.*, 2012)

A gene expression study aimed at investigating the genetic basis of leukaemia. There are two cell lines, the normal (usually referred to as the *wild type*) and mutant. The cell lines can be compared with no delay or after 24 hours of delay, so there are six possible comparisons. These comparisons are enumerated in the following table in which the first variable, usually referred to as a factor in the context of designed experiments and which is the cell line in this case, is denoted by either 0 for wild type or a for mutant, and the second factor, which is delay, is denoted by either 0 for no delay or b for a 24 hour delay.

Comparison (column 2 vs. column 3)	Factor combination	Factor combination
1	00	a0
2	00	0b
3	00	ab
4	a0	0b
5	a0	ab
6	0b	ab

For example, the final row represents a comparison of wild type after a 24-hour delay against a mutant after a 24-hour delay. The comparisons are made by: dyeing one of the factor combinations green and the other red; mixing the coloured factor combinations and applying the mixture to a specially prepared slide, known as a microarray slide; and measuring the response (y), which is the relative logarithm of expression which ranges from green through yellow to red. The microarray slide consists of many cells, each of which contains a specific gene, and the relative logarithm of expression is measured for each gene.

The optimisation is described in terms of a single gene. The model for logarithm of expression for a particular factor combination (x_j) on a typical slide (j) is

$$x_j = \mu + \alpha + \beta + \alpha\beta + E_j$$

where the means, in the hypothetical population of all such cell lines, for the factor combinations are given in the following table:

Factor combination	Mean
00	μ
a0	$\mu + \alpha$
0b	$\mu + \beta$
ab	$\mu + \alpha + \beta + \alpha\beta$

E_j is random variation with mean 0. The standard assumptions are that the errors are independent with the same standard deviation regardless of the factor combination. The deterministic part of the model is that each factor combination has a different mean, but it is convenient to describe this in terms of the parameters $\mu, \alpha, \beta, \alpha\beta$, which are referred to as, respectively, mean of wild type with no delay, effect of cell line (mutant compared with wild type) when no delay, effect of delay (24-hour delay compared with no delay) for wild type, and interaction. Note that the interaction, which is usually denoted by $\alpha\beta$ in the statistics literature, is a single parameter in its own right and not the product of the other two parameters – this notational convention does reduce the proliferation of Greek letters in more complex models. The interaction allows for the effect of both mutant and time delay to differ from the sum of the effects of mutant and time delay, as, for example, the effect of the time delay might be different for mutant and wild type. The main effect of the cell line is defined as:

$$\frac{\alpha + (\alpha + \alpha\beta)}{2}$$

and the main effect of delay is defined as:

$$\kappa \frac{\beta + (\beta + \alpha\beta)}{\oint}$$

The response from a single slide (j) will be the difference between two (x_j), and the six comparisons are (y_j) modelled as shown.

Comparison (column 2 vs. column 3)	Factor combination	Factor combination	Model for response (y_j)
1	00	a0	$\alpha + E_{j,1} - E_{j,2}$
2	00	0b	$\beta + E_{j,1} - E_{j,2}$
3	00	ab	$\alpha + \beta + \alpha\beta + E_{j,1} - E_{j,2}$
4	a0	0b	$\beta - \alpha + E_{j,1} - E_{j,2}$
5	a0	ab	$\beta + \alpha\beta + E_{j,1} - E_{j,2}$
6	0b	ab	$\alpha + \alpha\beta + E_{j,1} - E_{j,2}$

The researchers wished to estimate the main effects and the interaction, though not necessarily with the same precision (reciprocal of variance) because they are particularly interested in a possible interaction. Microarray slides and analyses are expensive, and they were restricted to $n = 160$ slides. A design $\mathbf{d}$ of the experiment is an allocation of d_i slides to comparison i subject to:

$$d_1 + d_2 + d_3 + d_4 + d_5 + d_6 = n$$

For any design, the variance of the estimators can be calculated by a standard calculation (for this application and any statistics text that includes the linear model, see Sanchez *et al.*, 2012; for the general result, see e.g. Montgomery, 2009). There are millions of possible designs, and checking them all is infeasible. Simulated annealing offers a practical means of obtaining a good design even if it is suboptimal.

When using simulated annealing the objective function, or objective functions, can be set in any sensible manner, and the following is a description of how they chose to set up the optimisation. The optimisation is relative to this subjective choice, so different research groups could come up with quite different optimal designs. The researchers set up three objectives to be minimised:

$$O_1 = (1 - w)(\text{var}(\hat{\alpha}) + \text{var}(\hat{\alpha} + \hat{\alpha}\beta)) + wD$$
$$O_2 = (1 - w)(\text{var}(\hat{\beta}) + \text{var}(\hat{\beta} + \hat{\alpha}\beta)) + wD$$
$$O_3 = (1 - w)\text{var}(\hat{\alpha}\beta) + wD$$

where $D = (\text{var}(\hat{\alpha}) - \text{var}(\hat{\alpha} + \hat{\alpha}\beta))^2 + (\text{var}(\hat{\beta}) - \text{var}(\hat{\beta} + \hat{\alpha}\beta))^2$ is a penalty imposed so that the designs considered estimate the effect of cell line at the two delays with equal precision and the effect of delay for the two cell lines with equal precision, and w is a weight which was set to 0.99 in order to keep the constraint satisfied, or at least almost satisfied. The researchers set three objectives, rather than take a weighted sum of the three objectives to give a single objective function, because they wished to consider a range of possibilities before selecting a design that suited them.

They set up a simulated annealing algorithm as follows.
Initialisation:

1. Select an initial set of m generating designs at random.
2. Initialise the set of Pareto optimal designs L to be those designs that are Pareto optimal among the initial set of generating designs. The Pareto optimal designs among the initial set are designs such that no other design in the set is as good or better on all three objectives.
3. Set an initial temperature, T_0.

Iterative steps:

1. For each generating design $\mathbf{d}_j$, construct a neighbouring design $\mathbf{d}_j^p$ by randomly selecting one slide from $\mathbf{d}_j$ and randomly assigning that slide to a different comparison.

2. If $\mathbf{d}_j^p$ is Pareto optimal with respect to L, update L by adding $\mathbf{d}_j^p$ and removing any designs that are no longer Pareto optimal.
3. With probability P, replace the generating design $\mathbf{d}_j$ with $\mathbf{d}_j^p$, where

$$ P = \min \left\{ 1, \exp \left(\sum_{i=1}^{3} \left(O_i(\mathbf{d}_j) - O_i(\mathbf{d}_j^p) \right) / T \right) \right\} $$

4. After N neighbouring designs have been tested at this temperature, reduce the temperature by a factor of κ.
5. After a given run time, stop and plot the Pareto front.

The crucial simulated annealing steps are 3 and 4. If $\mathbf{d}_j^p$ is better than $\mathbf{d}_j$ (the objectives are all smaller), the former is certain to replace the latter. There is also a probability of replacing $\mathbf{d}_j$ with $\mathbf{d}_j^p$ even if the latter is worse, but this probability decreases as the temperature decreases, and this feature is the rationale for the name of the technique.

The final results are shown in the figure and are known as a Pareto front. The axes represent precision, so larger values are good. If the algorithm is run longer, the Pareto front will tend to move forwards and cannot move backwards; also, gaps will tend to fill in. The front shows that the precision of the interaction can be increased with only slight decreases in the precisions of the main effects and is a valuable indicator of suitable designs. The algorithm as described has various parameters which are m, T_0, κ and N, and these were initially set at 40, 40, 0.7 and 100,000, respectively. Sanchez *et al.* (2012) used a hill-climbing algorithm (e.g. Greenfield and Metcalfe, 2007; Montgomery, 2009) to optimise the parameter values.

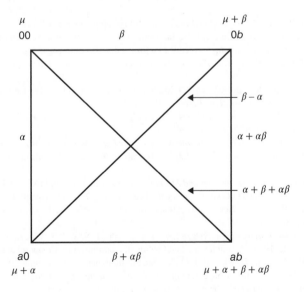

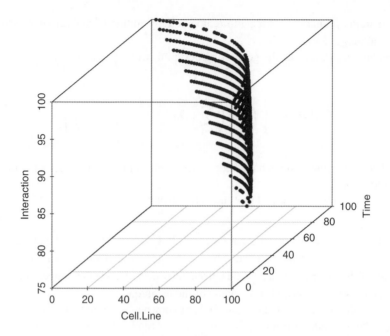

References

Barnett, S., and Cameron, R.G. (1990). *Introduction to Mathematical Control Theory*, 2nd ed. Oxford: Oxford University Press.

Bean, N.G., and Jewell, N.D. (2013). Value-driven river management: a Murray River case study. *Australian Journal of Water Resources* 17(1): 47–63. http://search.informit.com.au/documentSummary;dn=590350984388426; res=IELENG

Bellman, R.E. (1957). *Dynamic Programming*. Princeton, NJ: Princeton University Press.

Chadès, I., Chapron, G., Cros, M-J., Garcia, F. & Sabbadin, R. (2014). MDPtoolbox: a multi-platform toolbox to solve stochastic dynamic programming problems. *Ecography* doi:10.1111/ecog.00888

Chen, D-S., Batson, R.G., and Dang, Y. (2011). *Applied Integer Programming: Modeling and Solution*. New York: Wiley.

Fisher A.J., Green D.A., and Metcalfe A.V. (2012). Modelling of hydrological persistence for hidden state Markov decision processes. *Annals of Operations Research* 199(1): 215–224.

Fisher A.J., Green, D.A., Metcalfe, A.V., and Akande, K. (2014). First-passage time criteria for the operation of reservoirs. *Journal of Hydrology* 519: 1836–1847.

Greenfield, T., and Metcalfe, A. (2007). *Design and Analyze Your Experiment Using Minitab*. Chichester: Wiley.

Griewank, A.O. (1981). Generalized descent for global optimization. *Journal of Optimization Theory and Applications* 34(1): 11–39.

Howard, R.A. (1960). *Dynamic Programming and Markov Processes*. Cambridge, MA: MIT Press.

Jacobs, O.L.R. (1993). *Introduction to Control Theory*, 2nd ed. Oxford: Oxford University Press.

Kalman, R.E., and Bucy, R.S. (1961). New results in linear filtering and prediction theory. *Journal Basic Engineering, Transactions ASME Series D* 83: 95–108.

Montgomery, D.C. (2009). *Design and Analysts of Experiments*, 7th ed. New York: Wiley.

Nelder, J.A., and Mead, R. (1965). A simplex method for function optimisation. *Computer Journal* 7: 308–313.

Piantadosi, J., Metcalfe, A.V., and Howlett, P.G. (2008). Stochastic dynamic programming (SDP) with a conditional value-at-risk (CVaR) criterion for management of storm water. *Journal of Hydrology* 348(3): 320–329.

Puterman, M.L. (1994). *Markov Decision Processes*. New York: John Wiley & Sons.

Rockafellar, R.T., and Uryasev, S. (2002). Conditional value-at-risk for general loss distributions. *Journal of Banking and Finance* 26: 1443–1471.

Sanchez, P., Glonek, G., and Metcalfe, A. (2012). Pareto simulated annealing for the design of experiments: illustrated by a gene expression study. *Foundations of Computing and Decision Sciences* 37(3): 199–221.

Simon, D. (2005). *Optimal State Estimation: Kalman, H Infinity and Non-linear Approaches*. Chichester: Wiley.

Skogestad, S., and Postlethwaite, I. (2005). *Multivariate Feedback Control*. Chichester: Wiley

Webby, R.B., Adamson, P.T., Boland, J., Howlett, P.G., Metcalfe, A.V., and Piantadosi, J. (2007). The Mekong – applications of value at risk (VaR) and conditional value at risk (CVaR) simulation to the benefits, costs and consequences of water resources development in a large river basin. *Ecological Modelling* 201: 89–96.

Williams, H.P. (1999). *Model Building in Mathematical Programming*, 4th ed. Chichester: Wiley.

Part VI
Presentation

35

Writing the Thesis

Tony Greenfield

Introduction

A thesis or a dissertation is a work of scholarship that is published and made available for others to read. Even if you do not publish your work, or parts of it, as papers in academic journals, your thesis will be published by being placed in the university library, provided your examiners are satisfied with it. You may believe that the final stage of your work is the presentation of your thesis to your examiners, but that is not so. If they grant your higher degree, the thesis will be available for anybody to read for all time. Through your work, any reader may judge the scholarship of your department, of your university and of the examiners. It is in the university's own interests that your work should withstand criticism. You will fail the university if in any way your work lacks quality and, if it does, the examiners will be right to refuse or defer your degree.

You must therefore ensure that in every respect your thesis or dissertation is of the highest achievable quality. There are several good texts which may help you, including other chapters in this book. In this chapter, I summarise a few points which I, as an examiner, see as important. These will include a suggestion, but not a rule, for overall *structure*, guidance on *style* and advice about the *presentation of statistics*. But first I shall distinguish between a thesis and a dissertation.

A *dissertation* is a formal treatment of a subject, usually in writing. When the word is applied to undergraduate or post-graduate research, it usually refers to work done, either as a review of a subject or as the application of established methods to the study of a specified problem. A dissertation is usually submitted as part of an examination for a master's degree such as an MSc, MA or MPhil; for a post-graduate certificate; and sometimes for a first degree.

A *thesis* is a dissertation resulting from original research. It is usually submitted as the only document for examination for a doctorate such as a PhD, a DPhil or an MD. It may be submitted for a master's degree. Originality is the essential feature of a thesis and if you are submitting a thesis you must make clear which ideas were *yours*, which original work you did and which was done by others.

Research Methods for Postgraduates, Third Edition. Edited by Tony Greenfield with Sue Greener.
© 2016 John Wiley & Sons, Ltd. Published 2016 by John Wiley & Sons, Ltd.

In the rest of this chapter, I shall refer to the thesis but most of my advice applies equally to the dissertation.

Structure

The obvious structure is: introduction, background, materials and methods, results and conclusions. This suggests that there need be no more than five chapters, and indeed one academic supervisor I know insists that six should be the limit. But there is *no* limit to the number of chapters nor to the number of words in your thesis unless it is imposed by your department. Brevity is better than prolixity provided you do not skimp on information.

You must interest your readers and not bore them. You must catch and keep their interest so that they will read every word of your thesis with understanding, approval and pleasure. You must convince them that you know your subject and that you have contributed new ideas and knowledge. You must not deter or offend them with bad spelling, bad grammar, poor printing, uninformative diagrams, disorderly presentation, false or incomplete information, inadequate explanation or weak argument.

Here in more detail, with some comments, is my guidance for structure with the headings:

- Title
- Summary
- Keywords
- Contents
- Introduction
- Background and choice of subject
- Methods, results and analysis
- Discussion
- Conclusion and recommendations
- References and bibliography
- Appendices
- Glossary
- Notations
- Diary
- Acknowledgements.

Title

It is not easy to devise a title for a thesis. It must be short. It must identify your work so that anybody working in your subject area will immediately recognise it. Beware of the possibility of misclassification. You may be amused to find a book about traction (for treatment of spinal injuries) in the transport section of a library, but you wouldn't want that to happen to your work. Do not use abbreviations or words that may be difficult to translate into other languages without misinterpretation.

On the title page, you should add:

- your full name and existing qualifications;
- the name of the degree for which you are a candidate;

- the names of your department, faculty and university; and
- the date of submission.

Summary

You must tell your reader what your work is about, why you did it, how you did it and what conclusions you reached. This is also your first opportunity to declare your originality. Can you do this briefly, in no more than 200 words, and simply with no technical terms? If you can, you will encourage your reader to read on.

Keywords

Keywords are needed for classification and for reference. Somebody sometime will be interested in your subject but will not know of your work. You want to help him to discover it through your library's online inquiry service. What words would his mind conjure if he wants to find the best literature on the subject? Those must be your keywords.

Contents

In many theses that I have read, the contents list has been no more than an abbreviated synopsis: a list of chapter headings. The contents list must tell me where to find the contents, so it must include page numbers. This is easy with word processing. But do not restrict this to the chapter headings. I want to be able to find each of the appendices by name and by page number. It is frustrating to be told that a table of data can be found in the appendices. Even if you direct me to Appendix 9, my time will be wasted and my temper frayed if I have to search.

Please, in your contents list, give the page number of every chapter, of every appendix (which should have a name), of the index, of the references, of the bibliography, of the glossary, of your diary and of your acknowledgements. If you do that, your readers, including your examiners, will be able to find their way through your work.

Include a contents list of illustrations. For example:

Figure one   Flowchart of materials   page 23

and so on.

Introduction

Your introduction should be an expansion of your summary and your contents list combined. Your opening sentence should be a statement of the purpose of your research followed by the briefest possible statement of your conclusions and recommendations. Describe the path you followed to go from one to the other in terms of the contents of your thesis. Then your readers will know what to expect. Say who you are and in what department you are working; describe the resources that were available and what difficulties you met.

Review the contents of your thesis, perhaps devoting a paragraph to a summary of each subsequent chapter.

Background and Choice of Subject

Why did you choose the specific subject? Did you discover an interest in it because of something you had done earlier, or from your reading? Or was it suggested to you by your supervisor? What is the context of your subject: in what broader realm of knowledge does it fit? Is there a history of discovery? Who were the early workers, and what did they publish? Describe the narrowing of the field, related problems that others have studied, the opportunities for further study that they have revealed and the opportunities created by scientific and technical advances. Will any human needs, other than intellectual, be met through your research?

This is where you demonstrate your ability to study the literature with balanced and critical perspective. Name the most important of the references you found, but beware of boring your reader with a long and unnecessary account of every document that you found in your search just because it seemed slightly related. Recount the different viewpoints you may have found in the literature. State your opinion about the reliability of the evidence supporting the different viewpoints, and lead into your own. You should state this in terms of questions to be answered and hypotheses to be tested.

Methods, Results and Analysis

In a short paper, you would write a section on materials and methods followed by a section on results. In a longer study, you will probably write several chapters each dealing with different aspects of the whole. It may not be easy to separate the methods from the results. Possibly each of several chapters will have a description of methods and results. In some subjects, the most important part of the method will be a theoretical development. This must be rigorous and explained so clearly that any intelligent reader will understand it. There is a danger, in referring to theoretical development by others, that your readers will not know it. They will be frustrated if you simply give the result and a reference. That reference is unlikely to be easily at hand. They will appreciate your skill if you can explain the theory in a nutshell without re-enacting the full theoretical development.

Describe any technical equipment and techniques for observation and measurement. Compare the costs, reliability and ease of alternative techniques. Did you design and administer questionnaires? Did you have constructive or investigative discussions with individuals or groups? Did you use creative techniques? Be honest about the practical work that you did and what was done by others. Describe the methods of experimental design and data analysis that you used and what computer software you used. Examiners will not be impressed by a bland statement like 'the data were analysed by standard statistical techniques'. In describing the research design, you should refer to the questions and hypotheses stated in your introduction, perhaps repeating them in detail.

Results are usually best displayed in tables of data with summary statistics. If there are just a few measurements keep them in this section, but if they run into several pages put them in an appendix with clear references. Sometimes graphs will be suitable to illustrate the data but it is not always true that 'a picture is worth a thousand words'. Especially avoid bar charts and pie charts in a formal thesis. These should be reserved for platform presentations. Simple data tables will do. Whatever figures you use, make sure that you have explained clearly what they represent. We don't all see the same meaning in any results.

A chapter on methods and results is not the place to discuss results fully unless discussion of results at one stage of your research is part of your argument for developing further theory and methods for a later stage. If this is so, be sure to keep your discussion separate from the results in a section of the chapter.

Discussion

There is no need to repeat any results, but only to refer to them, interpreting and commenting on them with reference to your aims, your hypotheses, the work and opinion of others and the suitability of your theory, methods, experimental design and analysis.

Conclusions

Conclusions follow naturally from your discussion. Have you met your objectives? Have you found or done something new? Have you added to the store of human knowledge, even if you have only demonstrated a negative result? What did *you* contribute? Do you have any recommendations for further study?

Was the course of your research smooth? Did you follow any blind alleys, any deviations from your intended study design? You should report them so that later workers don't repeat your mistakes. Similarly, if you found any negative results, you should report them. It is unethical not to do so. Also be clear about what is meant by a negative result. It is rarely a demonstration that 'there is no effect'. Usually it is a statement that 'there is insufficient evidence to conclude that there is an effect', which is not the same.

References and Bibliography

References and bibliography are usually lumped together, but you should separate them.

In the bibliography, list those books which will be useful for background reading and further study and which describe methods and theory that are widely used and accepted, such as statistical methods, survey design and analysis, general mathematics, computing, communications and history.

In the references, list those books and papers to which specific reference is needed for the development of your argument within the thesis.

In both lists, you should consistently follow a standard form for each reference. The most usual standard is the Harvard system.

The standard for a paper is:

authors, date, title (in quotes), journal (in italics), volume (in bold), issue, pages.

The standard for a book is:

authors, date, title (in italics), publisher (including city).

Be sure that every reference is necessary for the development of the argument and that you have truly read every one. Your examiners will not be impressed by a long list of references of which some are hardly relevant. They may even ask you about individual ones. Be sure too that every reference in your list is actually referred to in the text of your thesis and, similarly, every reference in the text is in your list. A careful examiner will check this. You will help her, and may convince her, if against every reference in your list you put, in square brackets, the pages of your thesis where the references are made.

Please don't try to impress me, if I'm your external examiner, by including references to my papers unless they are truly relevant.

Appendices

Some people collect masses of material (data forms, memoranda, correspondence, company histories and handwritten notes) and fill hundreds of pages, which they describe as appendices.

Appendices are valuable only if they comprise material that is germane to your thesis and for which there is some possibility that your reader will make reference. You might describe special equipment, a detailed specification of a computer program that you used, a table of costs, a theoretical argument from other research (but any theory that you have developed should be in the main text), data collection forms, printed instructions to surveyors, maps or a large table of raw data.

Be sure to give each appendix a name as well as an appendix number, just as you would if it were a chapter of the main text. Number the pages consecutively with the main text so that the reader can easily find any appendix when it is referenced.

Glossary

In your thesis, you should highlight the first occurrence of every word and acronym that may be unfamiliar to your reader: underline it or put it in italics or bold or in a sans serif type. Define the word in your main text, but then enter the word into your glossary where you should repeat the definition. If you introduce a lot of technical language, your reader may quickly forget a definition and want to be reminded. Give page numbers to the glossary so that it can be found easily.

Notation

You may define notation as you like anywhere in your thesis but, as with technical words, you must keep to your definitions. The symbol π, for example, is normally used to denote the ratio of a plane circle's circumference to its radius. It is also used occasionally to denote probability. If you use it as such, define it in the text and again in a table of notations. Whenever in the text you use a symbol that may puzzle your reader, let him check its meaning in this table of notations.

Diary

How did you spend your time? Were you always at your desk or at the laboratory bench? Or were you in the library, at seminars and conferences, visiting equipment suppliers, interviewing, discussing your research with your supervisor or other researchers or lying on the beach (and thinking)?

A diary of your work will help your examiner to understand the complexity and difficulties of your research and will help those who follow to plan their own projects.

Acknowledgements

I mention acknowledgements last, not necessarily because they should go last but because they are not part of the substantial research report. They are important, however, and you should put them immediately after the title page.

You did not work on your own. You were guided and helped by your lecturers and demonstrators, by your supervisor, by technicians, by suppliers of equipment, by surveyors and assistants, by other research students with whom you discussed your work and by the secretaries and administrators, and you were sustained by grant awarders and by your family.

You should thank them all.

Style

At the end of this chapter is a list of some good books about scientific and technical writing. Here are a few points to ponder.

Personal Pronoun, I

There is continuing debate on whether to use the first-person, singular, active I. One professional institute banned its use in all reports some years ago. My daughter, as an undergraduate, assured me that she would fail if she used it in her dissertation. I was shocked but didn't interfere.

It is not simply a matter of taste. It is a question of honesty and of the credit that you deserve and need to justify the award of a higher degree. *You* are responsible for ensuring that others recognise what *you* have done, what ideas *you* have had, what theories *you* have created, what experiments *you* have run, what analyses and interpretations *you* have made and what conclusions *you* have reached. You will not succeed in any of this if you coyly state:

- It was considered (*write*: I proposed *or* I thought)
- It was believed (*write*: I believed)
- It was concluded (*write*: I concluded)
- There is no doubt that (*write*: I am convinced)
- It is evident that (*write*: I think)
- It seems to the present writer (*write*: I think)
- The author decided (*write*: I decided)

But be careful: too much use may seem conceited and arrogant.

Spelling

All word-processing software now has a spell checker, yet many published papers and theses are littered with misspellings and misprints. One reason is that the writers are too lazy or ignorant to use the spell checker.

But another reason is that many words, when misspelled, are other correctly spelled words. The spell checker will not identify these. If there are any in your submission, then you are telling the examiners that you have not read carefully what you have written. Here is a recent example:

'Improved understanding of these matters should acid cogent presentation ... '

The typist had misread 'aid' in poor handwriting as 'acid'.

The most common (anecdotally) of scientific misprints is the change from 'causal relationships' to 'casual relationships'. Other common ones, which the spell checker won't find, are:

- fro *for* from *or* for
- lead *for* led (past tense of 'to lead')
- gibe *for* give
- correspondents *for* correspondence *or* corresponding ('changes in watershed conditions can result in correspondents changes in stream flow').

My favourite malapropism is: 'Sir Francis Drake circumcised the world with a one hundred foot clipper'.

If you are writing for an English audience, use English rather than American spelling. Words like 'color' and 'modeling' look wrong to an English eye, as do words ending in -ize instead of -ise. On the other hand, if you write for an American audience, use American spelling. Fortunately, word-processing packages offer the choice of the appropriate spell checker.

Abbreviations and Points

Do *not* use points except as full stops at the ends of sentences or as decimal points.

- Write BSc PhD, *not* B.Sc. Ph.D.
- Write A Brown or Albert Brown, *not* A. Brown.
- Write UK UNO WHO ICI, *not* U.K. U.N.O. W.H.O. I.C.I.

Do *not* contract:

- department into dept.
- institute into inst.
- government into gov't.
- professor into prof.

Do *not* use:

- e.g. (write 'such as' or 'for example');
- i.e. (write 'that is');
- *et al.* (except in references); or

- etc. (put a full stop instead, otherwise the reader will wonder 'What are the etceteras?' and if you can't be bothered to tell her, she may not want to read further).

The plurals of acronyms may be written with the initials followed by a lowercase 's' without an apostrophe (e.g. ROMs).

Avoid possessives with acronyms (do not write NORWEB's) by recasting the sentence to eliminate the possessive.

Prefer single quotes to double quotes, but single and double are used when there is nesting of quotes.

Units

Use SI units with standard abbreviations without points. The base units are: m kg s A K cd mol. Note that, by international agreement, including France and the United States, correct spellings are *gram* and *metre*.

Do not use a dash to denote an interval.

> *Write* 'between 20 and 25°C', *not* '20–25°C'.

> *Write* 'from 10 to 15 September', *not* '10–15 September'.

Capitals

Like *The Times*, resist a tendency to a Germanic capitalisation of nouns by avoiding capitals wherever possible. Too many of them break the flow of the eye across a sentence. They also make pompous what need not be. The general rule is that proper names, titles and institutions require capitals, but descriptive appellations do not. Thus *government* needs no capital letter, nor does *committee* or *department*. The same goes for jobs that are obviously descriptive, such as prime minister, foreign secretary or even president unless it is used as a personal title ('President Washington' but 'the president'). There are a few exceptions such as Black Rod, The Queen and God.

> **Laws** are lowercase (second law of thermodynamics) unless they are named after somebody (Murphy's law).

> **Integers**: In text, write out 0 to 10 (as zero, one, …, ten) but for greater integers use figures (21). If an integer starts a sentence, write it in words.

Things to Avoid

- **Ornate words and phrases**, such as convey (take), pay tribute to (praise), seating accommodation (seats), utilize (use), Fred underwent an operation (Fred had an operation), we carried out an experiment (we did an experiment).
- **Needless prepositions** tacked on verbs: check up, try out, face up to.
- **Vague words** like considerable, substantial, quite, very, somewhat, relatively, situation (crisis situation), condition (weather conditions) and system.

- **Clichés** such as last but not least, as a matter of fact.
- **Passive voice**
  - As is shown in Figure 1 … (*write*: Figure 1 shows …)
  - It was decided to …
- **Obfuscation** – I found the following on the World Wide Web. It is attributed to Mark P Friedlander.

> *Learn to obfuscate*
> Children, children, if you please
> Learn to write in legalese,
> Learn to write in muddled diction,
> Use choice words of contradiction.
> Sentences must breed confusion,
> Redundancy and base obtusion,
> With a special concentration
> On those words of obfuscation.
> When you write, as well you should,
>
> You must not be understood.
> Sentences concise and clear
> Will destroy a law career.
> And so, my children, if you please,
> Learn to write in legalese.
> So that, my dears, you each can be
> A fine attorney, just like me.

If you have questions about specific points of style not covered here, you may like to use one of the many style guides available on the web. One example is the government website style guide at https://www.gov.uk/guidance/style-guide/a-to-z-of-gov-uk-style. Major news media also offer style guides, for example the BBC style guide at http://www.bbc.co.uk/academy/journalism/news-style-guide.

Statistics

Several textbooks offer guidance about the presentation of data and of data analysis. In '*A Primer in Data Reduction*, Andrew Ehrenberg (1982) wrote four chapters on communicating data: Rounding, Tables, Graphs and Words. These are worth reading before you write any papers or your report.

Rarely are measurements made to more than two or three digits of precision. Yet results of analysis are often shown to many more digits. Finney (1995) gives an example: $2.39758632 \pm 0.03245019$ 'computed with great numerical accuracy from data at best correct to the nearest 0.1%'. Such numbers are crass and meaningless, but computers automatically produce them. Would you then report them, pretending scientific precision, or would you round them to an understandable level that means something?

In his discussion of tables, Ehrenberg says that:

- rows and columns should be ordered by size;
- numbers are easier to read downwards than across;

- table layout should make it easier to compare relevant figures; and
- a brief verbal summary should be given for every table.

The briefest and best (in my view) of guides about the presentation of results is reprinted as an article from the *British Medical Journal*: 'Statistical guidelines for contributors to medical journals' (Altman *et al.*, 1983). This has good advice for all research workers, not just those in the medical world, and I suggest that you obtain a copy. Here are a few of its points:

- Mean values should not be quoted without some measure of variability or precision. The standard deviation (SD) should be used to show the variability among individuals, and the standard error of the mean (SE or SEM) to show the precision of the sample mean. You must make clear which is presented.
- The use of the symbol $\pm$ to attach the standard error or standard deviation to the mean (as in 14.2 $\pm$ 1.9) causes confusion and should be avoided. The presentation of means as, for example, 14.2 (SE 1.9) or 14.2 (SD 7.4) is preferable.
- Confidence intervals are a good way to present means together with reasonable limits of uncertainty and are more clearly presented when the limits are given, for example '95% confidence interval (10.4, 18.0)', than with the $\pm$ symbol.
- Spurious precision adds no value to a paper and even detracts from its readability and credibility. It is sufficient to quote values of t, χ^2 and r to two decimal places.
- A statistically significant association does not itself provide direct evidence of a causal relationship between the variables concerned.

References

Altman, D.G., Gore, S.M., Gardner, M.J., and Pocock, S.J. (1983). Statistical guidelines for contributors to medical journals. *BMJ* 286: 1489–1493.
Ehrenberg, A.S.C. (1982). *A Primer in Data Reduction*. London: Wiley.
Finney, D.J. (1995). Statistical science and effective scientific communication. *Journal of Applied Statistics*, 22(2): 193–308.

Further Reading

Barrass, R. (1978). *Scientists Must Write*. London: Chapman and Hall.
Cooper, B.M. (1975). *Writing Technical Reports*. London: Penguin.
Kirkman, J. (1992). *Good Style: Writing for Science*. London: E & FN Spon.
O'Connor, M., and Woodford, F.P. (1978). *Writing Scientific Papers in English*. London: Pitman Medical.
Partridge, E. (1962). *A Dictionary of Clichés*. London: Routledge and Keegan Paul.

36

The Logic of a Research Report

Tom Bourner and Linda Heath

Introduction

This chapter is about reporting research. What makes it different from other writing about research reports is that it looks at the underlying logic behind the structure of the research report. The focus of the chapter is on the content and structure most commonly adopted by articles published in academic journals that report research results. Later in the chapter, however, we show that the conclusions apply to most other kinds of research reports as well.

The large majority of papers published in academic journals follow a fairly similar pattern. The main aim of this chapter is to answer the question of whether this is just an academic convention or is there a defensible rationale (i.e. an implicit logic that can be discerned). Its subsidiary aim is to make the reader more aware of the structure of the papers they read, including spotting what's missing.

Why is this an important issue? It is important for the underlying structure of a research report to be clear to the novice researcher for at least three reasons. Firstly, it will make it easier to write a research report. Secondly, it will increase the likelihood of getting papers published. Thirdly, understanding the reasons for the underlying pattern of a research report helps to clarify the research process itself.

People who are new to research may believe that the form of a research report is a matter of academic convention, which implies rules to remember and follow. In fact, there are reasons behind that form. At least some new researchers are likely to find it is easier to remember a single coherent rationale than apparently arbitrary rules.

Understanding the logic of the research report is also likely to improve the reporting of research. Poorly reported research is of less value than well-reported research, so an improvement in research reporting will increase the value of the research in question. Also, better research reporting reduces the likelihood that good research will be lost due to failure to publish and thereby share the results of the research. And at the level of the individual researcher, well-reported research is easier to publish then poorly reported research.

Research Methods for Postgraduates, Third Edition. Edited by Tony Greenfield with Sue Greener.
© 2016 John Wiley & Sons, Ltd. Published 2016 by John Wiley & Sons, Ltd.

In addition, understanding the logic of the research report helps to improve the actual quality of the research itself, because (i) the research report is the immediate goal of a research project and it helps to know where the goal is located, and (ii) the principles of a good research report overlap to a large extent with the principles of good research.

As an added bonus, it turns out that understanding the logic of a research report also throws light on the content and structure of research *proposals*, including proposals for research funding. So a really good understanding of the underlying logic of a research report will help the novice researcher to write good research proposals with increased likelihood of successful funding.

In this chapter, we start by looking at the purpose of a research project and the role of the research report within that purpose. We then look at what must be contained within a research report for that role to be realised. Finally, we look at how those elements fit together in the resulting sequence. In other words, this chapter is largely based on reason: it is an attempt to deduce the structure of a research report from the presence of a starting point and the presence of a goal.

Background

There are many definitions of research, ranging from the loose and vernacular 'finding stuff out' to dictionary definitions such as 'systematic investigation in order to establish facts and reach new conclusions' (OED, 1998). In this chapter, our definition of research is the *intentional creation of shared new knowledge*. Each element of this definition is important:

- **Intentional**, because we recognise that knowledge can be created by accident, by serendipity or as a by-product of some other process.
- **Creation** is used in preference to the word *discovery* because we want a definition of research that includes not only discovery but also *invention*. In other words, we want to include not only new knowledge of that which already exists in the world but also knowledge of that which could exist, but does not yet exist. So our definition needs to be broad enough to encompass not only finding out about what currently exists in the world but also creating new components of the world.
- **Shared**, because new knowledge that is not shared does not add to the stock of knowledge from which we can all draw.
- **New**, because 're-inventing the wheel' doesn't add to the stock of knowledge.
- **Knowledge**, because research goes beyond providing more than mere data[1].

---

[1] There is a nice ditty by James Autry which goes:

> Data are not facts,
> Facts are not information,
> Information is not knowledge,
> Knowledge is not truth,
> Truth is not wisdom.
> (Autry, 1991: p. 89)

Those words are worth a little reflection by most post-graduate researchers, and perhaps by all researchers.

The key premise that underpins this chapter is that the purpose of research is to make a significant original contribution to shared new knowledge. To be worthy of publication, the new knowledge contributed by the research must be not only original but also significant. In other words, it must also pass the 'So what?' test. Research that contributes knowledge of a precise figure for the number of pebbles on Brighton Beach would not be worthy of publication as it would fail the 'So what?' test.

The proximate purpose of research is the advancement of knowledge. Its underlying social purpose is to contribute to the pool of knowledge from which we can all benefit. This means that sharing the results is a significant part of the research process itself.

The importance of sharing the results of research is recognised by Research Councils when they require proposals for research funding to include plans for the dissemination of the research findings. The importance of sharing the results of research is recognised by the emphasis given to publication by universities when conferring the title of 'professor'. And it is emphasised in the doctoral training of new researchers in Britain by the requirement that a new doctoral thesis should be worthy, at least in part, of publication and also by the requirement to deposit a copy of each successful doctorate dissertation in the British Library. This places the new knowledge contributed by the doctorate in the public domain. It symbolises its contribution to the pool of knowledge available to all. It conveys an expectation that new knowledge is to be shared.

There have however, been practical problems in realising that expectation in practice, and often a doctoral thesis has sat, gathering dust, within closed library stacks. As a result, a vast quantity of potentially significant research has hitherto been unread and unused. However, relatively new national initiatives such as EThOS (Electronic Thesis Online Service) give access to anyone seeking theses from institutions across the United Kingdom which are free at the point of use for anyone who wants to access them. The major advantage of making a thesis available online is that it becomes accessible worldwide to anyone with an interest in your area of research and facilitates wider availability of research material.

Research can be shared in various ways, including journals, books, presentations at conferences, research seminars, teaching and increasingly the internet. The traditional means, however, has been the academic journal, which uses peer review as a means of quality control to test claims of new knowledge.

The importance of sharing the results of research is so important that perhaps we should find a new name for that part of the research process which omits the stage of sharing the findings (i.e. 'research' which contributes only to private learning but not to public knowledge).

The challenge for the researcher is to make a significant original contribution to shared knowledge. The challenge of this chapter is to find the form of a research report that would best serve that goal.

The Argument

Imagine that your research has produced what you believe to be new knowledge and you now wish to share that new knowledge with the rest of us. To make this more concrete, imagine also that you wish to publish your research in an academic journal. You know this is the traditional way by which researchers disseminate the fruits of their research (i.e. their putative new knowledge). As a new researcher, this is probably the first time you've submitted an article to an academic journal. What should you include, and what should you leave out? How should you organise what you decide to include?

A sensible approach would be to look at some other articles that have been published in academic journals to find answers to these questions. If you do this, you'll see that articles are normally divided into sections, certain section headings appear with great regularity and the sequence in which the sections appear also seems to form a common pattern.

The following sections in the following order are pretty typical:

1. Introduction
2. Background, including previous work
3. Method
4. Findings
5. Discussion
6. Conclusions.

Sometimes, of course, slightly different words are used instead. For example, the word 'Results' might be used instead of 'Findings'. And sometimes a couple of the sections are conflated. For example, it is not uncommon to find an article with the section heading 'Discussion and Conclusion'. However, one way or another, all of these elements are usually found in an academic article.

To what extent is this mere academic convention? You've probably noticed a few academic conventions in your reading and your research so far, such as the use of terms like *ibid.* and *op cit.* and the particular way that references are listed at the end of published articles in academic journals.

It is the contention of this chapter that the format of the typical journal article is more than just academic convention. The format has a logic that helps to clarify exactly what should go in each of the sections. Moreover, understanding that logic also throws additional light on the research process itself.

Let's look at the basic rationale for the various sections, starting with the 'Findings' because that is really the raison d'être of the whole article:

> **Findings** – The driving force behind producing the article is to disseminate research findings, so this section obviously has to be included.
>
> **Method** – In order for readers to decide how much credence to give your findings, they need to know how you arrived at them.
>
> **Background, including previous work** – Before doing your own primary research, you'll have reviewed what is already known about the subject. There's no point in re-inventing the wheel. The endpoint of this initial literature review is to reveal a gap in the published knowledge of this topic. A section on 'Previous Work' therefore provides the knowledge context of the study and clarifies the goal of your own research and its rationale. It tells the reader what gap(s) in the literature your own research is directed at.
>
> **Discussion** – Having presented your own findings, you need to review your key results in the light of the existing knowledge in the field. This enables you to show exactly how your results relate to existing knowledge and indicates where your own contribution to knowledge fits into the emerging map of knowledge in the

field. In terms of the latter, there are two key questions: (1) 'What light does the literature throw on your findings?' and, even more important, (2) 'What light do your findings throw on the literature?'

Conclusion – Your conclusions will clarify and underline the nature of your contribution to knowledge, and this is also a good place to identify further questions that your research findings pose for the further development of knowledge in its field.

Introduction – The main purpose of the introduction is to give the reader the information they need to make a good decision about whether it is worth their time reading your article. This is the first section of the article, but it is often the last section to be written because it can be difficult to be entirely clear about what you are introducing until you've written at least a first draft of the other sections.

So that's a brief rationale for including each of the main sections of a research report in the form of an article for an academic journal. Now we'll return to each of the sections and look at what the rationale implies for the content of each of the sections. This time, however, we'll look at the sections in the order in which they are most likely to appear in the article you submit to the journal.

1. Introduction

We've seen that the main purpose of the introduction is to provide potential readers with information to help them decide whether or not to read the article. The following information would do this:

- What the article is about, including the main aim of the article.
- Why the main issue addressed by the article is important. Sometimes this is self-evident, but it always helps to provide your perspective on the significance of the issue in question. You have decided to research this issue, so you probably have a particularly acute awareness, or a particular understanding, of its significance. You'll also be much more aware of its significance than the average reader because you've researched the issue in depth. Moreover, this is a way you can start to establish that your research contributes a *significant* addition to our knowledge.
- Main conclusion(s). This also helps potential readers decide whether or not to read your article as it tells them what the potential payoff is likely to be.

2. Background, including Previous Work

This section provides the context for your own research. In our experience, it normally helps to include the following:

- Historical context of the issue/problem in question (e.g. previous attempts to address the issue).

- Definition of any key terms, as this aids clarity. Also, if there is any ambiguity in the use of any of the terms, or any disagreement about their meaning, it lets the reader know your own understanding of each term and the sense in which you are using it in this article.
- Review of the literature to locate the gap that your research is intended to fill. You normally can't afford to provide a comprehensive review of the literature in an academic article as there will be a maximum limit on the number of words allowed for articles submitted to the journal you've chosen. So how much detail to include? Enough to establish that there is a genuine gap in the literature on the topic to which your research is directed.

3. Method

How much detail should you include about the method? There is an almost infinite amount you could write about the method(s) you employed, including such details as the dates and times at which you carried out your research, the personalities of anyone involved and how they were dressed. None of this is likely to be worth including. Remember that the main reason for including a 'Method' section is to help your readers decide how much credence to place on your research findings, and that will give you a good guide on what to include and what to leave out. On that basis, it would be worth including:

- The reason you chose the method(s) you did choose. This may include your understanding of 'knowledge' in this area.
- Any plausible alternative method(s), with your reason(s) for their rejection.
- Enough detail about how you implemented the method(s) you used, to enable other researchers to replicate your research if they were in a position to do so. This is important because it helps other researchers to test your findings and puts them in a good position to challenge your claims to the discovery of new knowledge. It is through the process of researchers making claims to the discovery of new knowledge and others trying to challenge such claims that the pool of shared knowledge expands.

4. Findings

A fairly bald statement of your findings will put readers in a position to make their own assessment of them. If you present an *interpretation* of your findings, including only the particular results that support your interpretation, then it can impair the credibility of your work. This is because that would make it more difficult for readers to make an alternative interpretation, or interpretations, of your findings.

5. Discussion

This is where you get a chance to focus on the results that you regard as most interesting and significant. This is where you can interpret your findings in the way you find most persuasive. And this is where you discuss how your findings fit into the existing map of knowledge in the field. It is this which enables you to position your new knowledge within the latest literature in the field and thereby clarify exactly how your findings make a contribution to new knowledge.

6. Conclusion

Your conclusions are your interpretation of your key results and their significance. They follow on from your discussion of your key findings in the 'Discussion' section. Having established that you have made an *original* contribution to knowledge, you'll now want to establish that it is a *significant* original contribution to knowledge. So this is the place to indicate the significance of your conclusions for theory and/or application and practice. This is also the place to identify further questions that your research poses for the accumulation of further knowledge in the field or more widely. In summary, the 'Conclusions' section should include at least the following:

- Conclusions that embody your interpretation of your key finding(s). This is different from just a summary of what you have already said.
- Implications of your findings and conclusions for theory and/or application.
- Emergent questions leading to suggestions for further research.

Discussion

This section looks more closely at some of the issues that have emerged in this chapter so far, in particular: (i) how understanding the logic of a research report can increase understanding of the research process itself, (ii) the extent to which the argument generalises to other kinds of research report and (iii) the relevance of the argument for the development of proposals for research funding.

The Logic of the Research Process

At the start of this chapter, we asserted that understanding the logic behind the structure of a research report can throw additional light on the research process itself. So now we'll review the research process in the light of what we've said about the logic of a research report.

The starting point of research can be characterised as a 'presenting problem'. It may be some real-world problem, it may be some problem with the theory or it may be the problem of understanding a topic better, but one way or another it can be represented as a problem or question. The first step in the research process is to identify the presenting problem and its significance.

If, on reflection, the presenting problem turns out to be insignificant there would be no point in proceeding to the second step, which is to ask 'What is already known about this issue?' and then search the literature for the answer to that question. Not everything that is written on a topic constitutes 'knowledge'. It is therefore important to critically examine the results of the literature search in seeking to identify the knowledge. In other words, it is important to test the results of the literature search and, if necessary, challenge them. The result is a critical review of the published literature or, for short, literature review. At the end of this process, the researcher(s) will have a map of what is already 'known' about the topic and what is not known. The term *gap in the literature* refers to the latter. If the researcher discovers that what they want to find out is already 'known' in the published literature, then there is no point in proceeding further. More likely, they find that some aspects are already known and some aspects lie in the 'gap'.

It is now possible for the researcher to identify the 'research problem'. In contrast to the 'presenting problem', the 'research problem' is to fill the identified gap in the published knowledge (or some part of it).

The primary role of the literature review at this stage in the research is to enable the transition from the presenting problem to the research problem. The research problem is sometimes termed the *research question*, and it specifies what the research aims to find out. Only at the end of this stage can the researcher be confident that if the research is successful then it will lead to an original contribution to knowledge.

Having established what the research aims to find out (i.e. the research objective), then the next logical step is to find a method or methods fit for that purpose. This is step 3: choosing the research method(s). It makes sense to adopt the classic problem-solving process here: identify the range of options, and then evaluate the options to identify the one(s) most fitted for the purpose of the research. Having selected the research method(s), the next step is implementation. The end results of this step are the research findings themselves.

It is tempting to think of the discovery of the research findings as the end of the research process, but that is far from the case. You now have to make sense of your research results and identify where they fit in the existing map of knowledge on the subject. This means some interpretation as you make sense of your results, and that means discussing your findings. At the end of that process, you will have reached some conclusions.

Your conclusions are likely to extend beyond the location of your results in the emerging map of knowledge. They are likely to include your judgement of the significance of your results, which means looking at their contribution to application(s) and practice as well as theory. And it is likely that your research results and your conclusions will surface new problems and questions. Such problems and questions are the starting point of new research enquiries, so it is very likely that your conclusions will include suggestions for further research. It is sometimes said that *Homo sapiens* is a problem-solving animal; it is at least true to say that *Homo sapiens* is a 'problem-*finding*' animal, not to say a 'problem-*generating*' animal. It is this questioning attitude that underpins the problem-finding behaviour that is the force behind the accumulation of human knowledge.

You might think that the research process is really over now, and, if so, you would be wrong again. So far, what you've done could be described as personal learning. You've contributed to your own individual knowledge. But when we talk about a contribution to knowledge, we are implicitly referring to *shared* knowledge. So the final step in the research process is to add what you've discovered or invented to the shared pool of knowledge. This means producing a research report of some kind to place your research findings in that larger domain. 'Research' that does not do this falls at the final hurdle. Such 'research' fails to satisfy our definition of research ('the intentional creation of *shared* new knowledge').

It is clear that the logic behind the research report mirrors the logic behind the research process. The better you understand the logic behind the research process, the easier it will be to write research reports as articles that are accepted by academic journals. And the converse is also true: the better you understand the logic behind the research report, the more successful you are likely to be with your research.

Other Kinds of Research Reports

So far, this chapter has focussed on the logic of the content and structure of a 'typical' article for an academic journal. It has adopted the perspective of a research student preparing to share her research findings by writing her first article for publication in an academic journal.

Although this is the time-honoured way that academics have shared the results of their research and contributed to the advancement of knowledge, it is not the only way that it can be done. Whether they publish in an academic journal or not, research students are normally required to produce a thesis[2]. This is also a report of research, but it is much larger. Thus, for example, the literature review in a thesis is likely to be very much larger which means that it will normally be more comprehensive, and doctoral candidates will be expected to display the full extent of their finely honed critical faculties in evaluating the published work in their field.

Nevertheless, the doctorate is, in essence, just another research report, and 'the logic of a research report' applies equally to a doctoral thesis as to an article prepared for submission to an academic journal. The main difference, apart from its size, is that the *section* headings are likely to appear as *chapter* headings. So a 'typical' doctoral thesis will contain the following chapters:

Chapter 1: Introduction

Chapter 2: Critical review of the literature

Chapter 3: Method

Chapter 4: Findings

Chapter 5: Discussion

Chapter 6: Conclusions.

If your thesis omits any of these elements, you'll need to have a very convincing explanation. Likewise, if any of these elements appears in a different sequence, this is likely to be questioned at your viva.

So the logic of a research report applies equally to a doctoral thesis as to a published article. In fact, the first published article produced by many research degree students is likely to be a summary of their doctoral thesis. This is reasonably straightforward because each of the chapters in the doctoral thesis is summarised to form a section in the draft article. At this point, it is worth noting that the research degree regulations of many universities state that a criterion for the successful doctoral dissertation is that it should be 'worthy, at least in part, of publication'. There are few more convincing ways of persuading your examiners that your thesis is 'worthy, at least in part, of publication' than including the letter of acceptance from an academic journal of an article based on your research as an appendix in your thesis together with a copy of the article itself.

We've seen that the logic underpinning the content and structure of a typical article in an academic journal generalises to the content and structure of a doctoral dissertation. Does it generalise further?

---

[2] It is, however, possible in some universities to get a research degree on the basis of published work alone.

It certainly generalises to the sort of end-of-project research report required by most bodies that fund research projects. In fact, it generalises to any form of academic research report. Master's degrees and undergraduate degrees were once called 'taught' degree courses to distinguish them from research degree courses. There was a time when they were precisely that; they were made up entirely of taught components. Then master's degrees courses increasingly adopted the practice of including a final unit that involved research and the production of a dissertation as a research report. So gaining some training in research became the hallmark of postgraduate courses. After that, the practice spread to some undergraduate programmes, and then it became increasingly common to find a final-year module based on research and the production of a research report as a dissertation. This is often considered the 'capstone' unit of the undergraduate course and seen, therefore, as particularly important, sometimes carrying twice the weighting of other modules in the student's final year. Increasingly, therefore, developing the students' capacity 'to plan and manage a research project' seems to be emerging as an expected outcome of a higher education for undergraduates as well as postgraduates. In these days of accelerating change in the world and accelerating accumulation of knowledge, it seems to be higher education's newest 'transferable' skill. What we have said about the logic of a research report applies with equal force to all these forms of research report at all these levels of higher education.

Are there any forms of research report to which it *doesn't* apply or to which it might only partially apply? The most important exception is probably the business report. Research undertaken for a particular organisation is usually not intended to add to the pool of knowledge from which we may all freely draw. On the contrary, it is often the source of competitive advantage. Since the logic of the research report that we have presented in this chapter was driven by the aim of making a contribution to *shared* knowledge, it is likely that the content and structure of a research-based business report will differ from those of an academic research report. The main differences are likely to be:

- much less emphasis on a critical review of the literature,
- much less emphasis on method, and
- much *more* emphasis on recommendations for action as a consequence of the findings.

Research Proposals, including Proposals for Research Funding

A proposal for funding research is at the other end of the research process from the production of a report of the research, so it might appear to be quite disconnected. This appearance, however, is misleading. When writing a research proposal, it is helpful to keep an eye on the end result, which you now know is not the research findings, nor the research conclusions, but the research report. Any funding body is likely to want answers to the following questions:

- What is the presenting problem?
- Why is it a significant issue?
- What is known about the problem already?
- What is the gap in the knowledge you intend to fill?
- What method(s) will you use to fill it?

- What new knowledge does your research aim to produce?
- What do you expect your conclusion(s) to be, and what are the possible implications for theory, application or practice?
- How do you intend to disseminate your research results?

You'll see there is a pretty close correspondence between these questions and the sections of a research report. It is reasonable, therefore, to infer that a clear idea of the content and form of a research report provides a good guide to information that is important to include when writing a research proposal, including one to apply for research funding.

The bottom line seems to be that the logic of the research report casts light on the logic of the research process itself, and it also casts light on the logic of a research proposal.

Summary and Conclusions

Academic writing comes in many forms including, for example, books, journal articles, theses and reports to funding bodies at the end of research projects. This chapter has focussed predominantly on reports of research submitted to academic journals to disseminate research findings, but it is equally relevant to research-based dissertations and research reports submitted to funding bodies.

The starting point for this chapter was the point at which a researcher believes they have made a significant, original contribution to new knowledge and wants to share that knowledge by means of a research report. In communicating a significant original contribution to new knowledge, you need to ensure it is recognised by others, so you need to demonstrate:

- that it is knowledge;
- that it is original (i.e. *new* knowledge); and
- that it is significant.

In order for the readers of your research to recognise your contribution, they need to believe that it is true. So you need to provide grounds for that belief. This is why it is important to explain the method or process by which you reached your conclusion. In those cases where the research is replicable, this involves providing an account of the research process in sufficient detail to allow the reader to replicate the research to check on their findings. In other cases, it involves providing an account of the research in sufficient detail that the soundness of the steps taken can be assessed by the reader.

Since research comes in all shapes and sizes, it might seem that there is little that different research reports have in common. But there is at least one key element that they all have in common; all research seeks to make a significant original contribution to knowledge. To be more precise, this chapter applies to those research projects that aim to make a significant original contribution to shared knowledge.

Starting with the 'typical' content and structure of a research report, we can infer the steps taken in a 'typical' research project:

- Start with identifying the 'presenting problem' and its significance.
- Explore what is already known about the issue in the published literature on the subject.

- Recognise that not everything written on a topic constitutes 'knowledge'; therefore, it is important to critically review the literature. Identify gaps in the literature – otherwise, there is no point in going further.
- Translate the 'presenting problem' into the 'research question' (which involves filling the identified gap in the literature).
- Choose the research method(s) best fitted to answer the research question.
- Implement the method(s), and collate the research findings.
- Interpret your results, and identify how they relate to the existing map of knowledge on the subject to make a contribution to new knowledge.
- Recognise that your conclusions will go further than placing your new knowledge into the emerging map of knowledge on the subject; they will include your judgement of the significance of your results.
- Identify any new problems and questions that you have surfaced for further study.
- Share your new knowledge so that it enters the public domain and is not merely new personal knowledge.

This chapter has shown that there is an implicit logic to the pattern of research-based articles published in academic journals. We tried to simplify the process of writing a research report by making that logic explicit, and so save the new researcher from having to get to grips with apparently arbitrary academic rules of engagement. By so doing, we hope to have clarified the research process itself and helped the new researcher write better research reports and more successful research proposals.

References

Autry, J. (1991). *Love and Profit: The Art of Caring Leadership*. London: Chapman Publishers.
Oxford English Dictionary (OED). (1998). Oxford: Clarendon Press

37

Presenting Data

Sue Greener

Introduction

Today's research environment is alive with pictures, both static ones and video. We use pictures and videos to draw attention to tweets on Twitter, we use pictures to share on Instagram and Pinterest and that is just in social media. We can learn from the marketing media that pictures catch people's interest and are easier to 'consume' in a rushing world than a large amount of text. Yet you are producing a research report or thesis which is very heavy in text indeed. So how can you bring some user-friendly pictures into your work? And should you do so? In this chapter, we take a brief look at how data and ideas can be presented using a visual approach.

The last thing you want to do is to fill your thesis or research report with pretty pictures. Sadly this is quite a common approach in undergraduate studies, where students frequently offer themed word-processed reports using off-the-shelf templates and clipart. As someone who has to mark most reports online, the download time and unnecessary images do not put me in the best frame of mind to assess the actual work. So you want to find a happy medium. We use visual representations when we want to tell a particular story about data or ideas. These representations will still be explained in the text, but they give the reader a quick way of following your logic.

As mentioned in earlier chapters, a scientific thesis is unlikely to need much graphical illustration where tables of data may be sufficient to summarise results and provide a starting point for detailed analysis. However, a thesis more dependent on qualitative data, and using mixed-methods analysis for example, may use visualisations of data to good effect in order to help the reader notice key points within the context of the data.

Then, of course, you will often need to share your developing ideas and your conclusions with others; and at this stage, visualising your data in some form can be of great benefit. Today's over-dependence on slide-sets often seems to lead to endless bullet points of equal weight, making it difficult to structure ideas in the listener's or reader's mind (a common problem for university students faced with slide-sets at every session). While the slide-set which only contains images will be initially interesting but soon fail to provide enough detail

Research Methods for Postgraduates, Third Edition. Edited by Tony Greenfield with Sue Greener.
© 2016 John Wiley & Sons, Ltd. Published 2016 by John Wiley & Sons, Ltd.

to keep the listening mind focussed, using images in moderation can be stimulating. All this comment on slide-sets is of course dependent on them being accompanied and led by a strong presenter as, at best, they are only a visual aid, not a visual alternative to the presentation. More on this in Chapter 38 on presenting.

Other forms of written report, such as reports and summaries to research funders, colleagues and collaborators, can be enlivened and made more accessible through visualisations of different kinds. This may include charts, graphs, plots, maps and pictures, including photographs, provided they are used where they have a clear function and are not just offered without explanation or introduction. Placing illustrations carefully with an eye to meaning is one thing, but we also need to consider the overall impact of the page or slide. In reports and in research theses, the presentation as a whole should be easy on the eye and, most importantly, very easy to follow.

Not many readers of research documentation will start at the beginning and go through faithfully to the last full stop or appendix. While these documents are usually structured in a linear way (see Chapter 35 on writing your thesis), they are not always read this way. It is more likely, particularly for an examiner or assessor, that they will use the contents page and drop into areas of immediate interest or concern. This does mean that the pagination and contents page must be easy to follow and accurate: it will help if there is sufficient 'white space' between items to allow the eye to search quickly. When I examine a thesis or report, I look first at the contents page to get my bearings within the work, then the references to find out the range and quality of ideas used by the author, then possibly the conclusions, before starting out at the beginning to look for the logic of argument and data analysis which should lead to those conclusions. Others will have different approaches to evaluating a paper or report, but mine can certainly come unstuck if the contents page is unclear, uninformative or inaccurate.

The rest of this chapter will look in more detail at three approaches to the presentation of data distinguished by purpose. Firstly, we look at the inevitable charts and graphs or visualisations which have the purpose of guiding the reader to key points before explaining the thinking behind these choices. Secondly, we examine the use of data visualisations with the purpose of explaining your research to a non-expert audience. Thirdly there is a discussion of the use of data visualisations to help you, the researcher, to summarise and organise your thinking along the way.

Key Points Visualisation

In taught courses, universities and colleges consider the quality of graphics such as charts, graphs and boxplots as one of the assessment criteria for the quality of research work. Students are therefore encouraged to figure out how to visualise data in order to offer meaningful analysis and interpretations of those data. Unfortunately, that often results in a plethora of pie charts. If the research questions are simplistic, often featuring yes/no answers or proportions of a single variable distributed among a population, the pie chart offers a simple way to summarise this one-to-many relationship. It will help with relative magnitudes, frequencies or percentages and can come in different visual forms: slices can be raised and separated from each other, and colours and three-dimensional versions abound.

In research reports, pie charts are generally to be avoided. Such simple relationships can be better shown in a table if they are required. If you have to use them (and they do have a role in presentations to non-expert audiences), then make sure they are clearly labelled with data (e.g. percentages), that the computer-generated colours will be seen in final form (if the report is to be copied, it is unlikely the colours will be distinguishable in the intended way) and that three-dimensional pie charts do not overly distort the reader's sense of the proportions displayed just in order to make the pie look more impressive visually.

If you are trying to describe the behaviour of particular variables over time, their range, their dispersion around the mean or their central tendency, then a chart or graph will help. This can give an immediate impression of the behaviour of the variables which you can then describe and analyse in more detail in the text.

The behaviour of one variable against another or several other variables can be shown visually in simple charts using lines and bars. There are many illustrations of such charts in packages such as Excel (Microsoft) and Numbers (iOS). Again, such simple charts may be of more use in summarising a large quantity of data in a thesis or when presenting results to non-expert audiences. Here are a few examples.

A box plot is helpful to show the simple distribution of a dataset. This is sometimes called a whisker plot because the lines that are shown to extend vertically from the boxes, which show variability outside upper and lower quartiles of the datatset, are known as *whiskers*. Here is an example of a box plot (whisker chart):

Stacked column and stacked bar or area charts can show the relationship of individual items to the whole, so you can compare the contribution of different data items to a total across categories of data. This is a stacked area chart:

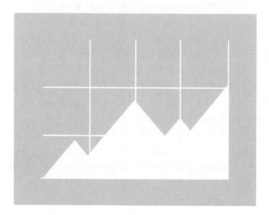

Radar charts or polar charts can display multivariate observations from a number of variables in two-dimensional format:

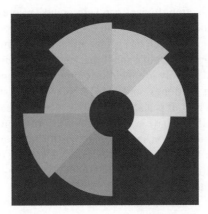

Bubble charts can show three points of data in a series. These are generally used to show correlations of *x* values, *y* values and sizes. The Gapminder website (www.gapminder.org) produced by Hans Rosling has some animated charts using bubbles to great effect, for example looking at the trends in birth rates relative to country populations over time.

For certain types of organisational, technological or geographical research, heat maps can demonstrate intensity or focus against a contextual background, such as a map or website screenshot. Individual values are usually shown as colours on such a map.

For reports to be presented online or delivered with projection, animated visualisations can be powerful. They allow the presenter or reader to focus on a detail within a graphic, which can be built in several layers and hyperlinked.

Continuing with the technology theme, social media interactions are now often visualised in network map form. For example, the range and strength of connections made across the population using a particular Twitter hashtag can be visualised using NodexL. An example of the use of #brightsoc over the period of a research conference on social media is at this URL, courtesy of Connected Action: www.sueg1.wordpress.com/2014/07/21/nodexl-visualisation-of-ecsm2014-twitter-hashtag/. Such network analysis can not only produce immediate visual impact but also derive top hashtags, URLs, words and relationships between words within the map.

The vital thing with all such visualisations, charts and graphics is to label clearly the relevant data, axes or shapes (unlike the generalised images offered above); to offer a self-explanatory title for the figure within your report; and to explain what meaning you derive from the image. Never assume that your reader will understand the image in the same light as you, the author.

Presenting Visualisations to Lay Audiences

During your research period, you will often be called upon to present progress to supervisors or funders, and to other academic colleagues or students. These presentations may be formal, as in funding stages when more funds have to be negotiated, or at research conferences, or they may be relatively informal occasions with colleagues or students in your own department.

The level of formality may well affect the state of your nerves at the time, but it should not affect the care you take to produce a clear explanation of your research.

Often a slide-set is called for on these occasions, and I have set out my views on bullet-point-only slide-sets in the Introduction. Images can help greatly in presentations for a number of reasons: they relieve the tedium of short text bullets; they offer opportunities for interpretation and conjecture to the audience, which can offer variety of pace in a good presentation; they can summarise much descriptive text in a small space; and they can force the presenter to look up from their text notes and actively use the illustration for the benefit of the audience.

Moderation is needed in all things, and a good balance of text, data and figures in both reports and presentations is best. It is also tempting to over-engineer illustrations in slide-sets. This is never a good idea. If you have a great deal of information to convey, consider using a build-up series of slides to beckon the audience gently into your complex figure, or use animation in some shape or form. Slide graphics can give you the opportunity to start with an overview image and then drill down or hyperlink within the slide-set to different areas of the initial image to offer more detail.

Presenting software such as Prezi, which offers a dynamic 'zooming' format for presentation, can be used to good effect where you, the author, have a strong sense of visual design and where there is a range of graphics or images available. If you do not wish to use many images, Prezi is not advisable as that is its particular strength. You begin with a blank canvas rather than a limited A4 slide space and can build the structure of your presentation in a non-linear way, which can be helpful when you know your content very well and are prepared to respond to audience questions to follow a path through the material which is appropriate to them. You can use the flexibility of this visual space to drill down to detail and then pull back to an overview of a section or the whole presentation, in a dynamic format. Do remember not to include too much movement across and within the canvas; it can make the audience feel a little queasy.

The other major tool in your armoury when presenting your work to a non-expert audience is video. Again, moderation will help; too many short clips will confuse and bewilder an audience, but a well-chosen short video can transform the illustration of a case study, a context or a problem. The caveat here will be checking the available technology well before your presentation and immediately before you present to ensure all will run smoothly (not just picture but also sound). A minor point, but try not to include videos of yourself in action; this really does look a little arrogant.

But suppose we are not talking about a personal presentation here; perhaps you wish to disseminate your research through a blog or a conference poster. Micro-blogs such as Twitter have much more impact when they include an illustration, attracting more retweets, which in turn widens the potential audience for your work. That illustration could be the one graphic or figure from your research which makes a big impact, or it could be something a little less strictly academic, such as a cartoon (there are many packages around to help you produce these if you are not personally artistic) or an infographic. The latter often uses pictograms instead of straightforward graphs or charts. The use of a relevant symbol, such as the outline of a body to express relative numbers of people or a factory building to express relative production output, can offer rapid impact, which is what infographics and blogs are all about. Provided you always offer a URL, shortened link or QR (quick response) code to a more detailed and scholarly account of your research, the use of a hook such as a cartoon or pictogram can be justified.

Using Visualisations in the Process of Research

Usually you have a clear idea of what you are looking for in your data: you may be attempting to prove or disprove a hypothesis, or checking the application of a concept in your specific data for example. In this case statistical analysis is your first port of call, and you may wish simply to present the results of this in a visual format as we have discussed. However, sometimes we can use visuals to look at the data in different ways, and you may find this visual manipulation can trigger new ideas.

Sparklines in Excel or Numbers spreadsheets are simple line charts created on the fly from series data and can give an instant impression of the direction or variability of travel of the variable you are looking at. This is particularly useful with a large number of data readings over time. They do not show axes or data labels, simply the shape that the variation of data creates in a condensed format, so they are useful for picking up trends as data change.

Scatter diagrams can show the initial outline of a possible correlation, which can then be tested through statistical analysis. Hand-drawn scatter diagrams are simple enough to use as you are working through your data, as are mindmaps or spider graphs, where the central idea is linked to other ideas as they arise through your relationship with the data. Some people find mindmaps (the original idea from Tony Buzan) a natural response to brainstorming or developing ideas. Others find them unnecessary and artificial, so much will depend on your approach and preference. Certainly, as your research becomes more complex and data-heavy, you will need to find some support for the widening scope of your study. Visual illustrations can really help here if you can find the ones you can use quickly and easily. First explanations of hazy ideas to your supervisor can often be supported by a quick hand-drawn diagram which clarifies the issues for him and for you as you create it. Of course, you can use digital technology here too, if you have a strong facility for it; you can produce graphs, charts and well-crafted figures swiftly, which may even get included in the final version of your research. But don't spend too much time on producing illustrations until you are sure they will make the final 'cut'. A quick sketch using pen and paper may do just as well, especially if your ideas are changing rapidly. I have seen research students waste a great amount of their time in producing beautiful images for me as their supervisor, only for us to agree quite quickly that that image does not add any value to the study.

A Final Word on Good Practice

Which brings us to a summary of this chapter. We have looked at illustrations and visualisations of data for three main purposes – explanation, dissemination and thinking. It should go without saying that any visualisation must be clear, since its usual purpose is to help someone derive meaning. This means that simplicity is an important feature of data illustrations. It also should be assumed that any data illustration has been double checked for accuracy. Particularly where spreadsheet packages are used to produce charts automatically, it is so easy to pick up an extra line of data or omit one, and this can be difficult to spot in a final graph.

The next critical point is where the illustration comes from and what right you have to use it. You can source photographs and images easily from the web, but either you must have explicit written permission from the author to use them, or they must be offered through Creative Commons licences in which case you follow the relevant guidance for citation. Similarly, of course, any secondary data used in your own illustrations must be correctly source referenced.

Should you reference yourself? Not if you have simply created a chart or other image and any others, which are not your work, are clearly source referenced. What about images you have used in other work, especially published work? If you have the luxury of other publications, beware too much self-citation. Be parsimonious with the references to other work of yours. Self-promotion has its place but can easily be overdone.

The final point? Ask yourself why you are using an illustration. If it clarifies a point or a dataset; makes a complex process understandable; helps to show relationships between ideas, data or arguments; or genuinely improves the overall presentation of your report, that is fine. In all other cases, if no particular value is added, leave it out.

38

Presenting and Sharing Your Research

Sue Greener

Introduction

There are other chapters in this book which deal with the presentation of your written research report, developing its logic and the illustration of your work with appropriate graphics or images. There is also a helpful chapter on the use of social media, in particular focussing on the technical issues related to blogging and websites and social networks. So this chapter looks at the personal ways in which we are likely to present or share our work. It will touch on the use of social media from a presentation and networking perspective, but it will start with you as presenter – horror of horrors – the one thing most people dread: facing an audience in person.

Your supervisor has just told you it is time to present your work in progress at an internal research conference in your university or organisation. Your logical reaction is to feel flattered in the confidence they are showing in your research study so far. You readily agree to a suitable event, time and place, putting it in the diary as if it were a pleasant engagement. However, it doesn't take most of us long to feel differently about presenting to an audience. Even if teaching has been part of your role for some time, it is quite different to face your peers and other experts in your field with your results so far.

What are you to do? Here are some simple ways to prepare and develop your presentation so that you make the best of this opportunity to share your enthusiasm and your results with others. The first step is to let go of the idea that this is all going to be about you. This is not about you. This is about your audience. This is about what they will go away with after your presentation, as well as how much they follow and how they engage with your material during the presentation. They are the whole point of the presentation, and there may be two of them or two hundred of them, or more, but their importance is the same. Your aim is to involve them in your ideas, possibly to gain feedback from them about your ideas, and definitely to leave them with a good lasting impression of those ideas. The more you focus on your audience

Research Methods for Postgraduates, Third Edition. Edited by Tony Greenfield with Sue Greener.
© 2016 John Wiley & Sons, Ltd. Published 2016 by John Wiley & Sons, Ltd.

and the need to involve and communicate with them, the less time there is to dwell on your feelings about the process of presentation. There is no benefit to be had from worrying that they will disapprove or be critical of your presentation; put yourself in their place. As part of the audience, you don't want to go away feeling you have wasted time listening to a talk which you cannot understand or follow; you really want to be able to say you chose to go to a strong presentation, as it justifies your action in attending. So the audience will usually be on your side from the start, willing you to do well. Let's look at how you can fulfil their hopes and give them a little love in return.

What Kind of Event?

You may be asked to deliver an internal presentation, as just suggested. In this case, the audience may be known to you and may be quite small. This is a great way to gain feedback in a relatively informal environment. Talk to colleagues about what is normal timing, content and practice in this situation. Best of all, go and attend as many others as you can before you have to do this yourself. You are likely to find that many of these internal seminars are in front of small audiences, and, provided you treat your audience well, they are likely to be constructive and helpful to you.

But perhaps you are required to deliver a presentation at a formal research conference? In fact, this is something to be enthusiastic about, because it is part of the developmental process, which will lead to sharing your research through academic publication. Again, you are encouraged to attend a conference in your field before you have to deliver at one yourself. Becoming familiar with relevant conferences, developing professional and informal networks at these conferences and getting used to conference programme formats are all really helpful ways in which to find your feet as an academic presenter. Your research supervisor(s) may be able to take you along to such conferences at a reduced fee and introduce you to other academics in the field. All this can ease your way into the academic conference routine.

Depending on your discipline and research topic, you may be asked to present at a commercial or professional conference or event. This will be organised very differently from an academic conference, so you will need to find out as much as possible about the event format, presentation requirements, technology available, any history of these events and, most of all, what kind of audience to expect. While the principles of good presentation will remain valid, the details of the day will be very different from those of a standard academic event, so do your research well.

Other kinds of event for presenting your research will include presentations directly to potential funders, where you may be able to assume a certain degree of familiarity with your subject, although this is no substitute for personal enquiry and research before the event. It is extraordinary how audiences seem to be able to speak socially about a topic for hours without really understanding what they are talking about. They will have heard of the popular buzzwords, but do they understand them in context? They may be coming with a clear set of criteria for accepting your presentation; try to find out about those criteria and hit them explicitly in your presentation. Find out as much as you can about what they know already, and tailor the starting point of your presentation to that level of knowledge.

The other major presentation event will be a *viva voce* for your degree, which will be given to internal and external examiners and usually a chair of the panel. For this event, it is less likely that you will be able to use visual aids, so the preparation will focus on your intimate knowledge and memory of the thesis you are presenting. Good practice here will be to annotate your personal thesis copy with flags or bookmarks to skip quickly to key points and sections under discussion, and, advice which certainly helped me in my viva, to summarise each page in a word or two written in pencil at the top of each page of your own copy. While this probably sounds crazy now, when you are in a viva you will probably be a little anxious and your ability to read and locate information will be somewhat reduced. These pencilled notes help enormously if an examiner says 'Now on page xxx, I want to ask …': as soon as you have turned to the page, the summary is easy to read and you are able to respond swiftly to the query. But let's not get ahead of ourselves. What are the principles of good personal presentations? Can you adapt them and communicate your ideas to people who may not know much about your research field? Can you win their hearts and engage their brains?

The Keys to Great Personal Presentations

All over the web, you will find guidelines on giving presentations. The popularity of TED talks via video (www.ted.com) has increased our awareness of presenters' capabilities and the power to communicate. You can watch TED talks on a vast range of topics and usually find compelling speakers with a powerful message. Perhaps this seems a long way away from presenting your own research? But you can learn by example. TED talks are rarely more than 18 minutes long. They are well rehearsed, and they are often backed up by very strong visual aids. That is something from which we can all learn.

Presentations require planning, preparation and practice. There is no shortcut here. Looking at each of these areas in turn, we can identify some of the keys to great presentations.

Planning

We start, as I stated in this chapter, with the audience. Identifying their needs is vital to a good presentation. What can you find out in advance about their perceptions and background? How widely has the invitation to your presentation been sent? Of course, it will be impossible to gain robust information well in advance about your audience, but every little piece of information will help you to work out how to put your message across to them. Here are some questions to ask about your potential audience.

- Are they hostile or friendly? It is very unlikely they will be hostile in any way, but if you happen to know someone is coming who disagrees with your research or approach, it is better to be forewarned to prepare your arguments.
- How knowledgeable are they? Perhaps they are in the same research group or department as yourself, in which case you will have clues as to their understanding of your field. Or perhaps the invitation has gone across the institution to a wider unknown audience, in which case you must plan to cover basics first and not make assumptions.
- Why are they there? Do they want to be there? Are some required to attend, in which case you need to give them reasons to enjoy the experience, or is this a purely voluntary audience?
- Have they paid to be there? This is unlikely in an internal event, but likely at a conference. They will want to find value in their attendance; what value can you offer them?

- How many of them are there? This may be limited by venue or room size, or simply by inviting specific numbers. This is important so that you can visualise the scale of the audience in advance. However, always be prepared to have fewer than expected. This is nothing personal! People often want to attend research presentations but are prevented by other commitments.
- How do they perceive you? Is this the first time they have heard you? What do they know about you? How much do you need or want to tell them to build your credibility? If your institution is unknown to them, do you need to introduce that first?
- How should you appear to them? This is related to the last question but may affect the way you can build rapport with your audience. It is not just about what to wear, but also the tone of your presentation. Be aware that anxiety often makes presenters appear either arrogant, where they put themselves above the audience, or humble, where they put themselves below the audience. Don't succumb to either behaviour: you are equal to them and have every right to talk to them as equals with neither excessive humility nor excessive arrogance.

And while we are considering your appearance and credibility, we should ask the question of you – why are you speaking? Are you intending to:

- Teach
- Inform
- Sell or persuade
- Inspire
- Convince
- Entertain
- Or a combination of these aims?

You may feel you are just there to tell them about your research, but think a little more deeply about your purpose. It could be that their feedback is vital to you, in which case you will need to build their understanding of your topic or message. Perhaps they could help you with funding or data access, in which case some persuasion by logical argument will be important. Perhaps you don't require any action from them as a result of your presentation, but you still want to make a strong impression for your own academic credibility and, possibly, career.

It is also a good idea to work out in advance on whose behalf you are speaking. Is it only you as sole researcher? Or perhaps your supervisors' names are on your slides and their reputations are also in your hands? Or perhaps you are representing your institution to others, locally, nationally or internationally. All these questions are designed to get you thinking about how to plan your presentation – what purpose you have, who they think you are and what results or outcomes you want from the presentation. Once these objectives are clear to you, you can start to think about the planned structure of your presentation.

So far, we have suggested an overall length not exceeding 18 minutes. This is not by accident, but stretches an audience's concentration, so to ramble on for longer is unlikely to help you achieve your aims. That is not to say the whole presentation slot should always finish after 18 minutes. On the contrary, most slots will be at least 20 minutes at a conference, often 30 or even 60 minutes. The remaining time is for dialogue with the audience. This is why it is so important to find out something about them, so that you can encourage their contributions and ask them some questions if they don't ask any of you. Once your audience starts asking questions about your presentation, time can go very quickly, so it is usually helpful to suggest early on that question time is left to the end of, or specific points in, the presentation.

Consider the structure of that part which you are presenting. Can you make it so simple that neither you nor your audience can forget it? For example, you may wish to offer a structure of three, four or five parts. More than that is harder to recall. You may find it helpful to imagine a physical shape for your presentation, such as a triangle of three points, a square or a five-pointed star. You could use this shape in a visual aid, or verbally, but whether you tell the audience about the shape or not, it may help you to remember where you are at any point of the presentation. The classic structure is, of course, a beginning, a middle and an end. Don't discard this idea; it can be very helpful for the listener, and some presenters seem to think they can do without one of these sections, which makes the logic of the argument hard to follow. It can also be great for an audience to be told very clearly how many key points you will be making. They can see how far along you are by how many points are still to come. They can therefore pace their attention. If you are to love and take care of your audience, things like this matter; help them any way you can to know how much more talking there will be, when you are getting to the most vital part and when you are preparing to close. These are cues which help your audience maintain attention.

Preparation

Once you have the basics planned, then you can start the detailed preparation of a presentation. What medium will you use for visual aids? Most people use slide-sets, often Microsoft PowerPoint, but fail to use them effectively. There is much help on the web on how to get the best out of presentation software, but here is a key piece of advice: a full set of slides without a single image or graphic will be hard for the audience to swallow, particularly if you are using a bland standard theme. Don't go to the other extreme and just use images, either; however good those images are, the audience will follow up their own thoughts from them and you may lose their attention. Another problem can be to overlay images with text, or to put an image at one end of the slide and text on the other end. While these things are possible technically, they increase the cognitive load on the audience as attention switches from image to text and back again, as they try to make connections. Don't make them work so hard; use images which are meaningful and explain their relevance to the point you are trying to make. And don't use too many!

Do also consider alternatives to slide-sets. Academic presentations often use zooming presentation software such as Prezi; it is free to use and can be a very powerful presentation tool. However, beware: if you are not great at visual design, Prezi or similar can be overdone and distracting – a simple slide-set may present your message more clearly. If, on the other hand, you are good with visual design, Prezi may give you the additional excitement and flexibility which a standard slide-set does not.

Other media will be important too: you may use web links or video extracts or an audio recording with the speaker's photo on projection. Use multiple media sparingly, to keep the essential structure and message clear for your audience. And the golden rule with multimedia is to access the venue early and check that everything works. Bringing in a live link via Skype or similar is also sometimes helpful, but the potential dangers of the internet not functioning at the crucial time offer risks which you may not be prepared to take. Always have a backup plan if you hope to use live internet interaction. Have an appropriate screenshot in your presentation to talk about if the link goes down. Or have an alternative way of interacting. Live polling is increasingly useful to find out what an audience is thinking and what they understand about your topic. The technology (e.g. Poll Everywhere or Socrative which are free to use or freemium cloud-based audience response systems) can transfix and excite an audience when there are enough in the audience with smartphones or web-enabled mobile devices, yet they are also dependent on various technologies functioning in the venue at the right time.

Let's not forget that it is you who are the presenter, not the technology. The technology, be it slides, video, polling and so on, is only a visual aid to your message: you are more important than the slides. Never simply read from the slides, and particularly do not turn your back on the audience to read from a screen behind you; this is rude and ineffective presentation. If you are happier simply speaking to your audience without visual aids, that is fine, but do remember the 18-minute rule. Longer than that, and your audience may find it hard to concentrate. One way to support a simple verbal presentation is to offer a summary of your presentation (text or slides) in a paper document to each of your audience. This is something they can use to maintain concentration as you speak, and of course make notes on if they wish. Once the 18 minutes is up, make sure you get some interaction from your audience to keep them with you.

A Little Extra Research

So you have decided on the media you may or may not use. The content of your research may be in a very long text version, several chapters perhaps, or may just be in your head. The next step will be to interpret and shape the content to suit the audience and the objective of your presentation. You may need to do some additional research here, to tailor your content to the audience or to update facts and figures. Consider, for example, the geographical context of your work. You may be researching in a wide global field, but your audience may live in a specific region. Make sure you can offer them something about that region. And perhaps the literature you wish to present was reviewed two years ago; it is likely that there will be more now, and up-to-date references will be helpful to your audience, as well as showing how you are up to speed with current developments in the field. Today's audiences are likely to have mobile devices with them and may wish to follow up on references, websites or terminology which you mention. Make sure it can be easily found – perhaps offer them current web links for later review. The key is to understand your audience enough to find common ground with them and build bridges from their understanding to yours.

Preparing the Content

We are really back to structure here, so we can now decide what background or opening statement will be useful to set the scene of your presentation. You know the key points you wish to get across; think about how you will link them so that the audience follows your argument without getting distracted. Check for logical gaps. It is easy when you are very familiar with a subject area to produce slides or paragraphs which, in themselves, are beautifully formed, but which do not fit together in a logical sequence.

Consider the ABCD of your introduction:

- Attention
- Benefits
- Credentials
- Design.

Aim for getting attention from the very start: you may say something unexpected, something that appears unconventional or shocking until you explain it: a challenge, a puzzle, a statistic. Once you have their attention, spell out the benefits for the audience of listening to you (what you have to offer them). Then deliver your credentials: why are you the right person to offer these benefits? And close your introduction with the shape or design of the presentation, to interest them in the key points to come.

Through the main 'middle' section of your presentation, aim to involve and engage that audience. Give them subject matter with which they can identify, recognise their level of understanding and take them on a journey with you towards the key message. Think about using questions in your presentation to encourage interaction. You may not want to stop presenting and get the audience speaking, but you can still get a show of hands, or ask a rhetorical question to get them thinking. A great presentation technique can be to design the whole thing as a series of questions, questions you have asked yourself about the topic, to which the audience also wants the answer. This really generates involvement: have them thinking 'Yes, I really want to know that' and they are with you all the way.

Should you ever use humour in your academic presentation? The answer will entirely depend on you. Never attempt set piece humour, the kind of jokes with a story and a punchline, because they generally backfire. Also, academic audiences may be culturally diverse and may not consider your jokes funny, and they may even give offence. However, if your normal social talk contains an element of humour, it would be reasonable to include some in your presentation. This really only works if you are comfortable with it. You may consider anecdotes or stories to illustrate a theme, involving your audience with laughter, but if the idea is worrying for you, do not even attempt it – stick to what you know you do well.

What about the close of your presentation? This is the most often neglected part. You are beginning to feel relieved that you have covered the key points and are coming to the end of your time, so you relax and suddenly it's all over. This is really not a good idea. Audiences will usually remember the start and the end of your presentation better than the middle. Even if the middle was well presented, the close must be memorable and helpful to the audience. What we do not want is either for the presentation to fizzle out as you murmur, 'That's all really' (!) or for the chair of the session to stop you mid-sentence because your time is up and you have ignored warnings to come to a close. Either way, you look unprofessional and unprepared. So the first rule is to time your presentation through rehearsal (more on this in this chapter). The second rule is to watch the time and respond calmly by concluding, even omitting some slides, to bring your presentation to a proper close. The third rule is to prepare that proper close carefully in advance to regain attention 'and finally…' and to use this increased audience attention to ask for action, feedback or questions and to repeat your key message.

One more point about preparation. We began this section suggesting that you should consider carefully the media you wanted to use to support your presentation. We end this section by asking whether you need notes. Inexperienced presenters often create verbatim scripts for a presentation, then find that nerves make it hard to follow the detailed script, lose their place and, as a consequence, lose their audience. This is not to say that writing out a verbatim script is a bad idea. It can be a very helpful idea, as it gives you the chance to perfect some phrases and clarify the logical structure. However to read from it is rarely sensible. You would only do this if you were broadcasting your presentation and had a very tight time frame to keep to, or had to follow a particular 'party line'.

The ideal is to create your structure, perhaps use a mind map or similar to ensure all points are included and you have worked out a sensible order of points. Then, if you wish, write out the script, or you may only need to do this for the beginning and end if you know the material extremely well and are happy speaking about it to time constraints. Practise delivering your presentation, and then aim to reduce to keywords and headlines, either on cards (tagged together so they cannot be dropped and fall out of order) or on a tablet device which is clearly readable. Try to avoid using A4 paper scripts as they can develop a life of their own if you are at all nervous: A5 cards or smaller are easier to handle. While I am discouraging you from using a full and detailed script, I would equally discourage you from having no notes at all. Even if you don't use them on the day, headings and keywords in order are a helpful reference, giving you confidence, and the flexibility to switch points if needed. If you have no notes, there is a strong chance you will speak for too long, and fail to stick to the key points, spending ages on one part and then having insufficient time for the rest.

Delivering Your Presentation

We have looked at planning, preparation, research and content, so now to the all-important event itself. Your presentation slides may be immaculate and inspirational, but we want the audience to follow you, not your slides, so your body language, the way you deliver the ideas, matters.

The biggest key of all must be rehearsal. It is true that you can over-rehearse. You are not trying to memorise every sentence you prepared and produce a presentation which is so slick it is unbelievable. What you are aiming to do is to get the audience to see who you are and to engage with what you have to say. You are not trying to be someone else who is good at presentations; you are trying to be you, putting across your research to equals in the room.

It is perfectly possible to understand everything this chapter has to offer, yet still not deliver a good presentation. The difference is made by practice. You should rehearse your presentation sufficiently so that you know its key points really well, and the order they are presented, but have not reached the stage where even you are bored with the content. In your rehearsals, if you find a phrase which works particularly well, note it down and try to keep it in there. But also try to express the same point in different ways in practice, so that if you can't recall exactly what you meant to say at the time, you can still explain the point. It is important to remember that your audience have not seen your rehearsals or your verbatim script if you had one. They only experience what happens in the moment. So if you forget something you meant to say, if you change the order by mistake, don't panic. They don't know unless you confess. This is where you become the show and the show must go on, without it looking as if you made a mistake. (Unless of course you missed out a vital part, in which case you simply apologise without fuss and go back to it.)

It would be very odd if you were not nervous on the day. When you have no nerves before a presentation, you should start to worry! The effect of nerves is usually to speed up your responses, and that helps you to think faster than your audience. If you are still worried, try building a 'horror floor' – this means imagining the very, very worst that can happen in your presentation. The very worst, when made explicit, is often not very bad (such as having to go back to a previous slide, or correct audience misunderstanding in question time, or forgetting what comes next and having to look at your notes). The earth will not stand still, and you will survive the presentation. Clearly visualise your success, the audience engagement, the applause. They want you to succeed, so focus on them, not on you.

We have already mentioned the importance of time management. Whatever time slot you have been given, stick to it, and practise sticking to it. Remember that if you have to allow for audience questions, this will eat up time quickly. Use a stopwatch (on most mobile phones and laptops now) to check and double check your average presentation time as you rehearse. It is not an option just to go faster at the end to fit more in; instead, omit a point and move smoothly to your prepared close and key message.

A little on non-verbal communication: consider eyes, stance, gesture and voice. Be generous in giving eye contact to the audience. Share the love, and don't focus on just one person; it can make them so uncomfortable. Lift your gaze to the back and sides of the room, not just the front row. Consider making eye contact with the audience before you start speaking; this engages their attention and helps them to 'hear' you. Above all, do not look backwards at a screen, except momentarily to check it if needed, and do not stare endlessly at your notes or your laptop! The point of your presentation is your audience, not you and your slides.

Allow your stance to demonstrate ease and balance. Try to stand for presentations if possible, and if you are shaky, stand beside a table or chair to give yourself some support without being seen to lean on the furniture. Stand tall, relaxing into a good posture. This looks both confident and authoritative, even if you don't feel it. Let your message dictate your facial expression; by all means smile, but not if you are delivering a serious point! Keep your hands relaxed by thinking not of your hands but of your shoulders which should be relaxed down, letting your hands drop to your sides unless you are using a gesture to make a point. Keep hands out of pockets.

Finally, your voice. We all have different ways of speaking, different sounds and accents, and different languages. All are good, none is bad, unless it stops the audience from hearing the message clearly. Sometimes anxiety makes the voice tighten and rise, the pace quickening so that the presentation becomes a gabble. Focus on speaking 'slow and low'. Slower means it sounds more important; lower gives you more authority. However, there is a good reason for varying the pace and pitch occasionally, because it wakes up the audience, so you might want to give an example in a lighter tone, compared to the conclusion it is designed to demonstrate. If you have the chance to speak out loud in the presentation venue before you start, then take it; hearing yourself and your sound in the room can help you decide how much you will need to project your voice, which should be focussed as far as the end wall so that all can hear. If you are offered a microphone in a larger venue, always accept gracefully. You do not prove anything by laughing it away, saying you are quite capable of making yourself heard without the mike. In fact, you prove that you are inconsiderate: the microphone is there to offer an easy way of hearing for every member of the audience, rather than those you are looking at at the time. It is also sometimes there to help those with hearing issues. Be generous and accept the mike.

A good sound will support your presentation, especially if you enunciate clearly, avoiding hesitation and repeated ummms and errrrs. When you introduce technical terms or jargon related to your field, take care to explain each term or give examples, search the audience for signs of misunderstanding and respond if you see this. Generally it is best to use simple words, as in academic writing, because that means that you are not hiding behind obfuscation. And remember that silence, or a pause, is particularly effective in a presentation. Pause before saying difficult words, pause to emphasise key phrases and pause to refocus the audience. If you lose your place, or have trouble changing a slide, don't panic. The pause will seem longer to you than to your audience. There is usually no need to chat in this pause; just sort out the issue, or ask for help, and return to your presentation.

Questions

It is easy to feel that at the end of your delivered presentation, everything is over and you can relax. You can't. You will now need your sharpest concentration to focus on listening to questions from the audience. Always empathise with questioners, even if they are asking something which you have already answered in the presentation. Never bluff if you do not know how to answer. You might ask them to explain the question further, or ask to speak with them offline after the presentation in more detail, or you could simply say that you don't know the answer and tell them what you will do about that (how you will find out and contact them perhaps).

Suppose you are greeted with a silence when no one has a question? You haven't failed; you may have been very clear, or perhaps they are reflecting on what to say. If the silence goes on for a while, you may have a session chair to pose a question to you if you are in a conference. If not, you may wish to explain how you can be contacted if questions arise later, and repeat the key point of your presentation to close.

Sharing Your Presentation

Before you give a presentation, consider carefully how it should be shared. If you are presenting a work in progress, you may not wish to share more than slides with your audience on the day, or a summary sheet for them to follow. It may go no further because there is much that may change. However, you may want to share your presentation with a wider audience; this is increasingly the case for academic researchers as they build social networks and contribute to repositories of research. If you have slides which you are able to share (e.g. any images you have used are correctly attributed and ideas you have found elsewhere are correctly referenced), then you could use a network such as Slideshare, Dropbox or Google docs with a public link, or if you use the free version of Prezi you will automatically be sharing this with others unless you opt out. Or you may have your own or your organisation's website to which the presentation is uploaded. You could go further and record a video of the presentation, ideally a short summary video, and put this on YouTube or a similar site on the web for disseminating your ideas. All the software mentioned here is free to use.

You could also consider blogging about your presentation. This is not the same as bragging about your presentation (though some do!). The idea is to give a context to the presentation, perhaps the event at which it was presented, its purpose and audience and so on. Or you could use micro-blogging, such as Twitter and tagging with keywords, to develop interest in your ideas. This is not just about self-promotion. Academic research is there to be shared, and developing a social network of academics in your field will help you gain feedback on your work, as well as expose your ideas to a wider audience: that audience being the whole point of presentation.

39

Reporting Research

Iveta Simera and Douglas G. Altman

Medical research makes prominent headlines because it explicitly affects people's lives. Even if we are not involved in the research as participants, we are all its potential consumers. As patients, we want to be treated according to the best evidence from well-designed and well-conducted research. However, in order to identify the 'best' evidence, research studies need to be published in an accurate, complete and timely way. Only such papers can be further processed into systematic reviews and clinical practice guidelines and so truly inform clinical decision making. If the published primary research is not reported well, or not published at all, available evidence becomes incomplete and unreliable and precious resources invested into the research are wasted.

This chapter provides a brief overview of general principles of reporting medical research studies with a particular focus on the following study designs: randomised controlled trials, analytical observational studies, and systematic reviews and meta-analyses. We will introduce the main reporting guidelines that have been developed to help document all necessary methodological aspects and findings in research manuscripts. Although the reporting guideline activity has been largely focussed on medical research, the broad principles apply to research in many other fields, especially biological sciences.

Ethical Imperative of Responsible Publication of Research Studies

Authors, editors and publishers all have ethical obligations with regard to the publication of the results of research. Authors have a duty to make publicly available the results of their research on human subjects and are accountable for the completeness and accuracy of their reports. They should adhere to accepted guidelines for ethical reporting. Negative and inconclusive as well as positive results should be published or otherwise made publicly available. Sources of funding, institutional affiliations and conflicts of interest should be declared in the publication. Reports

Research Methods for Postgraduates, Third Edition. Edited by Tony Greenfield with Sue Greener.
© 2016 John Wiley & Sons, Ltd. Published 2016 by John Wiley & Sons, Ltd.

of research not in accordance with the principles of this Declaration should not be accepted for publication.

Declaration of Helsinki – Ethical Principles for Medical Research
(World Medical Association, 2008)

Publication is the final stage of research, and scientists have moral responsibility to ensure their studies are communicated honestly to provide a clear, accurate, complete and balanced account of what was done and what was found. Researchers should avoid misleading, selective or ambiguous reporting in their papers.

International standards for authors of scholarly publications have been proposed at the 2nd World Conference on Research Integrity in Singapore 2010 and are freely available online (Wager and Kleinert, 2011). The key principles for responsible practice in research reporting are summarised in Box 39.1.

Box 39.1   Key principles for responsible research reporting.

The research being reported should have been conducted in an ethical and responsible manner and should comply with all relevant legislation.

Researchers should present their results clearly, honestly and without fabrication, falsification or inappropriate data manipulation.

Researchers should strive to describe their methods clearly and unambiguously so that their findings can be confirmed by others.

Researchers should adhere to publication requirements that submitted work is original, is not plagiarised and has not been published elsewhere.

Authors should take collective responsibility for submitted and published work.

The authorship of research publications should accurately reflect individuals' contributions to the work and its reporting.

Funding sources and relevant conflicts of interest should be disclosed.

Reproduced from the International Standards for Authors of Scholarly Publications (Wager and Kleinert, 2011).

General Principles of Reporting Medical Research

A good research manuscript combines a well thought out, logical structure, complete and accurate description of the key study elements, and a clear and concise writing style.

In principle, health-related research can be divided into two broad groups: experimental and observational. In experimental studiesl the investigators identify eligible participants and determine how these participants are allocated to different intervention (exposure) groups. A typical example of experimental research design is a randomised controlled trial (RCT), in which people (often patients) are randomised to receive one of two or more treatment options (see Chapter 18). In observational studies, the investigators observe exposures (e.g. smoking status) and outcomes (e.g. cancer) in study participants but do not influence what happens. Each type of research has its strengths and weaknesses, and each is more or less suitable for answering a particular research question. One of the fundamental points in medical research is to choose a research design that is appropriate to answer the given research objectives. Study reports should then provide enough details about the chosen research methodology to allow readers to judge how well the study was designed and carried out, how reliable the presented findings are, and how relevant and generalizable the reported findings are to the readers' specific situation.

Medical research papers are usually subdivided into several sections, each with a specific purpose:

Title – attracts readers to the key aspects of the study

Abstracts – summarises the paper and together with the title helps reader identify the nature of the study and briefly summarises what was found

Introduction – explains why we did the study and what question we asked

Methods – describes how we did it

Results – presents what we found

Discussion – considers what the findings mean

Conclusions – if present, offers a succinct summary of the major findings.

Box 39.2 outlines the key general content of a research paper in more detail.

Box 39.2   Structure of the medical research paper: describing the key content elements.

Title	Indicate the focus of the paper and include important relevant 'keywords' allowing the study to be identified by electronic searches. Be concise, precise and informative.
Abstract	Most journals require a structured abstract, typically providing key information on: – Objectives – Methods (setting, participants, interventions or exposures, and outcomes) – Results – Conclusions.
Introduction	Clearly explain why you have undertaken the study and what question(s) you were trying to answer. Be brief and relevant to the study: start from the broad context of what is already known, proceed to the specific unknown problems and finish with clearly stated study objectives.

Methods	Describe in a logical sequence how the study was designed, carried out and analysed. A typical methods section provides key information on the following:
	– Study design
	– Setting and location
	– Participants (or objects)
	– Intervention (exposure)
	– Outcomes (variables)
	– Sample size
	– Statistical methods
	– Ethical issues (e.g. consent).
	The text should be clear, accurate and complete (providing enough details to repeat, assess and compare with other studies).
	Content should correspond to the Results section.
Results	Report results of the investigations described in the Methods section (in same order) using text, tables, figures and statistics.
	Be as brief and clear as possible (but don't leave out 'disappointing' results).
	A typical structure and chronology include:
	– Description of study participants (separate for important subgroups)
	– Presentation of answers to the main questions (starting with primary outcomes, then secondary outcomes and any other analyses).
	Report on harms (adverse effects) as well as benefits.
	Pay special attention to the design of your tables and graphs.
Discussion	Discuss what your findings mean and where they stand in the context of other studies.
	A typical Discussion section structure and chronology include:
	– Brief presentation of the main findings
	– Assessment of study strengths and weaknesses
	– Comparison of findings with previous studies
	– Consideration of clinical and scientific implications
	– If relevant, suggestions for future research.
Conclusions	(This section is not always present separately in a research article.)
	Any conclusions should be fully supported by the study findings.
Acknowledgements	State source of funding and any relevant conflict of interest.
	Acknowledge all who contributed to the study but who do not qualify as authors.

Rigorous reporting of the scientific details specific to your research study is of the utmost importance. Recently, the expansion in the development of systematic reviews and clinical guidelines has triggered more intensive scrutiny of the available medical research literature. Systematic assessment of studies highlights frequent serious shortcomings (Simera *et al.*,

2012). These include but are not limited to:

- omissions of crucial information in the description of study methodology, inclusion and exclusion criteria, and intervention details
- inadequate reporting of statistical methods and analyses
- selective reporting of only some results (outcomes)
- inadequate reporting of harms
- confusing or misleading presentation of data and graphs
- incomplete numerical presentation of data precluding inclusion in subsequent meta-analysis
- misinterpretation of study findings in the main article and abstract ('spin').

It is also common that whole studies never get published or are published with a major delay (Simera *et al.*, 2012). Poor reporting practices distort the available body of research evidence and compromise its utility and reliability. Such practices are unacceptable, whether deliberate or resulting from lack of knowledge of what to report.

To avoid the above-mentioned shortcomings, a number of reporting guidelines have been developed to help in preparation of research manuscripts. However, it is important to know what will need to be reported before initiating the practical part of the research. We strongly advise researchers to become familiar with the relevant guidelines when designing a study and working on its protocol. The EQUATOR online resources (www.equator-network.org) described in this chapter should be of considerable help (Simera *et al.*, 2010).

Reporting Guidelines

Reporting guidelines provide structured advice on the minimum information to be included in an article reporting a particular type of medical research. They focus on the scientific content of an article and thus complement journals' instructions to authors which mostly deal with the technicalities of submitted manuscripts. Some are generic for defined study designs (e.g. RCTs) and should always be observed when reporting this type of study. Most published guidelines are more specific, however, providing guidance relevant to a particular medical speciality or a particular aspect of research (e.g. reporting adverse events and economic evaluations). The content of each of these guidelines was very carefully considered by multidisciplinary groups of relevant experts, and there is a strong rationale for each requested information item. Items range from 'simple' requests such as the identification of study design in titles and abstracts to items focussing on aspects of the study methods that might introduce bias into the research.

Following internationally accepted generic reporting guidelines (e.g. those introduced further in this chapter) helps to ensure that the study manuscript contains all necessary information that readers need to assess the study's relevance, its methodology and the validity of its findings. In the following paragraphs, we introduce the most frequently used generic reporting guidelines: the CONSORT, STROBE and PRISMA Statements, and the EQUATOR online resources that support preparation of high-quality research manuscripts. Many journals publishing medical research require adherence to these guidelines.

Reporting Randomised Controlled Trials: The Consolidated Standards of Reporting Trials (CONSORT) Statement

Well-designed and well-conducted RCTs are the best available methodology to evaluate the effects of healthcare interventions. In general, they deliver reliable results that can inform future research or clinical practice. Poorly executed trials with inadequate methodology are associated with bias and may produce exaggerated intervention effects (Schulz and Grimes, 2006). Thorough assessment of a trial's design, conduct and analysis is vital to assess its reliability. However, such assessment is only possible if the trial report presents the critical information needed for such an appraisal.

The CONSORT Statement is a reporting guideline that consists of a checklist of 25 essential items to be included in an RCT report (Table 39.1) and a flow diagram designed to help

Table 39.1　CONSORT 2010 checklist of information to include when reporting a randomised trial*

Section/topic	Item no.	Checklist item	Reported on page no.
Title and abstract			
	1a	Identification as a randomised trial in the title	_____
	1b	Structured summary of trial design, methods, results and conclusions (for specific guidance, see CONSORT for abstracts)	_____
Introduction			
Background	2a	Scientific background and explanation of rationale	_____
and objectives	2b	Specific objectives or hypotheses	_____
Methods			
Trial design	3a	Description of trial design (e.g. parallel, factorial), including allocation ratio	_____
	3b	Important changes to methods after trial commencement (e.g. eligibility criteria), with reasons	_____
Participants	4a	Eligibility criteria for participants	_____
	4b	Settings and locations where the data were collected	_____
Interventions	5	The interventions for each group with sufficient details to allow replication, including how and when they were actually administered	_____
Outcomes	6a	Completely defined pre-specified primary and secondary outcome measures, including how and when they were assessed	_____
	6b	Any changes to trial outcomes after the trial commenced, with reasons	_____
Sample size	7a	How sample size was determined	_____
	7b	When applicable, explanation of any interim analyses and stopping guidelines	_____

Table 39.1   (*Continued*)

Section/topic	Item no.	Checklist item	Reported on page no.
Randomisation			
Sequence generation	8a	Method used to generate the random allocation sequence	_____
	8b	Type of randomisation and details of any restriction (e.g. blocking and block size)	_____
Allocation concealment mechanism	9	Mechanism used to implement the random allocation sequence (e.g. sequentially numbered containers), describing any steps taken to conceal the sequence until interventions were assigned	_____
Implementation	10	Who generated the random allocation sequence, who enrolled participants and who assigned participants to interventions	_____
Blinding	11a	If done, who was blinded after assignment to interventions (e.g. participants, care providers or those assessing outcomes) and how	_____
	11b	If relevant, description of the similarity of interventions	_____
Statistical methods	12a	Statistical methods used to compare groups for primary and secondary outcomes	_____
	12b	Methods for additional analyses, such as subgroup analyses and adjusted analyses	_____
Results			
Participant flow (a diagram is strongly recommended)	13a	For each group, the numbers of participants who were randomly assigned, received intended treatment and were analysed for the primary outcome	_____
	13b	For each group, losses and exclusions after randomisation, together with reasons	_____
Recruitment	14a	Dates defining the periods of recruitment and follow-up	_____
	14b	Why the trial ended or was stopped	_____
Baseline data	15	A table showing baseline demographic and clinical characteristics for each group	_____
Numbers analysed	16	For each group, the number of participants (denominator) included in each analysis and whether the analysis was by original assigned groups	_____
Outcomes and estimation	17a	For each primary and secondary outcome, results for each group, and the estimated effect size and its precision (e.g. a 95% confidence interval)	_____
	17b	For binary outcomes, presentation of both absolute and relative effect sizes is recommended.	_____

(*continued*)

Table 39.1 (*Continued*)

Section/topic	Item no.	Checklist item	Reported on page no.
Ancillary analyses	18	Results of any other analyses performed, including subgroup analyses and adjusted analyses, distinguishing pre-specified from exploratory	_____
Harms	19	All important harms or unintended effects in each group (for specific guidance, see CONSORT for harms)	_____
Discussion			
Limitations	20	Trial limitations, addressing sources of potential bias, imprecision and, if relevant, multiplicity of analyses	_____
Generalisability	21	Generalisability (external validity and applicability) of the trial findings	_____
Interpretation	22	Interpretation consistent with results, balancing benefits and harms and considering other relevant evidence	_____
Other information			
Registration	23	Registration number and name of trial registry	_____
Protocol	24	Where the full trial protocol can be accessed, if available	_____
Funding	25	Sources of funding and other support (e.g. supply of drugs), and role of funders	_____

*The CONSORT Group strongly recommends reading this statement in conjunction with the CONSORT 2010 Explanation and Elaboration (Moher *et al.*, 2010) for important clarifications on all items. If relevant, we also recommend reading CONSORT extensions for cluster randomised trials, non-inferiority and equivalence trials, non-pharmacological treatments, herbal interventions and pragmatic trials. Additional extensions are forthcoming: for those and for up-to-date references relevant to this checklist, see www.consort-statement.org.

document the flow of participants through a trial. The CONSORT website (www.consort-statement.org) provides the most up-to-date guideline version and lists all CONSORT publications. Of particular value is a long explanatory paper that outlines the rationale for each requested checklist item supported by examples of good reporting (Moher *et al.*, 2010). The key methodological details that should never be omitted in trial reports include the method of generation of the random allocation sequence and its concealment at the point of enrolment, any blinding, sample size calculation, a definition of primary outcomes and the results of pre-specified analyses.

The CONSORT Statement provides generic recommendations for reporting parallel group RCTs. The CONSORT group have subsequently extended the main guideline to accommodate particular trial designs (cluster, non-inferiority and pragmatic trials), interventions (non-pharmacological, herbal and acupuncture) and types of data (abstracts and harms). All these guidelines are available from the CONSORT website. The EQUATOR Library (discussed further in this chapter) also lists some additional extensions developed by groups independent from CONSORT (e.g. for reporting trials in allergy and occupational medicine).

Reporting Analytical Observational Studies: The Strengthening the Reporting of Observational Studies in Epidemiology (STROBE) Statement

Observational studies represent a large proportion of published health research. They provide valuable information about potential causes of diseases or their associations with various factors. Long-term observations of patients exposed to experimental treatments provide valuable data about potential harms associated with these interventions.

There are three main types of observational design: cohort studies, case–control studies and cross-sectional surveys. Unfortunately, the usefulness of many observational studies can be limited by potential confounding and bias. A minimum set of recommendations for reporting these studies is specified in the STROBE Statement (www.strobe-statement.org). Similar to the CONSORT Statement, the STROBE guideline consists of a 22-item checklist, of which 18 items apply to all three study designs and four are design-specific. The long explanatory paper explains all checklist items in detail and is a very helpful resource to aid understanding of the methodology of observational studies generally (Vandenbroucke et al., 2007).

The STROBE Statement has several extensions: STREGA recommendations for genetic association studies, STROBE-ME for molecular epidemiology (biomarker) studies and a draft proposal for STROBE for preparation of conference and journal abstracts (all available from the STROBE or EQUATOR websites).

Reporting Systematic Reviews: The Preferred Reporting Items for Systematic Reviews and Meta-Analyses (PRISMA) Statement

With constantly increasing numbers of research studies, well-conducted and well-reported systematic reviews and meta-analyses are essential for informing further research and for clinical practice. Mastering the methodology for performing a systematic review is an extremely valuable skill for postgraduate research students. Unfortunately, as with other types of research, considerable shortcomings have been documented in the reporting of systematic reviews. The PRISMA Statement (http://www.prisma-statement.org/statement.htm) is a guideline facilitating reporting of systematic reviews that assess the benefits and harms of a healthcare intervention. The PRISMA Statement consists of a 27-item checklist and a four-phase flow diagram for documenting the numbers of studies considered for the review, from first identification to those finally included (Liberati et al., 2009). As with randomised trials, good reporting helps to determine how well a systematic review was conducted. Detailed information about inclusion criteria, the searching process, assessment of risk of bias in primary studies and funding sources is of particular value. Although the guideline's primary focus is on systematic reviews of randomised trials, most of its content is of generic applicability.

The Enhancing the Quality and Transparency of Health Research (EQUATOR) Library for Health Research Reporting: A Free Online Resource

The EQUATOR Network is an international initiative that aims to improve the reliability and value of the medical research literature by promoting transparent and accurate reporting of research studies. The EQUATOR online Library for Health Research Reporting (www.equator-network.org) provides extensive resources to help scientists prepare high-quality research manuscripts. These include links to over 300 reporting guidelines, guidance on scientific writing, guidelines for responsible research conduct and publications, and other useful information such as resources facilitating the design of research studies. The resources are also grouped together according to the relevance to the main groups of users: authors of research articles, editors and peer reviewers, and developers of reporting guidelines. The portal is regularly updated to ensure the availability of the latest guidance.

Concluding Remarks

Accurate, complete and timely reporting of research studies is an integral part of responsible research conduct. We strongly encourage researchers to explore the available resources and learn about basic reporting principles relevant to their studies and any appropriate reporting guidelines before starting their research. Following reporting guidelines helps in preparing high-quality research manuscripts, facilitates peer review and increases the chances of manuscript acceptance by a suitable journal.

References

Liberati, A., Altman, D.G., Tetzlaff, J., Mulrow, C., Gøtzsche, P.C., *et al.* (2009). The PRISMA Statement for reporting systematic reviews and meta-analyses of studies that evaluate health care interventions: explanation and elaboration. *PLoS Medicine* 6: e1000100.

Moher, D., Hopewell, S., Schulz, K.F., Montori, V., Gøtzsche, P.C., *et al.*, for the CONSORT Group. (2010). CONSORT 2010 explanation and elaboration: updated guidelines for reporting parallel group randomised trial. *British Medical Journal* 340: c869.

Schulz, K.F., and Grimes, D.A. (2006). *The Lancet Handbook of Essential Concepts in Clinical Research.* Amsterdam: Elsevier

Simera, I., Kirtley, S., and Altman, D. (2012). Reporting clinical research: guidance to encourage accurate and transparent research reporting. *Maturitas* 72: 84–87.

Simera, I., Moher, D., Hirst, A., Hoey, J., Schulz, K.F., and Altman, D.G. (2010). Transparent and accurate reporting increases reliability, utility, and impact of your research: reporting guidelines and the EQUATOR Network. *BMC Medicine* 8: 24.

Vandenbroucke, J.P., von Elm, E., Altman, D.G., Gøtzsche, P.C., Mulrow, C.D., *et al.* (2007). Strengthening the reporting of observational studies in epidemiology (STROBE): explanation and elaboration. *PLoS Medicine* 4: e297.

Wager, E., and Kleinert S. (2011). Responsible research publication: international standards for authors – a position statement developed at the 2nd World Conference on Research Integrity, Singapore, 22–24 July 2010. In Mayer, T., and Steneck, N. (eds.), *Promoting Research Integrity in a Global Environment.* Singapore: Imperial College Press/World Scientific Publishing, pp. 309–316. http://publicationethics.org/international-standards-editors-and-authors

World Medical Association. (2008). Declaration of Helsinki – Ethical Principles for Medical Research Involving Human Subjects. 2008 update. http://www.wma.net/en/30publications/10policies/b3/index.html

Further Reading

American Psychological Association. (2009). *Publication Manual of the American Psychological Association*, 6th ed. Washington, DC: American Psychological Association.

Hall, G.M. (ed.). (2008). *How to Write a Paper*, 4th ed. Chichester: Wiley-Blackwell.

Note

A journal article based on this chapter is Simera, I., and Altman, D.G. (2013). Reporting medical research. *International Journal of Clinical Practice* 67: 710–716.

40

Social Media – How to Make It Work for You as a Post-Graduate

Suzanne Fraser-Martin and Catherine Fraser-Martin

Most of you by now will likely have access to any or all of a personal computer, laptop, notebook computer, tablet computer or smartphone. These may be your own, or have been provided to you by your university. Any of these could represent your gateway to the wonders of the internet. Hopefully you'll have heard of this: there are few subjects whose study can't be significantly assisted by net access. The internet is a fabulous place filled with experts, news, amazing stories, friends and family and all the other fantastic things which you probably already find totally indispensible. (It's also full of opinions, misinformation, biased sources and speculation, and part of the challenge of net-aided research is learning to filter the wheat from the chaff. More on this later.)

Assuming all the foregoing, you probably already know where your favourite websites are and how to check your email, but you may be unfamiliar with, or even totally mystified by, blogs and micro-blogs, the etiquette of social networking sites, the possibility of creating and maintaining your very own personal website or just precisely why the Birds are Angry. Fear not! We hope to assist by answering at least some of your questions as to the pitfalls and perils of the electronic mysteries; and aside from exactly why the Birds are Angry, hopefully we can give you a clearer picture of what all these things are and how you can use them to your advantage as a post-graduate.

Before we begin there are a couple of things we'd like to mention. Firstly, before you go charging off and sign up to everything, read the terms and conditions for each service first. Some have strict policies on the use of your real name, and not a pen-name, nickname or whatever-username-was-available-at-the-time name. So don't be surprised when you sign up, connect to all your friends and your account gets suddenly deleted because you've breached the clear terms and conditions of the site. Secondly, other services and sites simply won't be of use to you. Have a good idea of what you want it to do for you before you go find a service which matches it. Finally, for usernames and passwords: choose something memorable and

Research Methods for Postgraduates, Third Edition. Edited by Tony Greenfield with Sue Greener.
© 2016 John Wiley & Sons, Ltd. Published 2016 by John Wiley & Sons, Ltd.

write it down somewhere safe. It would be of no use to spend considerable time setting up a blog and email account only to lose the passwords just as the virtual invite for the Annual Conference comes out. General warnings over; on with the tour.

Email

It's quite probable that you'll already be familiar with emailing. This is a system whereby you have a unique electronic address (which will generally look something@like.this) and are able to write and send electronic letters, messages and missives to all those you want to contact at the click of a button. At 3 AM. Or from a rollercoaster. From a festival, the top of a mountain or the belly of a whale. As a practical system of communication, especially in an academic setting, email's been around since the 1970s, and has been more or less ubiquitous since the mid-1990s. So this section isn't intended to teach you how to use email, or in any other way to teach you how to suck eggs. On the other hand, it may be that you've had little cause to use email outside of your studies; or it may be that you've generally used it only for relatively informal exchanges with friends and family. In that latter case, you may be used to treating email as a reasonably casual form of communication, peppered liberally with LOLs, ROFLs, OMGs, and of course these pesky things: **:o) :o(:o/** and their vastly over-complex friends that have emerged over recent years.

Rather than showing you how email works, then, this section is directed more towards the best way to use the technology for the maximum benefit to you and your colleagues – and that mostly involves avoiding some of the pitfalls that have claimed many who have gone before.

As a post-graduate you should have, at the very least, one email address through which people can contact you directly. Your university will probably already have provided one, or would likely be able to do so if you request it; or, failing that, there are various online companies, such as (at the time of writing) Yahoo!, Google, Outlook and innumerable others who provide a basic free email service – and do bear in mind that 'basic' here would have read as 'mind-bogglingly sophisticated' to someone used to email in the 1990s).

Let's touch on some of the finer points of email etiquette:

1. Keep a personal, individual email address by all means, but if you're still at the stage of deciding what address you're going to use, do make sure that you're not going to end up feeling embarrassed to distribute it to others – especially professionals and colleagues you may be setting out to impress. Whatever else you may be putting on an application form to join your professional body, a personal email address of 'bigboy99' or 'sweetasl33t' will speak volumes – but possibly not precisely the volumes you'd like to convey. It's the same when applying for jobs at the end of your studies. Any job. Your friends may consider a whimsical address amusing because they know you, but a professional body, senior lecturer or leading name in your chosen field will not – guaranteed. Think of a couple of the big heavyweight names in your area of expertise. Now imagine them asking you for your email address. If you hesitate about handing it out just because of the unique address you've chosen, it's probably time to set up a new one, even if it's just one to use for professional contacts.

2. If you do end up using multiple email addresses for different things (one for social contacts, one for professional contacts, one for family members, one to contact your research

assistants/subjects, a posh one for your application forms and the like), consider using a third-party email reader program – also called a mail client or sometimes a 'mail user agent' – which will allow you to set up and download messages from multiple addresses at once. Popular choices for this type of mail reader include Mozilla Thunderbird, Microsoft Outlook, Pegasus Mail or Apple Mail for Mac users. Decent mail readers are available as free downloads – a quick internet search will find you several. Once installed, an email client will enable you to see the inboxes, sent items and so on for several email accounts at once; send mail from those addresses; and manage the contents without having to go log in to several different sites every day. If you'd like to use a client of this type, check with each email provider whether they support the use of such to download messages from their servers. Some companies don't allow this, or they enable the facility only for certain types of account.

3. Your email address may be held on one or more distribution or mailing lists. This is a group of people who are sent copies of the same message all in one go. It could be a list of everyone in your department, your company or your social group, or it could be an optional list you've joined. These latter can allow you to receive regular updates on a collaborative project, to follow topics of interest in discussion groups or to pick up news from particular sources. Whatever the list might be for, when you receive messages from it, there will likely be a number of different people's addresses on there. If you're going to reply, *check to ensure you're sending to the right address.* Be aware that there's a world of difference between the buttons marked 'Reply' and 'Reply All'. The first will send back just to the person who sent you the message. The second, however, will *reply to the whole list* – or everyone whose address was included in the original message. Take note: this mistake has already scuttled careers for corporate employees, civil servants and politicians…. At the very least, confusing these two can cause a lot of embarrassment. Take a moment to check you're clicking correctly. This will help ensure that you send your enthusiastic slagging-off of your mutual lecturer just to your selected peers, or your sympathies and Grandma's recipe for chicken soup to your ill friend, rather than sending all of these to your every fellow student, lecturers and Head of Department. Important buttons, then. Check and double check your address fields at the top of the email (To, BCC, CC and the like): if you want to send your reply to just one person, make sure there's just one address listed.

4. Once you're sure you've got the right number of addresses listed, make sure those addresses are correct. The same potential for embarrassment applies, only now your message is going off to a 58-year-old woman from Missouri called Cindy – who will forward it on to all her friends, who'll forward it to their friends if they find it funny and/or outrageous enough. They'll send it to all their friends, and you'll be amazed how few steps it takes before everyone has a copy (those of you whose studies have ever involved the word *exponential* might not be that surprised). And then you'll start getting the messages, offers and invitations from people who want to greet you in Jesus' name and offer you a once-in-a-lifetime opportunity to make millions of dollars, or just get to know you, you know, *real well.*

5. If someone, say for example your Least Favourite Lecturer In The Whole Universe™, sends you a not-altogether-complimentary critique of your work, resist the temptation to reply back immediately. Switch the computer right off and go away for a bit. No, further than that. Right away. Go outside, or something. Look at the flowers and make patterns

in the clouds. Write some poetry, maybe. Typing and sending off an email when All Cheesed Off is *not a good idea*. The very technology which allows you to send messages painlessly at the click of a button will cheerfully work against you and heartily laugh as your righteously incensed reply goes viral. So instead, once you get back in from your nature ramble, re-read the message, bite your tongue, cheerfully thank them and say that you'll take their points under consideration. There might be times when it doesn't feel like it, but they are acting in your best interests. They may be a leader in their chosen field, and whoever they are, they'll likely respond better to a civilised and adult response from you. (And if absolutely nothing else will convince you, look at it this way: if you view it as someone trying to wind you up, you can gain enormous satisfaction from showing them how staunchly unflappable you are.)

6. DON'T TYPE IN CAPITALS. This doesn't just go for emails. In most internet text-based media, it's come to be viewed as SHOUTING! If you want to emphasise sections of text, then how you do it depends on what text format you're typing in. There are two basic text types in most modern applications. The most common is called *rich text*. Rich text is commonly used by word processing, email and publishing programs. It includes invisible codes that enable the user to mark sections of text as bold, italic or underlined, or use variable text sizes or fonts within one document. Most emailing systems will now handle rich text. And rich text offers many options for emphasis. It's better to use italics to emphasise a segment of text, bold for highlights and the like. But even then, *don't overdo it* (like that).

   The other text type is called *plain text*. This is letters, numbers and special characters only, and includes no codes to embed fancy text decorations. That means you can't use italics or bold or any of that other nice stuff, and you're limited to a single text size and font for your whole document. You might run into plain text if you're using a very basic text processor (e.g. the Notepad application included in Microsoft Windows), or if you're working in computer code of any sort. If you want – or need – to write an actual document in plain text, then there are preferred, non-shouty methods of emphasising certain sections, usually by surrounding the section with additional characters, *like this*, _like this_, or /like this/.

7. If you're on a mailing list or email discussion group of some sort, and someone posts a message that's essentially badly spelled, badly punctuated, badly ... erm ... grammarised (?) gibberish... Don't reply just to criticise their presentation mistakes. As fun as it no doubt can be, around most of the internet, this is considered petty and will overshadow whatever actual point you've been trying to make, no matter how valid. Internet convention has it that presentation isn't as important as the message you're trying to get across, and that anyone who's reduced to criticising spelling and the like has automatically lost the argument. So when judging other people's contributions, you should probably bear this in mind.

   When judging your own contributions, please bear in mind that presentation – spelling, grammar, punctuation – is *absolutely crucial*. We can't overemphasise – not even with bold text – the importance of this. Don't be fooled into believing the myth we've just told you about in the previous paragraph. That only applies to *other people*. You're trying to make an impression online, especially if you're involved in a debate of any sort. You want to show that you know what you're talking about. And just like sending in a CV or résumé, or creating marketing copy, the impression the reader will form is only partly based on what you're actually saying. If you clearly can't spell, or haven't taken the time to check your presentation, the reader will assume you either don't know or don't care that much about what you're saying. That will inevitably undermine the apparent authority with which you're saying it. If you turn in a piece of work online that you haven't checked properly, it might well be that no one will say anything about it, because they all know they'll be called petty if they do; but you can rest assured there will be derision going on behind all those silent screen names. In the real world, people are free to judge your skill, knowledge, intelligence and even your morality based on how you present yourself – and they won't hesitate to do so.

8. Avoid sending unsolicited emails to lists of people you don't have good reason to contact. Unsolicited bulk email is, for historical Monty Python–related reasons, referred to as 'spam', and those who send spam are 'spammers'. We'll cover this in more depth later, but in brief: being a spammer is a Very Bad Thing. If you are a spammer, you are widely considered the true lowlife of the Internet. Please don't be a spammer.

9. Don't send on chain letters. They can look like the paper versions you may remember you used to get (not so common these days, since email is so much easier, but they do still crop up occasionally): letters that promise you that your wish will come true/you'll meet the love of your life/you'll come into lots of money/you'll not die horribly tomorrow if you just send the message on to a hundred other people. All right, twenty. All right, ten. Five …? Then there are those wonderful 'make-money-fast' messages, usually from important folks in banking departments in certain African countries(*) with a spare million or so just hanging around with no legal owner, which they offer to send to you, *their most important and honourable friend*, if you'll just pay them a small admin fee…. *Delete all such messages.*

    (*Although these 'advance fee fraud' scam attempts can originate anywhere in the world, statistically the bulk have traditionally been found to be from Côte d'Ivoire and Nigeria, so we're not just picking on Africa here. It's for this reason that the fraud type is now commonly labelled as 'the Nigerian Scam' or '419' – the latter referring to Chapter 38, Section 419 of the Nigerian Criminal Code, which theoretically prohibits the scam.)

10. If you're in a situation where you get to critique the work of another, say a mailing list, discussion group or review site, don't send any message that attacks the person themselves, rather than critiquing their work. It can be easy to do, especially when emotions are running high. Ad hominem attacks are rightly looked down upon, and are sometimes referred to as *flames* or *flaming*. Once a discussion starts to get flame-ish, it tends to escalate, and the resulting match of fast-flying insults which get progressively more personal and degrading is called a *flame war*. Quite aside from the unpleasantness of being involved in them, they have been known to encompass entire mailing lists and wipe out discussion groups as people leave in the hope of finding more civilised company elsewhere.

11. Occasionally, you'll likely find in your Inbox some manner of Dire Warning or Outraged Appeal. A message that cautions you against some new, nasty virus that's going to wipe out your hard drive; urges you not to have contact with a particular person because they're a hacker, paedophile or someone who still uses an umbrella in last season's colours; expresses outrage at some example of government malpractice; or names a particular person as just the worst, most abhorrent and vicious criminal since criminals were invented. These are usually chain emails, floating around the internet and being forwarded on and on over and over again – and usually there's very little truth in them.

Some common examples of this at the time of writing are that John Venables – one of the killers of toddler James Bulger in 1993 – has been released and given a new identity/holiday in the Caribbean/mansion house in Scotland paid for by the Lottery Fund. Another fairly recent one claimed that Maxine Carr – girlfriend of Ian Huntley, who carried out the Soham murders in 2002 – has had face-changing surgery at the cost of the British taxpayer, been given a million pounds and been secretly relocated to live in her new home *under your floorboards.*

The vast majority of these messages are hoaxes or, at the very least, are massively out of date. Messages that are forwarded out into the 'wild' will tend to disappear from view for a few years and then resurface and be forwarded on all over again – so don't be surprised to see out-of-date warnings turn up again years down the line.

See the above points relating to the embarrassment factor when your entire mailing list of recipients come back to you one by one, complaining that the instructions you forwarded to them all to help them get rid of a nasty time-delayed virus has just sent their banking login details to someone of loose moral character, and wiped their entire home operating system in one fell swoop. They're currently having to do their personal business through a computer at the public library (NB: Don't ever use a public computer to do your banking!), whilst their shredded home data are being recovered and their bank is set to recovering their money. Mind you, you'll probably know all about it by this point, as in all likelihood you'll have forwarded that email on in good conscience and then followed the instructions yourself. So you'll be sat there right with them in the public library, with an empty bank account. And no backup of any of your work as it was all on that one home machine. And not on your university personal hard drive space, *which is where a copy should be at all times.* See also the 'Backups' section.

Whenever you receive a chain message of this type, especially one asking you to forward it on, it is worth checking it out on one of the urban legends sites online. Sites such as Snopes (www.snopes.com) record and analyse the current crop of email hoaxes and keep an encyclopaedic database of the old ones.

12. Whilst we're on the subject of hoaxes and such which can potentially damage your data, operating system, bank account contents and/or pride, we'd like to tell you about Trojan horses, or Trojans. A Trojan is a destructive computer program which some unscrupulous people will hide as an attachment on an email and send out as spam. These Trojan viruses can potentially do many different things, from the destruction of your data, to encrypting your computer and offering to sell you the decryption key for money (Ransomware), to allowing the writer of the program remote access to your computer. Remote access means that they can see everything you store on your computer, to seeing which websites you look at, and even using your computer for their own uses, such as using it like a puppet to attack other computers or to spread the virus further. The most popular uses for a Trojan at the time of writing are either to obtain login details for bank accounts, or to assimilate your computer into what's known as a 'botnet'. This is a group of machines being controlled remotely without their owner's knowledge and used to contribute processing power to large-scale online criminal activity. If you get an email from someone you don't recognise with an attachment that's titled as something odd, and especially if the attachment file type ends in '.exe.' don't open the email, or the attachment – just delete it.

13. As a side note, any email which purports to be a request or notice from your bank *isn't*. Even if it looks like it's from your bank. In those cases, it's just coincidence. Any email

asking you to 'verify your account details' – especially if it looks like it's from a bank – is an example of a type of fraud called 'phishing'. No bank will ever send you a request to log on from an email. Most banks won't send you any important messages by email at all (and, in general, our advice would be to never accept any contact from banks that isn't by proper paper letter – unless you're standing in a branch talking to staff, in which case it would seem crotchety to insist on a letter). If you do get an email message purporting to be from your bank, don't click on any link it provides; it will probably take you to a fake site which is just waiting to catch your bank login details to clean out your account, or, at the very least, the site may automatically download malicious software to your machine. If you use internet banking, go to your bank site through your own saved link: not much extra legwork, and a lot safer. A good tip is to go to your nearest branch and ask to be taken off their marketing distribution lists: the only communications you should then receive from them will now be your official account statements. On paper. Through the mail.

So, if you receive such an email, please do the electronic equivalent of shredding it and using it as bedding for a hamster to pee on. It is fit for no more. A bank will never contact you by email, just to make it absolutely clear. No exceptions. (As a side note, if you have a hamster, remember they need actual physical bedding....)

Hitting the Toenail on the Head: A Small Portion of Spam

At point 8 above, we briefly mentioned the phenomenon of 'spam'. It didn't take long, once email took off, for someone to work out that you could use the new communications revolution for advertising purposes by sending your advert to, well, everyone on the network. Eight hundred and seventy-one times. Per minute. Forever.

As this practice in turn sadly took off, someone somewhere coined the term *spam* to describe the sending of bulk messaging. The most common explanation for this term is that it was based on an old Monty Python sketch. A couple descend into a café from the ceiling (because Monty Python, that's why); the café is full of Vikings and has a menu consisting mainly of Spam, a budget canned meat product. The couple make a few attempts to order something that doesn't have Spam in it, but are repeatedly drowned out by Vikings chanting, 'Spam, Spam, Spam, Spam, Spam, Spam, Spam', endlessly, over them. It's this inability to hear content over noise that's led to spam being so named.

Spam is a handy tool for unscrupulous advertisers: because email isn't charged per message, spammers can send a million messages for no more cost than sending one. And, unfortunately, sometimes it works: people (very occasionally) do respond to spam messages, which not only exposes them to fraud but also muddies the waters for those who argue that spam isn't a legitimate form of advertising. In any case, the vast majority of internet users and providers detest spam and those who send it: it consumes internet bandwidth (the available capacity for data to be sent around the network) and clogs up inboxes everywhere. Estimates for the scale of the spam problem provide some sobering numbers: in August 2013, Kaspersky recorded 70.7% of global email traffic as spam (http://www.kaspersky.com/about/news/spam/2013/Spam_in_Q2_2013_More_offices_in_danger_from_targeted_plausible_fakes). The total number of messages that equates to is numbered in the *trillions*.

It's not just an email phenomenon, either. Spam can easily be automated to post messages to blogs and other websites. Because these are indiscriminate, they rarely make any sense for the blog they're posted to, and if they're allowed to post themselves unchecked, they can prove a serious embarrassment for the site owner. Not only because they make the place look untidy, but because they might well be pushing goods, services or opinions that may prove disagreeable or offensive to readers.

There are steps you can take to fire a few shots of your own in the ceaseless war against spam. Most immediately, you can train your email filters to deal with it. You have two filters. The first is installed in your email client software, or is part of your service provider's webmail. While perusing your email Inbox, you'll find an option to 'Mark as Spam' (or some similar command): use this wherever you can. It not only gets rid of annoyances this time, but also trains your mailing system's spam filter to recognise unwanted messages. (It also reduces the chance of future 'false positives' – that is, messages you actually want but that the filter wrongly thinks are spam. In the best traditions of internet humour, these are known as *ham*.)

Your second filter is installed in yourself, and it's just as important. It's vital that you train yourself to take a pragmatic, no-nonsense approach to email. If you are in any doubt about the sender of a message, delete the message. At the very least, don't open any files that may be attached to it: that's a sure-fire way of subjecting your computer to viruses, Trojans and all manner of other nasty malicious payloads. You'll recall we mentioned the hazards of emails purporting to be dire warnings from your bank: it's crucial that you develop a zero tolerance for unexpected messages. Unless it's clearly from someone you know, or otherwise something you're expecting, have no truck with it.

Websites

A few years ago, to set up your own website you needed to be able to communicate directly with a computer and tell it exactly what you wanted in a programming language. The most common of these for websites is HTML. Indeed, the previous edition of this book detailed common HTML coding instructions to be able to set up your own site from scratch. Thankfully, now there are website providers who do all the complicated coding at your click of a button. You probably already visit some websites on a regular basis: BBC News, Wildebeest Knitting X-treme, Czech Traffic Wardens Reunited.... The variety is endless. Some of you may even have your own website already.

Setting up your own is not difficult: there are plenty of free sites out there which will host a website for you – a basic one at least – though the more bells and whistles on offer, the more likely you are to have to pay out for them. The first thing you'll notice is the address – what's technically termed a URL (Unfeasible Radish Latitude – just kidding. It's actually a *Uniform Resource Locator*) or URI (Uniform Resource Identifier). Some common URLs you'll probably have seen are www.google.co.uk; http://news.bbc.co.uk or http://wildebeestknittingxtreme. com (and, by gods, we hope that last one doesn't actually link to anything ... but it can be yours for a small amount a year....). A site which hosts your website for you for free will most likely put something into the address to reflect the company's ownership of that web space. So, if you set up a free site with Example Provider Inc., you'll likely end up with a URL that says something like: http://epicwildebeest.exampleprovider.com.

You can pay for a unique address which just reflects the content of your site, such as yoursite.co.uk rather than yoursite.wordpress.com. The technical term for doing this is called 'registering a domain' (and has nothing to do with whips or chains – unless you're into that kind of thing.... Your research may be all about that, and then it's perfectly acceptable. Just sayin'....). But registering a domain costs money: usually an annual subscription fee, the amount of which can vary from not-an-awful-lot-at-all, to dear-sweet-lord-I-need-a-second-mortgage. (As an interesting aside, some people make a tidy living – or did, during the 1990s, before what's known as the dotcom crash – by buying up domains they thought might be in demand, and then selling on the registration for higher prices. A substantial cause of the dotcom crash was the massive over-inflation of prices in domain trading.) Still the majority of serious published individuals setting up as experts (even newly qualified ones) in their respective fields will have their own websites advertising their expertise and services. It's relatively easy to set one up for yourself. There are a number of providers offering free web-hosting services: be sure to have a look at some of the example sites first, check out the terms and conditions and be certain that the host provides the kind of service you want before you sign up.

Social Networking Sites

Now that we've mastered email and websites, we can move onto something a bit more complicated. Just kidding. Social networking sites have been some of the fastest growing areas in recent years: these are places in which you can store some personal information, photos and the like, and generally keep in frequent contact with those friends who you are unable to see day to day. It's a fab thing if your friends are all the way off in the distance (likely if you're at university), but some folks use it to keep in touch with those they see every day anyway, and some even use it as a replacement for actually physically talking. As a student, you're obviously (and quite rightly) thinking, 'I should be spending all my available time studying and writing my essays – I don't want to get bogged down in ultimately meaningless electronic social pleasantries with people I see at work/at the club/in the library/in lectures'. But think of it like this: it's a networking tool which allows you instant and unlimited access to professionals in your field, experts, research company directors, specialists and other researchers who can offer you support, encouragement, criticism or even a job contract or further paid research opportunities ... so maybe setting up a profile on one of these isn't such a bad idea for a post-graduate in a professional sense. The sites which offer social networking services are many and include Bebo, Google+, Myspace, LinkedIn and Facebook. The majority of social networking site providers have a specific focus; the majority of LinkedIn users are there as professionals, Myspace generally has casual users with the focus originally on music production and performance. Providers such as Google+ and Facebook fall somewhere in between and aim to cater to everyone.

Fair words of warning: social networking sites and the commentary which goes on in them are not considered public domain. They are, in fact, run by private companies which have an owner and are there to make money. Don't be surprised if, when someone says something nasty about you in words too rude for this book, the site admin doesn't respond particularly quickly to your incensed complaint. Just before you pick up the phone: it's no good ringing the police about it either. Before you start sniggering about the sheer stupidity of some folks: it happens. Every day. Too frequently to count. Remember, a social media site is not a public forum, it's technically a private one which you willingly signed up to join. If you want a truly public forum to air your views, go and stand on a street corner and tell people; there are only private virtual forums available. As an exception, the police will probably listen if say the abuse you receive is part of a long-term aggressive and ongoing case of domestic violence, but it's very unlikely that they'll be able to do anything if not. So don't waste their time by calling them at 3 AM on a Saturday and complaining that some unknown's called you a N00b in your comments. If you really can't stand it, delete your account and start again with a new username.

Again, with these sites, there's a code of etiquette: if you're looking to present an image of being a professional academic, or at least aspiring to be one, then:

1. Talk about your subject as a craft – including your recently mastered skills and knowledge, achievements and discussion points. You can include short courses, conferences or formal study evenings that you've been to or, better yet, that you've presented papers at.
2. Include a section which shows any future conferences you are planning on presenting at, so your adoring public can come see you in person and begin their stalking hobby. Okay, maybe not immediately, but they will, oh yes. Just kidding: there are far more people on the celebrity scale, and academics don't generally register. Most academics begin their lifelong specialisations around the point of post-graduate study, and specialisations are incredibly, well, specialised. You may end up as one of only a few people who have studied that particular area or, better still, the only one. In the whole world. Thusly, your work and papers will be in demand (eventually). A good aim is to have a paper published whilst you undertake post-graduate study, or present at a small conference. If you're on a Master's programme, and plan to study at the PhD level, then it will be looked on as a bonus. Start making contacts now.
3. Be careful of precisely what information you share on your profile. See above stalking point. 'Nuff said.
4. Include short extracts of your research, but not all of it! If you're going to turn professional, then you don't want all your work in the public domain free to begin with. It'll make it a tad difficult for people to want to contact you to discuss it if it's right there. By all means, post abstracts or extracts; it'll help other academics and research companies decide if your research fits their remit. And if you'd like to offer a few short discursive papers or the like to whet people's appetite for your work, you can consider it an investment of sorts.
5. Don't post pictures of yourself on a drunken night out. You and your friends might think that climbing up a town centre notable's statue at 4 AM to leave your underpants stretched over its head is hilariously funny, but a scout for a prospective publisher might not. A lot of social network users will tend to use the system as a means of broadcasting their every life event, warts and all. As a professional or aspiring academic, you're a business, and a business user will need to be more discriminating about what they post and what content they allow others to post on their profiles.
6. Include some details about your home life, but not a blow-by-blow account. Readers may be vaguely interested in whether you have a cat, but they don't care that it threw up a massively yucky hairball on the duvet this morning.

Blogs and Micro-blogs

Now that we're on a roll, we'll move swiftly onto blogs and micro-blogs. A simple definition is that a blog (from *web-log*) is an electronic journal, or diary, in which an entry gets time/date stamped as you post it up. People use blogs for all manner of subjects. Some are personal diaries, some provide reviews, some showcase artwork and some offer political commentary. There are news blogs, religion blogs, cookery and lifestyle blogs, advice blogs....

Most blogging services will allow control over who sees what. While you may want most entries to be public, some can be 'locked down' so that they're only visible to people you choose. A blog is in effect an even easier way of setting up a personal website. Some hosts allow you what are called *static pages* and some don't, so it depends on what you want your site to do. An example of a pure journal-based blog would be Livejournal, and one which allows multiple static pages alongside a journal would be Wordpress. A blog is easy to set up, and allows you to put up words, pictures or photographs, links to other sites and news items. A simple blog can be set up, with colour and presentation theme selections, and given basic words and pictures within an hour or so; it costs nothing to maintain or set up in the first place (aside your time and effort, of course). In addition, many blogs can interact with other social media sites such as Twitter and Facebook, allowing you to make an announcement on those media every time you add a new post to your blog. This can be extremely helpful in spreading the word to your millions of adoring fans.

For beginners, Wordpress is exceptionally easy to use. Other sites you could use are Google Blogger, TypePad or Livejournal. Many of these free services are ad-supported, which means that advertisers pay the host to have their adverts put onto your blog. Most hosts will offer paid versions which have extra bells and whistles and remove the adverts, but we would suggest the free versions work just fine to begin with.

Micro-blogging, also known as *social news*, is something similar but different, in that you are able to put up information in short bursts, for Twitter in particular with a limit of 140 characters, which are time- and date-stamped as you upload them. This can include location information, and links to pictures and information sites. Commonly you will also have the opportunity to upload personal details to a profile page. A good example of this is Twitter, but there are other sites such as Posterous, Instagram and Tumblr available. The etiquette for blogs and micro-blogs are just the same as for websites. No drunken underpant-abandoning photographs. See above.

Whether to go for a blog or a micro-blog service depends on what you want; if you want to discuss deep and lengthy philosophical questions – such as the nature of the mind, or whether you can make two cans of beans and half a loaf last three days – then you may want to consider starting a blog. If you want to give brief updates with links to articles or news, you're more likely to favour a micro-blog. You can always link a micro-blog to produce a notification every time you update your blog with a post. Different people read blogs and micro-blogs, and there's nothing stopping you from using both to reach a wider audience.

A Word about Trolls

There are trolls on the internet. Oh yes.

They come in all shapes and sizes, from the mildly annoying email troll to the full-blown ranting troll. Trolling, in an internet context, is the act of deliberately posting off-topic derogatory or inflammatory material in an online community with the primary intent of provoking an emotional response. At the time of writing, there have been recent incidents which have made the national news in which memorial websites set up in remembrance have had offensive content posted expressly to provoke a grieving family. The term *trolling* comes from a form of fishing in which a line is dragged behind a boat in the hope of luring fish to bite.

Internet trolls can be costly in several ways. A troll can disrupt the discussion on a newsgroup, disseminate bad advice and damage the feeling of trust in the community. Furthermore, in a group that has become sensitised to trolling – where the rate of deception is high – many honestly naïve questions may be quickly rejected as trolling. This can be quite off-putting to the new user who upon venturing a first posting is immediately bombarded with angry accusations. Even if the accusation is unfounded, being branded a troll is quite damaging to your online reputation. It is considered sensible to lurk (watch but not post on mailing lists, newsgroups etc.) for a period to avoid duplicating posts and get a good feel for the other people within the group.

Thankfully, in the United Kingdom we have the Communications Act 2003. The following are high-profile cases of trolling which have resulted in legal action:

- Sean McDuffy was jailed for 18 months in September 2011 for posts on social networking sites about 15-year-old Natasha MacBryde after she took her own life (http://www.bbc. co.uk/news/uk-scotland-north-east-orkney-shetland-17385042).
- MP Karen Bradley raised trolling in Parliament in 2011 after a Facebook page was set up mocking the death of 17-year-old Hayley Bates in a car crash (http://www.bbc. co.uk/news/uk-england-stoke-staffordshire-14162669).
- Colm Coss was jailed for 18 weeks in October 2010 after posting obscene messages on Facebook sites set up in memory of Big Brother star Jade Goody, and several other deceased individuals (http://www.bbc.co.uk/news/uk-england-manchester-11650593).
- In October 2014, Brenda Leyland committed suicide after being confronted by a reporter about her abusive messages to the family of missing child Madeline McCann (http://www. bbc.co.uk/news/uk-england-leicestershire-29501646).

Take a moment to think about that: these trolls went after people who have died and their grieving families for their own gratification. They will think nothing about going after an academic, especially a new one.

Put anything on the internet, and sooner or later you'll get a troll. It's that simple. The trick is telling the difference between a troll and a genuine someone new to your discussions. This is something to keep in mind if you start a news group or mailing list to discuss your specialist subject.

Words of sound advice: *do not feed the trolls*. If you identify a poster as someone who is simply trying to get a rise out of you – or others – don't give them what they want. Don't reply; don't attack them; don't attempt to reason with them. They want the attention, and so your best weapon is to deny them the attention they crave. End of advice.

Almost. One more thing, for an author venturing onto the internet: not everyone will like your work. Some will tell you so. Some will tell you so really quite comprehensively. You need to accept that people use far more aggressive language online than they do in real life. (For one commentary on the online disinhibition effect, see: http://www.wired.co.uk/news/ archive/2013-05/30/online-aggression.) What you get the angry 'UR BUK SUXXORS' (Translation: I say, old bean, your work is not the best one I've ever read on the subject) post, take a breath and try to discern the basis of the objection (if there isn't one, ignore or delete the comment). If you can work out what the person is complaining about, feel free to address the complaint – but don't lose any sleep if you don't manage to persuade them to like your work. And – important point – *don't get defensive*. Defensiveness can so easily lead to aggressiveness, and then you've blown your reputation straight away. Lose your rag with a critic, even an unfair one, and people will pounce on you. At the time of writing, although he may have been eclipsed by the time you read this, a good example of this is the software developer Derek Smart, whose reaction to criticism has become the stuff of internet legend, and the foundation of what is currently the largest and longest running flame war in the history of the internet. In case it looks like we're gloating at misfortune, we point him out because he's such a good lesson in the importance of accepting criticism gracefully.

Spell Checkers

Spell checkers, especially electronic ones, can spell beautifully but are not intelligent. *Dew knot trussed you're spell checker*. If you spell a word wrong, but it's still a valid word – as in the last sentence – your spell checker will cheerfully ignore it.

By preference, use your spell checker, then get a non-academic friend to read through your work for you. They'll stop at all the specialist words and question them, so are much more likely to flag up spelling errors for you. Another technique is to read your whole work backwards – reading forwards is a lot faster, and as you wrote it, you'll know what's in each sentence and paragraph. But by reading each word forwards, but the sentences backwards, you have to think about what you're doing and are more likely to see smelling pistakes.

Backups

A friendly warning when dealing with electronic copies: back up your work frequently. Print off a copy and keep it in your safe/bank deposit box/folder at work/fridge; burn a copy onto a CD; or copy it to a memory card or memory stick.... Just keep a copy somewhere other than your computer. Computers, and by extension the internet that's made up of them, are temperamental. They will, sooner or later, eat your work and leave no trace of it, especially if you're in a rush to send an email copy to a friend of a friend who just happens to be having dinner tonight with the leading names in your field and has offered to take your latest batch of precious research backing up your new theory along to show them, but needs an email copy *within the next five minutes*.... Put at least one complete copy elsewhere. Even have an agreement with a friend to email them your work for safekeeping at every draft revision. The longer your work, whether essay or thesis, the more places you should store it and the more frequently you should make backups. Just in case a miniature rabid badger breaks in and decides that your hard drive is the perfect place to dig an intricate sett for hibernation. Don't laugh: lots of people have lost lots of data by deciding a potential problem was too unlikely to hit them. Don't rely on redundancy either – a RAID disk will protect you in the event of one of your disks failing, but won't protect you from accidental deletion. Human error causes far more data loss than electronic failures.

Clouds

Not the fluffy white things which rain on you, but in terms of computing – specifically, *the* Cloud.

Originally, the programs that you used on your computer opened up inside your computer, allowing you to create, manage and update documents on your own hard drive, usually within a few physical feet of where you were typing. In short, you knew where your data were: inside the box with the flickering lights on the front, right there on your desk. Cloud computing, so-called because the network of computers is often depicted as a cloud in diagrams, is a further advancement in which your information is stored on a server somewhere … out there.

Just in case you're wondering if you've ever come across one, the entire internet is technically a cloud storage device.

This kind of storage can be incredibly useful for post-graduates. Your data is stored off-site; you can access your data if your research requires you travel around the country, or the world; you can use several different devices to update your material – computer, laptop, phone, tablet and so on; it can help you co-ordinate a large number of additional researchers or individuals; and it can allow several people to contribute at the editing and feedback stages.

Free Cloud storage services are provided by several companies. Popular examples at the time of writing include GoogleDocs or Dropbox.

So, now that you've had a browse round some of the different types of social media available to you and their benefits and drawbacks for a post-graduate, go have an explore and set up a strong electronic presence through which to network.

Good luck!

Part VII
Next Steps

41

Protection and Exploitation of Intellectual Property

Patrick Andrews

Introduction

To gain a research degree, universities require that you make an original contribution to knowledge in your chosen field. As well as representing an intellectual advance (no pressure!), this may also result in discoveries which are worth someone's money. That person is called a customer, and they are of central importance to what follows. Academe is beginning to recognise (if not embrace) the concept of customers, which should make commercialisation easier.

This chapter aims to equip you with some tools so that you understand the significance and the processes of turning research into money. Based on courses in entrepreneurship which have been run at universities from Cambridge to Tallinn, it is centred on UK practice, but the principles apply anywhere with a legal system (www.scotskills.com).

You have extra career options because your research training may allow your discoveries to be sufficiently fundamental as to be both widely applicable and hard to compete with. You will emerge at the end of reading this material, I hope, with enough understanding to (i) decide your personal approach to commercialisation, and (ii) avoid being taken advantage of by more experienced commercial players.

Note: I am not a lawyer, and this chapter isn't legal advice (nor is it costing you £400 an hour).

What Is Intellectual Property?

> Getting a patent is just a way to get sued.
>
> Arthur C. Clarke

The term *property* suggests ownership. Increasing numbers of people argue that intellectual property (IP) is qualitatively unlike other forms – such as your car or your laptop. IP is also

Research Methods for Postgraduates, Third Edition. Edited by Tony Greenfield with Sue Greener.
© 2016 John Wiley & Sons, Ltd. Published 2016 by John Wiley & Sons, Ltd.

an umbrella term for many very different things. We have, however, to deal with the system as it exists. This section discusses the processes by which ideas may come to be legally recognised as yours. Who owns your most brilliant ideas? According to the law, it may not be you.

You gain legal rights (intellectual property rights (IPR)) when you can prove that an idea is both novel and uniquely yours. As a postgraduate, you will most probably own whatever IP you create. The ideas of employees, including professors and postdocs, will in general be owned by their university. Getting your IP recognised as yours allows you to do some interesting things:

- You can discuss your ideas with everyone without losing any rights.
- You can allow other people to use your idea if they pay you to do so (licensing).
- You can make claims for compensation in court against anyone who may be using your ideas without permission (infringing).

There are several forms of intellectual property (see www.patent.gov.uk and www.uspto.gov). I will outline the main ones here.

Know-how

This is best described as the tricks of the trade. If, for example, your Auntie Bessie makes superlative jam, it might make sense to keep the recipe a secret. If someone independently thinks up her technique and makes jam which is just as widely liked, then she will have no legal case against them. Rather than attempt to spend large amounts patenting her method, she might be advised to register a trademark, as discussed further in this chapter. In a similar way, magicians innovate in the form of know-how which is maintained within societies like the Magic Circle (http://www.boingboing.net/2007/09/12/magicians-innovate-w.html).

Copyright*

You own copyright in, for example, documents you write. It's a free process which works pretty much everywhere, and these rights last for 70 years after you die (www.ia-centre.org.ukwriting). You don't have to make an application to own the writing, art, film, photographs, software, architectural designs, music, drama, choreography or sound recordings you have created. They are yours automatically. Software authors can claim copyright to their code, but if someone simply changes something (e.g. all the variable names, but not the algorithm), it's very hard to claim infringement against them.

The entertainment industry owns huge stockpiles of copyright and regularly sues people for infringing their rights. Digital rights management (DRM) technology has been implemented by many companies to stop pirating. The problem is that DRM often stops fair use as well. Piracy seems much more serious, to me, when it's connected with fake medicine or unregistered aircraft components (see http://www.webroot.com/gb/en/home/resources/tips/ethics-and-legal/the-societal-costs-of-digital-piracy).

Design Right and Registered Design*

Design right is similar to copyright. It automatically protects, for up to 15 years, visual features of a design that you have created, rather than its function. Design registration is more like patenting and provides protection for up to 25 years.

Unregistered or 'Intellectual' Assets

It's surprising to some that even the following are considered assets with a potentially significant value: contracts, goodwill, demonstrators, brand, your reputation, skills and so on.

See, for example, http://www.reydoog.com/intellectualassets.php.

Trademarks*

Even those who deny that they are influenced by branding will have a preference about the source of their laptop or their underwear which takes advantage of someone's carefully thought-out branding message. People buy because a brand represents certain values they want to be associated with (think Apple). Trademarks form part of a branding strategy and can, over time, come to be worth more than the rights to the actual products. These can include words, numerals, slogans, names (including signatures), logos, letters, shapes of goods or their packaging, smells or sounds – and they are indicated by adding 'TM' to each. To register a trademark, talk to the Patent Office's Search and Advisory Service.

Patents*

A patent gives you the legal right, in a certain country, to stop others from doing anything with your idea.

- The idea must refer to a thing or a way of doing something.
- The thing must be novel (i.e. unknown when filed – its 'priority date').
- It must involve an inventive step (i.e. it mustn't be just an obvious next step).
- It must be 'industrially applicable'.

In effect, you get 20 years' monopoly in return for public disclosure. This deal is based on governments' belief that giving everyone access to your ideas (rather than keeping them as trade secrets) will be good for the economy. To encourage you to do this, you get an unfettered first shot at the marketplace. The idea that inventors of all kinds would never invent without the prospect of a monopoly is clearly wrong, however (see e.g. http://iotd.patrickandrews.com).

There is huge emphasis placed on protecting one's IP. In the 'How to Turn Your IP into Money' section, I will question the value of that uniform mind-set when you want to make money. This whole IP business takes itself very seriously and can become so technical that, before that, I thought I'd throw in some amusing background material. Take a look at www.patentlysilly.com.

How to Turn Your IP into Money

> Don't worry about people stealing an idea.
>   If it's original, you will have to ram it down their throats.
>
>   Howard Aiken

I'm now going to tell you about:

- how to set up a personal IP strategy (i.e. decide about spending money on protection);
- how to get your inventions the best protection you can reasonably afford;
- The need to keep the whole process in perspective.

Let's assume that you work with an inspiring supervisor and challenging fellow researchers, so that coming up with ideas is not a problem.

Most post-graduate students are not university employees and thus own their project's IP. Some universities ask postgraduates to assign, in advance, any IP which they generate during their stay there. This entitles them to exactly the same commercial rights and responsibilities as a professor. This may sound good, except that the university will decide what IP to protect and how to exploit it (and you aren't being paid a prof's salary). They argue that since they bear all the costs, for example of IP protection, they should keep the lion's share of any revenue. Deals available to those who have assigned their IP to the university will probably involve revenue sharing (e.g. 75% of the first £20,000 income, and 25% of any more).

This may appeal to you if you want to avoid almost all the hassle and don't mind handing over the majority of the potential rewards associated with commercialisation. Your university will also require you to commit to some involvement in commercialising your work, if they are to take on the task of paying for patent applications.

Whilst you consider your options, don't discuss your business-sensitive stuff in public. In addition, do maintain an up-to-date, complete record of all your research activity. If you talk about your ideas with anyone, they may later have a claim on them. This is pretty ridiculous in any modern research lab because part of the value of the research process emerges from kicking ideas around together. You have to decide whether to keep silent about them until you have chosen whether or not to exploit each, or to share them with your research group – in which case, all members might consider themselves entitled to be named co-inventors. This includes your supervisor, who should be completely *au fait* with your latest results. If your work is sponsored or supported by somebody other than a research council (e.g. a company), they too may wish to benefit from any wealth which arises from it – indeed, they may try to insist that you assign your rights to the university before work even begins, as discussed in this chapter.

What you do in terms of protection depends on your personal motivations. Do you actually want money, fame, high-standing freedom from the 9 to 5 or some combination? The sad fact is that it is very difficult to get UK companies to invest in new product ideas. This is especially true of those which have been thought up by external people and which still require a great deal of costly development work to incorporate them into any kind of saleable product. Research outputs themselves are almost never easy to translate into product benefits.

Given that commercialisation is a potentially difficult and time-consuming activity, try to assess the level of involvement in it which you want to commit. Developing products and services is utterly different from undertaking research. I don't have space to describe the differences, but the main one is that everything you do is orientated around making something which fulfils a list of benefits which you know customers want, within a deadline. If you want to develop a new product based on a technology derived from laboratory work, it may take longer than you ever expect.

If you decide that one of your goals is to try to make money from your ideas and discoveries, it's vital that you:

- Talk to your supervisor to clarify any claims he or she may wish to make (e.g. to be named as a co-inventor on a patent application or to some share of any future revenue from your IP). Most supervisors tend to be keen to support the considered ambitions of their protégées. People do tend to get 'funny' around money and status, though. Academics are particularly careful to avoid embarrassment. Get agreement upfront with any co-inventors or 'stakeholders.'
- Talk to your technology transfer or commercialisation office.
- Seek out any sources of student entrepreneurship advice or support at your university.

These people will be employees and subject to a contractual requirement not to divulge your commercial secrets to anyone, but such rules do sometimes get broken, so try to explain only what is necessary and then listen carefully to any advice you may receive.

Most people will complete their work, considering the economic implications of publishing papers, as they progress, as well as the academic consequences of delaying publication whilst IP protection is being sought. Their research degree completed, they will be in a better position to devote the time necessary to make and sell products and services based on IP. (*Note*: Just because you achieve a PhD in electroquantumlinguistics doesn't mean that your only commercial option is to use that in business.)

Let's now imagine that you have finished your research, had some ideas which look like they may have commercial potential, gathered some compelling data and/or made a successful demonstration, talked to all the stakeholders confidentially and emerged with an agreement which asserts that you are at least a joint owner of the ideas and are taking responsibility for their commercial development. If, at any stage, problems become insurmountable, then you should feel free to move on to a new opportunity. Post-graduates are generally young enough to experiment with several such proto-ventures. Failure in academe is considered uniformly bad. Failure in business, if controlled and learned from, is a mark of valuable experience.

Make an Outline Business Plan

Do you have:

- An idea appealing to a defined, reachable market?
- Passion, skills, experience and resources? (Shaz Rasul, entrepreneur)

Don't get overly focussed on the minutiae of IP protection. The process itself won't make any money; in fact, it can be very costly. *What makes money is being able to apply the new knowledge your research has uncovered to solve an important problem for some identified customers*. You must also have the enthusiasm to make commercial things happen. The basic options for making money from your IP are:

1. consultancy (based on your know-how),
2. licensing your IP,
3. bootstrapping a new business,
4. external investment, and
5. assorted other approaches.

You need to choose the combination of these approaches which best fits your own requirements. Whatever the scale of your business venture, you will have to make a large number of *decisions*.

Firstly, talk in general terms to potential customers. Politely ask that they sign a non-disclosure agreement and restrict what you discuss to what the invention does, rather than how (if you don't, you will probably lose the ability to protect your IP, but it makes no sense to be too secretive). Then consider these questions as objectively as possible:

- Is there some really significant advance?
- What are the areas of technical uncertainty, and what future developments can you envisage?
- Will anyone be able to make a problem-solving product based on your inventions?
- Will it require IP protection? What is the best form, and how much will that cost?
- When do you expect the costs to start to be outweighed by the sales?

These questions are best answered by preparing a simple, first-pass business plan. It's a vital exercise, even if you are starting a back-bedroom enterprise. Remember that the map is not the territory.... The Sun Microsystems business plan was 12 pages long, so don't get hung up on planning. Do a coarse calculation or two first, and limit the resources you expend on this.

Here are the elements that need to be thought through and *written down*. The process of writing allows any team members to share their thinking explicitly and also generates new ideas (but you will know that, because you will have completed your thesis by this stage!).

Executive Summary

Possibly the only bit that ever gets read by anyone else.

The Business Opportunity

What are you going to sell to whom? 'We will create and sell V units of W by X (date) and make £Y profit per sale' – this is your *business model*. What makes your offering special? Don't get focussed on the wonders of your invention. You need to think about whether the business opportunity is correspondingly great. What problem gets solved for whom?

Marketing Strategy

How many of your target customers are out there? What are the current issues driving their purchases (e.g. legislation)? Why will they buy from you? Who are your competitors, and how powerful are they in terms of customers (even the best idea has competitors … a Nobel prize–winning antigravity device may have to compete against stepladders in certain applications)? What are your relative strengths and weaknesses? What developments are in the pipeline, and how will everyone in the market react to these? How will customers get to hear about your offerings? Who will actually do the selling?

Management

Who will be in charge of what activities? Have you and your team got the necessary skills and experience? (Very few individuals have the abilities and experience to start a successful business venture solely by themselves.)

Operations

Where will you be based, and how organised? Where will your customers be, and how will you communicate with and deliver to them?

Financial Forecasts

Make some educated guesses about V, W, X and Y above. Describe how you got these numbers and why they are reasonable (think about visiting business libraries and doing online searches for data from similar ventures). Consider each item of cost and when the sales will happen. This allows a timeline of incomings versus outgoings to be graphed, so that your profitability can be assessed. Perform these analyses based on the least and most attractive scenarios. If it looks only marginally attractive, maybe you can find a way to cut costs or boost sales. If you and your trusted advisors can't, then it's fine to back off; you have expended limited resources and gained some valuable experience in planning and decision making. The name of this game is limiting risk – to both yourself and any backers.

You will want to select a legal status for your new business activity. Here are the main options:

- Sole trader (unlimited liability for debt; all the profit is yours)
- Partnership (share the profits, costs and debts as well as some skills synergy)
- LLP (risk limited to your investment)
- Ltd company (risk limited to your shareholding, companies' house returns are required etc.)
- Social enterprise (profit is all reinvested in some social purpose; no dividends issued).

So far, we haven't talked about how you get to make money once the business is successful. Common approaches include:

- Salary + bonuses + expenses
- Shares which increase in value and can be sold (in a trade sale or to investors)
- Dividends on profit (paid to shareholders like you from profits)

You can help with this planning process by seeking out trustworthy friends and relatives, preferably those who have made some money themselves and who aren't emotionally involved (e.g. as investors) in the feasibility of the plan.

Let's look at the routes to commercialising your IP (which we mentioned in this chapter) in more detail now.

Consultancy

You can sell access to some of your specialist knowledge in the form of consultancy. Often start-ups do some of this work to make ends meet (whilst developing a secret product in the background). You can only charge for your time, so this kind of service is not as scalable as product sales. As well as technical know-\how, consultancy requires a keen awareness of wider business issues, and it can take a while to develop the required level of trust from clients.

Licensing (Basics)

In licensing your IP, you allow someone else to make use of your ideas in return for money (a royalty). This is normally some percentage of their sales to which your invention has contributed. A licence agreement will need to be drawn up detailing exactly what is being licensed to whom and in return for what. Unlike consultancy, you can sell the same IP simultaneously to numerous customers and your exposure to legal action may be limited if you have a giant corporation as a licensee (litigants will attempt to sue them for preference). If you don't fancy the risks of building your own company and you have a proto-product that merits large-scale international sales support from a global licensee, then licensing may be a good way to optimise your risk–reward ratio.

The biggest hurdle is often finding companies willing to invest their money in your invention. This process takes time to negotiate a mutually satisfactory deal. Another downside is that you have to share a lot of information and the proceeds with your licensee. Realistically, although you can license any form of IP, companies will really only buy into registered ideas, in order to give them some reassurance that they are dealing with the recognised owner. This means that you have to pursue at least the early stages of IP protection – which can be costly (see below). There are difficulties too in that even a detailed licence agreement can be broken – due perhaps to market downturns. You will therefore need to keep track of whether your licensee is paying what is owed and fulfilling their other obligations.

Bootstrapping

This just means that you are doing everything without external funding. This is the way that most businesses used to be founded. There has traditionally been an association between IP-based business and the need for large-scale investment. A certain cachet attaches to being a 'high-growth company'. This addiction to glamour and the search for the 'next Google' encourages people to think in the short term. It generally takes a long time, however, to build a successful, sustainable business – even in the internet era. Recently, IP-based businesses are becoming much more feasible without needing venture capital or large loans. (You can design, manufacture and market a real, physical product line from a garage easily these days with a 3D printer, some CAD software and an internet connection).

Admittedly, bootstrapping is really suitable only for certain types of business. If you are aiming to create a production facility for a new solar-powered car, then it's not really possible to use your back bedroom as a factory; you would *need* external funding. Bootstrapping can work well for businesses which are predicated on slower growth or part-time work or where you as inventor want to be in control of everything. If you need later to scale up your activities, investors will be very much keener to get involved if you have evidence that you were capable of launching a product that attracted customers.

External Investment

Venture Capital
VC may be required at some stage of your commercial activities. You might need to buy advanced equipment or invest in expert staff or marketing. VC investors are looking to see if your business can deliver a *massive* return on their money, whilst minimising their exposure to risk. You will need to have some IP protection at least in train as well as a business plan which paints a realistic, clear picture of how the VCs will make their money.

They will want this plan to detail the following elements:

* a compelling summary which 'sells' the business idea;
* management who have done it before;
* a business model they understand, in a market they are familiar with;
* massive and growing customer need, based on robust market research; and
* a strong competitive edge (i.e. a product which is 10 times faster, better and/or cheaper).

In return for the money they bring, VCs seek to own 20–40% of your company and will probably want to place someone on your company's Board of Directors. They provide development cash from a 'pool' (to which the real investors have promised payments). The money will be staged, so that only if you deliver what the plan said at a given point do you get the next round of cash. As more money is raised, and more shares issued, the value of your holding is diluted (temporarily, with luck).

You don't repay anything, even if the whole thing fails, but they will expect you to have 'skin in the game'. In other words, you need evidence that you have committed money, time and other resources of your own. They will aim to get their profits at 'exit' in perhaps 3–5 years by selling their shares or the entire company (i.e. a 'trade sale'). In practice, their business model acknowledges that of the deals they do, only a few will be really successful, but these will make enough money to more than cover all the others which struggle or fail. See:

www.artofthestart.com

http://www.ted.com/index.php/speakers/david_s_rose.html.

Business Angels
Business angels are individuals who are putting up their own cash. Although they will ask for a smaller slice of the business than VCs, they too will expect to take a role in helping to guide your business and they may also encourage other angels to get involved, thus spreading their

risk and accessing extra expertise. As with all investors, you need to assess whether the deal they offer is attractive to you and whether you can work with them.

Debt Finance

You could ask a bank to lend you some money. I wouldn't recommend it to start-ups, given the risk level involved. In view of the anti-entrepreneurial attitude of financial institutions, your access to borrowing may be extremely limited, irrespective of how good a plan you have (and even if you put your house up to guarantee the loan).

Joint Venture

Sometimes a company with whom you have a relationship will be interested in your IP from the point of view of using it to develop a new product line, for example. It may be that forming a new company or joint venture with them offers you some advantages such as a mutually cost-effective way to access large markets and build momentum faster than normal. Inevitably they will demand some control, given the risk they are taking. They will be a guaranteed customer for your deliverable (albeit the sole one, at least to start with). You can also expect to get some administrative support and help with legal costs. Joint ventures work best when both partners bring something unique and a combined effort emerges. Company personnel and plans which are stable over time also greatly help. Do be aware that once a large company has gained access to your invention(s), they may be inclined to simply take over the whole project and make you redundant. Possibilities like this are good to get written into the JV agreement, which you will be required to enter into with them (after negotiating the detail).

Assorted Other Approaches

To commercialise your IP, you may be inclined to simply pay someone else to do the legwork. If you have pots of cash, then you could talk to any of a number of technology development consultancies. By contrast, so-called invention promoters have a bad reputation, so do be hesitant about working with companies offering to market your invention for a big initial fee. Undertake an online search for 'invention scams' first.

One alternative to the IP protection approach is to enter your ideas into competitions or open innovation programmes. Big companies realised a while ago that the costs of running giant R&D labs and a costly patent portfolio might not be a maximally cost-effective way to innovate. Now, big players like P&G and IBM are offering cash in return for ideas from employees and the public alike. Apple, Starbucks and Lego, for example, have realised that their most passionate customers will avidly support their brand by contributing ideas – often for no reward. There are many opportunities online to get involved in such competitions (e.g. www.innocentive.com). Even governments are moving more towards offering prizes for strategically important solutions.

Open source businesses started in the software world but now can be found offering instructions for hardware too (www.arduino.com). The basic business model is to supply the code, instructions or data for free (preferably via download) and make money from selling, for example, printed manuals, special versions, extra upgrades and insights (http://en.wikipedia.org/wiki/Open_source).

Before you start writing your own plan, you need to know a bit more about IP, with an emphasis on patenting.

Patents – Pros and Cons

Owning a patent or trademark allows you to talk freely to potential idea buyers (licensees), investors and so on, and to consider legal action against infringers (although that is only possible if you have a substantial war chest for legal fees). These days, you can do an online search to check your idea is original and then submit an online patent application based on your back-of-the-envelope notes.

The problem with this approach is that it's very hard to be sure that nothing in the vast patent databases could be claimed to be similar to your idea. Patents can be written in deliberately obscure language. On the upside, there are ~50 million patents and these represent a huge resource (80% of which appear nowhere else). You might consider licensing someone else's patent to support your own business or, if it's more than 20 years old, you can make use of it for nothing.

It's also unlikely that your best attempt at drafting will be granted a patent; and even if it is, it will be vulnerable to attacks in court which experienced lawyers, acting on behalf of some company, are trained to make. This is why patent attorneys can charge highly for their ability to draft robust applications.

If you believe you have had your legal rights to any of the items above marked '*' infringed, it is up to you to start court proceedings against them. This may be a serious problem in some countries where the costs are *enormous* and the legal system unfamiliar. Applying for legal registration of your idea also exposes it to potential copyists everywhere.

The patent process is the same for businesses with totally different scales and models. Since you are unlikely to found, for example, a big-pharma outfit at once, smart tactics will be required to stay ahead of the IP game.

Some Smart IP Tactics

Even if someone tells you that your idea is 'registrable', that doesn't mean you should apply for a patent. Getting a patent doesn't imply the commercial value of your invention. Patents have been granted to the following:

- apparatus for facilitating the birth of a child by centrifugal force;
- a bird feeder shaped like a church; and
- a pillow with a retractable umbrella.

Also, don't rush to patent. If you get a filing date too early, you may not be so sure about the detail of how the invention works, and thus be picked apart in court by challengers or if you try to enforce it against infringers. You need to achieve a balance between priority and finding a licensee or joint venture partner who can help with the costs and share the benefits. It's always worth doing a free, initial, coarse search to check for obvious earlier versions of the idea (via e.g. http://gb.espacenet.com). But 'absence of evidence isn't evidence of absence', so if you want to get a patent, you will still need the services of a patent attorney who will undertake a proper search and draft an application on your behalf (www.cipa.org.uk). I'd always recommend getting patent applications (that you intend to pursue to grant) drafted by a professional (with whom you have a friendly relationship and who knows you have limited cash). Before you meet, for an initial free, confidential hour, write down your draft claims (i.e. the clever part about how your idea works) in detail. Don't expect the paperwork which emerges to include all the cleverest aspects of your invention. Some won't be original; some may not be legally defensible. Your first priority should be to avoid accusations of infringement – another reason to employ professionals (http://www.technologyreview.com/read_article.aspx?id=16280&ch=infotech).

Remember too that patent applications can make your little business look very clever. You can submit handwritten applications as long as you don't expect to pursue them to a granted patent (in the United Kingdom). These can be withdrawn within 12 months and, although you lose the registration date, the content remains secret. This allows a certain amount of low-cost boasting to take place in the interim: 'We have several active patent applications'.

Procedure: How to Get a Patent (UK)

If you decide that making a patent application is a good idea, the UK outline procedure is laid out here. A patent attorney will do this work for you, for a fee, if you actually aim to own a patent at the end of the process.

- Get form 1/77 etc. from the Patent Office website.
- Upload your description of the novel features to the Patent Office. This establishes an official priority date.
- The parts of the patent application of most importance are these:
  - your explanation of why your invention is more than the next logical step from earlier work ('prior art'); and
  - who has contributed exactly what in creating the invention.
- Once the application is on file, you can talk about your idea, but first get a nondisclosure agreement (NDA) signed by anyone to whom you describe the content. An NDA is not a guarantee of silence, but it helps people understand that you are serious and makes them think before gossiping.
- The Patent Office will do a search and get back to you with queries in connection with other ideas that may challenge your claim to be first.
- Assuming you can rebut any such issues, you have 12 months to find partners or withdraw and lose your priority date.
- You also need to decide whether it's worth extending your application internationally via something called the PCT (Patent Cooperation Treaty) process.
- 18 months: If not withdrawn, your application will be published.
- **48 months+: Your patent should be granted.**
- 60 months: Annual maintenance fees become due (see below).

Patent costs

Here are some broadly indicative costs. Your mileage may differ.

- Priority filings (UK) – £2500–5000
- PCT filings (12 months later) – £4000–7000

- National/regional phase costs – £2000–6000 per region/country
- Prosecution to grant – £10,000+ per country
- Five years after grant, renewal fees of £15,000–20,000 a year to keep a single patent going are not atypical.

IP Exploitation without Formal Protection

It's important to keep this in perspective and not get intimidated by the seemingly huge costs and apparent complexity. If you can cope, for example, with the details of a doctoral research project, then you are certainly capable of deciding and implementing a personal IP strategy. *People do make serious money via their research-based IP.* They often achieve this by a combination of legal rights and business nous (as well as wealthy backers). Here are some extra insights that are worth being aware of.

The United States has switched to a first-to-file patent regime. This means that if you can get a patent application for your idea submitted before anyone else, it gives you a big advantage if legal disputes about ownership arise. With this change, the US Patent Office also now offers a low-cost ($65) Provisional Patent application to microbusinesses (even outside the United States). Unlike the situation in the United Kingdom, this allows lone inventors to talk to large companies about licensing on an equal footing. US companies are culturally much more interested in buying ideas from small, external inventors than are UK companies. This is the niche in which my business now operates (www.hawkshawpd.com). Do feel free to contact us for more information.

Now that supply chains are getting complicated, and partners are often also competitors, anyone who starts legal action in haste can make themselves widely unpopular – as well as impoverished. Sometimes companies who are infringing will respond positively to an offer to license your IP for an attractive price. You get to name them as licensees, and they get to continue a business legitimately which they obviously think it's worth being in.

In certain cases, it's possible to be innovative and have your products disguise their own workings, so that legal protection can be de-emphasised. Software can often be effectively obfuscated internally, and electronics can have redundant circuitry so that the costs of reverse engineering outweigh the benefits of copying. Conditional access technologies, such as DRM, are designed to make content available only to those who have paid for a code or key. More generally, as product life cycles are shortening, companies are now opting to avoid the glacial IP protection mechanisms in favour of brand development, based on making quick sales and then moving on to the next product.

Conclusion

So now you know the basics about intellectual property. You still need to understand product development, publicity, presentations and negotiating deals. These are vital skills which are better learned from experience than a textbook – so I wish you good luck with the challenges of creating and profiting from your own IP!

42

Career Opportunities

Sara Shinton

A vast number of career paths are open to doctoral graduates, but it is easy to lose sight of this if you are based in a university and faced with nothing but academic careers. This chapter will help you explore the opportunities that are open to you and provide practical advice whatever your chosen career path.

In the last 15 years, there have been several first-destination and longitudinal surveys and reports (some highlighted in Box 41.1), so awareness and understanding of post-PhD destinations are better than ever. Reading these reports will give you a great insight into your options and dispel any perception that non-academic or non-research options are unusual.

Box 42.1   Key providers of information on doctoral career destinations, academic paths and the higher education sector.

Vitae: 'What Do Researchers Do?' Since 2004, Vitae have conducted extensive research into doctoral career paths and have produced a wide range of reports and supporting materials, all of which are available on their website, www.vitae.ac.uk.

Royal Society: With a focus on the sciences, the Royal Society have produced a range of reports, including 'The Scientific Century' (2010) which discusses the academic sector, careers and funding. Available from http://royalsociety.org.

RCUK: In the United Kingdom, the combined Research Councils (RCUK) contribute around £3 billion (in 2012) to fund research and have huge influence in the sector. They have commissioned a number of reports into the value of a research career, all available on their website: www.rcuk.ac.uk.

HEFCE: The Higher Education Funding Council for England (HEFCE) is another key supporter of UK academia. They regularly analyse the trends of staff employed in universities, giving insights in salaries, contracts and diversity. These are available from www.hefce.ac.uk.

Research Methods for Postgraduates, Third Edition. Edited by Tony Greenfield with Sue Greener.
© 2016 John Wiley & Sons, Ltd. Published 2016 by John Wiley & Sons, Ltd.

The data on post-PhD destinations allow us to challenge the view that academic roles are the only substantial careers open to doctoral graduates. The reality is that within a year of graduation, less than half of all PhD graduates work in academic research. Of those, only around 1 in 10 go on to become permanent academic research staff. Although precise figures vary across disciplines, it is a universal feature that competition in academia is fierce. Most of those who go into temporary research positions ultimately join the majority of PhD graduates in non-academic posts. Graduates of all disciplines build successful careers outside of the traditional view of research.

If you are reading this chapter with a clear preference to pursue an academic career, don't despair! But stop assuming that your doctoral qualification is anything other than a starting point. We will consider detailed advice for post-doctoral researchers later. But as a starting point, remember that successful academics have generally been strategic in their research choices, focussed on producing valuable outputs and ensured their research has national or international value.

Choosing the Right Career

Even with the wealth of information that now exists (or perhaps because of it), choosing the right career can be an intimidating decision. So it is important to know that research students have access to expert support and an ever-increasing amount of tailored information. It is an unfortunate truth that many researchers fail to make the most of the support available until the last days of their PhD or later. By then, they have missed many of the opportunities available and are disadvantaged by starting career transitions that could have begun years earlier.

This section aims to help you to get started with your career decision making, pointing you towards further help and trained advisers. If you start considering your options early, you can:

- develop a tailored skills profile, focussing on the skills needed for your preferred career;
- look for opportunities for work shadowing or to work on short internships (even as a research student), to gain an authentic insight into potential careers;
- extend your network into career areas, giving you access to insider information, news about opportunities and the chance to build your visibility; and
- identify the employers who are active in your preferred job area and understand how they recruit, what they are looking for and what they are offering.

Perhaps most importantly, you will be more motivated to complete your thesis and move on to your new career.

Even though university careers services are larger, more engaged with academic departments and well equipped, there are still many students whose decision to pursue a PhD is partly based on having never thought about careers as undergraduates. If this applies to you and you are waiting for a moment of inspiration to point you to the ideal career, it's unlikely to happen. There is, however, a good chance that you'll drift into postdoctoral research – the right first step for an academic path or if you are passionate about your research, but otherwise just a postponement of the decision.

Often, misconceptions about careers services stop researchers from visiting them. The messages from careers advisers presented in Box 41.2 might encourage you to look again at the support they offer.

Box 42.2   Messages from university careers advisers.

'We don't just know about undergrads. Most of us are over 21 and have had varied careers before joining a Careers Service.'

'Remember that we collect the data on the first jobs that PhD graduates enter and we've collected this data for years. We can give you a far more detailed picture of destinations from your department than the national data, excellent though that is.'

'We can give you a confidential space in which to figure out what to do next and ways to work out what's right for you. We can't tell you what to do, but we will help you make the decision for yourself and support you in finding vacancies.'

'Don't worry if we're not specialists in your subject. We're careers methodology experts: we live and breathe "how to find jobs".'

'We see research students leave the academic track every year and never see alternatives to academia as a second choice or an admission of failure.'

'Like you, we spend lots of time with academics. You can talk to us about how to get on in academia, as well as how to get out.'

'There are lots of careers resources aimed specifically at PhDs. Your Careers Service can point you towards these, and those for specific employment sectors, subjects and countries.'

'Many Careers Services run events aimed at PhDs. Come and ask us about our events and tell us what you need. The more we understand about the employers or jobs you'd like to know more about, the easier it is for us to develop our expertise and put on relevant events.'

So having made the case for using the support on your doorstep, let's start with some simple steps and point you to resources tailored for researchers.

It is my firm belief that a career choice based on personal enthusiasm is the most likely to lead to long-term satisfaction and success. Any discussion I have with researchers about career options always begins with the following question:

What Do You Enjoy Doing?

As a researcher, you will have a variety of roles and responsibilities. These could include project planner, writer, teacher, speaker, experimenter, interviewer, programmer, problem solver and thinker. Start the process by reflecting on which aspects of your research you feel most enthusiastic about. Don't stop with your research. Think about how you spend your free time. Are you involved in societies? Voluntary work? A small business? Start to build a picture of the things you enjoy doing.

My next question is a critical one, as it will be the foundation of convincing an employer to recruit you.

Which Skills Can You Demonstrate?

You can probably come up with a list for yourself, but it can be difficult to identify all your skills, particularly if feedback on your performance as a researcher focusses on *what* you are doing rather than *how* you are doing it. There is a substantial resource to help you analyse your skills called the Researcher Development Framework (RDF). This has been developed through interviews with successful researchers and analysis of studies into research skills. Available online through the Vitae website, it will give you an extensive list of the skills which you could bring to the labour market and, again, will help you reflect on what you enjoy about research.

One of the strengths of the RDF is its breadth. It doesn't just focus on research management skills, but encompasses creativity, networking, strategic thinking, team working and enterprise. There is a range of material on the Vitae website to help you to use the RDF to support your career development and broaden your awareness of the value of your PhD.

Is There Anyone Who Can Suggest Careers Ideas Based on the Strengths They See You Demonstrate?

Although personal reflection is an essential part of career planning, it is important to get the perspectives of others. Your supervisors should give you feedback on your performance as a researcher. If they tend to focus on your research, rather than you, the insights you gain from your own skills analysis should help you to ask direct questions to get this feedback. Your understanding of your supervisors should help you articulate this, but general questions might include:

> Have you seen an improvement in my communication skills / ability to work well with others / writing?

> Can you suggest ways in which I could be more organized / creative / collaborative?

> Am I prioritising the right things in my research?

And, most importantly:

> I'm starting to think about my career choices after my PhD. Do you have any suggestions for me to consider?

Remember that your supervisors are likely to have worked with many students over the course of their careers. They might have useful ideas or contacts you could exploit.

Are There Any Careers Resources Specific to Your Discipline?

Academics aren't the only experts with knowledge of the labour market related to your research area. Many disciplines are represented by professional bodies and learned societies, particularly in the sciences and engineering. Most professional bodies will use their network of members as a careers resource, producing case studies and guides to career options. These subject-specific resources can give far more detailed insights into the opportunities available after a PhD. Dr Sarah Blackford at the Society for Experimental Biology is the author of a careers book (Blackford, 2012) for bioscience researchers. The Institute of Physics is another of the many societies who employ a careers specialist and produce careers material, often tailored for researchers (Shinton, 2011) and free to institute members.

If you work in a field without this kind of organisation behind it, the expertise will still be there, but you'll have to dig a little deeper to find it. Academics and researchers will attend conferences in your field from a range of institutions and (if relevant) other sectors. With a little confidence and an understanding of what you are looking for from a career, you can use these networks as a source of careers information.

What Have Other Researchers Done after Their PhDs?

As mentioned in this chapter, the understanding of doctoral career destinations and motivations has dramatically improved in the last decade and gets better all the time. When the first analysis of the first destinations of UK-domiciled doctoral graduates was published in 2004 (Shinton, 2004), it surprised many by revealing that over half of these graduates immediately left the academic path. These data have been updated and analysed in increasingly sophisticated ways, giving insights into careers aspirations, regional trends and longer term career directions. The Vitae website (see Box 41.1) now hosts an invaluable guide to doctoral careers with information on popular sectors, common destinations and profiles of researchers who have made the transition from doctoral student to professional.

Common Destinations of Doctoral Graduates

Drawing on the destinations featured in these reports, this section will set out a few examples of careers open to researchers. Box 41.3 points to the best websites to help investigate these and other options. This research is essential for helping you to understand the demands of specific sectors and roles and for marketing yourself and your skills effectively.

Post-doctoral Research

Perhaps the most obvious career destination for doctoral graduates is to continue with academic work by taking on a post-doctoral research position (or *post-doc*). This can be a chance to continue your current research or to move into a new area. Most positions are held in universities, but there are opportunities to work as a postdoctoral researcher in research institutes or (less frequently) industry.

The transition from PhD to post-doc can be very straightforward – as with your doctoral research you are likely to be working on a specific project, under the management of an academic. The main differences are that you will be expected to take more initiative, and you may manage the day-to-day supervision of students in your group. In scientific and technical subjects, the funding for your research will probably have been secured by the lead academic (the principal investigator, or PI), whereas in the arts, humanities and social sciences you are more likely to have developed your own project.

Post-doctoral research is not a long-term career in itself. It is a transition between being a research student and becoming an academic, so you must be aware of the clock ticking from the point you start your contract. Spend time learning about the expectations in your field – what should your publication profile look like? Should you have secured your own funding before applying for a lectureship? Will you benefit from moving institutions, subject areas or even countries? How long should you spend as a post-doctoral researcher?

If you still have huge enthusiasm for research but aren't sure you want to continue indefinitely in academia, post-doctoral research can be an attractive option. Many post-doctoral researchers choose to leave the academic path after a few years, moving into industry, research related and many other roles.

If the post-doc route appeals to you, you should start investigating potential departments and groups as soon as possible. Conferences are a great way to get to know other researchers and potential supervisors and to learn about opportunities. If you do aspire to an academic career, moving institutions is likely to be beneficial, and in some fields international experience is highly valued.

Seek advice now from academics in your department, particularly those who have been recently appointed. Find out how they found their posts, what advice they have for you and what they are doing to progress their careers. Start to track recruitment of academics in your discipline and be clear on the demands of institutions. Understand what you need to demonstrate on your CV (publications, research funding and teaching) to improve your employability.

Also tell your supervisor as soon as possible – their network is likely to be stronger than yours, and they may be able to recommend you to key people.

Research in Other Sectors

Researchers can be found in many different kinds of organisations working on a wide variety of projects. The sector which will be most relevant to you will partly depend on your research area and partly on your personal values and motivations.

Industrial research is a common destination for scientists and engineers. There are many different sectors, such as the pharmaceutical, specialty chemical, agricultural, manufacturing and energy sectors. Most PhD graduates will be initially drawn to research and development roles (and many employers are keen to recruit doctoral-level graduates into these roles) where you can work at any stage between basic research and product launch. If you are drawn to the idea of working on projects which lead to applications and products, this sector may appeal more than academia.

Research institutes are traditionally publicly funded, although most now undertake work for industrial and commercial clients as well as conducting government-funded research. Most focus on particular fields of research, so your choices may be geographically restricted.

Government departments employ researchers from many disciplines. Some will work in local and national government where they support policy decisions by providing essential summaries and analyses of specific issues. Others will work on applied research in specific areas such as statistics, defence or social research.

The *National Health Service* (NHS) is the fifth biggest employer in the world and the largest in Europe, so unsurprisingly, amongst its 1.7 million employees there are a variety of researchers, including a range of clinical scientists. People in these roles support clinicians through specialist services to diagnose disease and develop new treatment methods. The NHS also undertakes a great deal of research into health policies, procedures and patient care. Again, these roles are very focussed on the objectives on the NHS and more applied than some academic research.

Commercial research organisations investigate and analyse data which relate to business operations, customer behaviours and markets. Some companies apply complex mathematical models, whereas others use social science research methods. Work is often project based, and you will need to be flexible to adapt to the demands of different clients.

If a non-academic research career appeals, you must be sure to market your PhD in a way which appeals to these employers. To do this, you must understand their particular requirements. Talk to researchers from these sectors who attend the conferences in your field. If the specific conferences you have attended are purely academic, discuss with your supervisor alternative meetings which might have a broader remit. Use the networks close at hand (doctoral graduates from your department, careers service connections and professional bodies) to build links in these areas and investigate how to present your background effectively. Remember that other sectors often place a greater emphasis on a broader set of skills including team working, leadership and project management, so take opportunities to develop these during your PhD.

Box 42.3   Starting points for researching careers and job sectors.

www.vitae.ac.uk – As mentioned in Box 41.1, this is a key site for doctoral researchers. It includes extensive information on the sectors and roles identified in its surveys, resources to audit and develop your skills and advice on managing and completing your PhD.

www.prospects.ac.uk – This is the official careers site for Higher Education in the United Kingdom. An exhaustive site includes careers advice, occupational and sector profiles, vacancies and case studies. The site is particularly relevant if you are interested in moving away from research roles.

www.academiccareers.manchester.ac.uk – A wonderful site filled with authentic academic insights and advice. Researchers and lecturers from many disciplines have contributed to the resources which gives you a full picture of the demands of an academic job and advice on achieving success.

www.jobs.ac.uk – The principal site advertising jobs in UK universities, this carries vacancies for *ALL* university roles (including administrative, research and managerial) along with tailored careers articles on many aspects of academic job hunting.

University Roles

Most universities are amongst the biggest employers in their home cities, and only a proportion of these jobs are directly in teaching and research. Each university relies on hundreds of staff to ensure it runs effectively and provides high-quality services to its students and staff. As a research student, you will have met and interacted with many of these people – careers advisers, researcher development staff, computer services, departmental administrators and librarians. They are the tip of the iceberg – there are many other challenging and rewarding careers in higher education. These include:

- research support roles, which ensure institutions are effective at identifying and securing research funding;

- staff development units, which train staff to fulfil their roles better;
- press and PR officers, who ensure that the interesting and important work happening in an institution is shared with the wider community and media;
- schools liaison staff, who perform outreach activities, help non-traditional applicants aspire to a university education and support their transition into study; and
- technology transfer units which promote the commercialisation of research through patents, licensing or forming spin-out companies.

If you start to look at your institution as an employer as well as a place of learning, you will start to see the many opportunities available. Use your careers services or researcher development unit as a starting point. The staff are likely to have good insights into the variety of internal careers and may have personal contacts you can talk to about their work.

Research-Related Careers

If you looked at the RDF (Researcher Development Framework) mentioned in this chapter, you'll be familiar with the wider range of skills that researchers develop. Some of these skills also relate to specific career areas related to the research process.

These skills include writing and editing, which may lead to potential careers with journals and publishers, working to help researchers disseminate their work through peer-reviewed articles, monographs and books. Similarly, if you are interested in research strategies and the processes by which funding is allocated, you might find a career working for a funding body interesting.

Some researchers are particularly skilled organisers, networkers or presenters. These skills could be the basis of careers in teaching, in public engagement or in a professional body or learned society. These organisations support and promote research and education. Their activities include publishing, organising conferences, supporting networks, promoting their subjects to the public, supporting teachers and academics, lobbying policy makers and engaging with the media.

Moving on from Research

For many researchers, the PhD is the final chapter in their research career. After graduation, they find new challenges in a huge range of roles. Some move into business roles, starting on a managerial path; others are drawn to specific areas such as finance. Self-employment appeals to some, and others move into charity and not-for-profit organisations.

The key message is that the next step of your career should be based on your interests, talents and motivations – your PhD is an important part of this foundation, but it shouldn't define you.

Key Messages

If you apply the skills you've developed as a researcher to the process of career choice, you have the potential to be a formidable job seeker. Use these skills to reflect on your own preferences, to study and understand the doctoral job market and to identify the best experts to support the transition from student to professional, in whichever field you chose.

Start early and think about ways to develop a tailored path to your next career. Look for opportunities to develop the skills your next employer will seek.

Finally, remember that the job you take immediately after your PhD is just the *first* step in a career. You don't need to decide what to do for the rest of your life, just to find something that takes you in the right direction, continues to build your employability and gives you personal and professional satisfaction.

Acknowledgements

Thanks to Elizabeth Wilkinson, University of Manchester Careers Service, and Sarah Blackford, Society for Experimental Biology for invaluable discussions about the support and relevance of careers services and learned societies to doctoral graduates and to Dr Keith Morgan, Shinton Consulting for insights into academic careers.

References

Blackford, S. (2012). *Career Planning for Research Bioscientists*. Oxford: Wiley-Blackwell.
Shinton, S. (2004). *What Do PhDs Do?* Vitae.
Shinton, S. (2011). *Moving On: The Physics PhD Student's Guide to Boosting Employability*. Institute of Physics.

Index